CW00536597

THE SCIENCE OF ROMAN HISTORY

The Science of Roman History

BIOLOGY, CLIMATE, AND THE FUTURE OF THE PAST

Edited by Walter Scheidel

PRINCETON UNIVERSITY PRESS

PRINCETON & OXFORD

Copyright © 2018 by Princeton University Press

Published by Princeton University Press,
41 William Street, Princeton, New Jersey 08540

In the United Kingdom: Princeton University Press,
6 Oxford Street, Woodstock, Oxfordshire OX20 1TR

press.princeton.edu

All Rights Reserved

ISBN 978-0-691-16256-0

Library of Congress Control Number 2017963022

British Library Cataloging-in-Publication Data is available

This book has been composed in Miller

Printed on acid-free paper. ∞

Printed in the United States of America

10 9 8 7 6 5 4 3 2 1

CONTENTS

Maps

Figures

Tables

LUCA BONDIOLI is director of the Sezione di Bioarcheologia at the Museo delle Civiltà—Museo Nazionale Preistorico Etnografico "L. Pigorini," in Rome, Italy. His recent research interests have focused mainly on advanced methodologies and techniques for information retrieval from fossil/ archaeological human bone in an evolutionary, functional, and population perspective. He has worked on a variety of paleoanthropological topics focusing on dental maturation, radiographic applications in the fossil record, and skeletal biology of Roman remains. He is involved in Early Pleistocene fieldwork in Buia (Eritrean Danakil).

MICHAEL G. CAMPANA is a computational genomicist at the Smithsonian National Zoological Park and Conservation Biology Institute. His research uses ancient DNA data for the study of disease and animal populations through time.

PROFESSOR OLIVER E. CRAIG specializes in biomolecular archaeology, specifically the recovery of proteins, lipids, and DNA from ancient skeletal remains and archaeological artifacts to provide insights into past human activities. His particular interests lie in temporal transitions and variability in human diets, cuisine and subsistence practices, and the impact that dietary changes had on social evolution, health, and the environment. Oliver is interested in combining a broad range of analytical techniques to study palaeodiet but particularly stable isotope analysis of human bone and organic residue analysis of food remains on ceramics. He directs the BioArCh centre at the University of York.

PETER GARNSEY is Professor Emeritus of the History of Classical Antiquity at Cambridge. His current interests include the history of political thought, comparative systems of penal law, and Roman society and culture, with special reference to skeletal evidence.

REBECCA GOWLAND is an Associate Professor in Bioarchaeology at Durham University. Her research interests include the inter-relationship between the human skeleton and social identity; health and the life course in the Roman World; palaeopathology; and social perceptions of the physically impaired. She has co-edited (with Christopher Knusel) *Social Archaeology of Funerary Remains* (2006) and (with Lindsay Powell and William Southwell-Wright) *Care in the Past: Archaeological and Interdisciplinary Perspectives* (2016), and has co-authored (with Tim Thompson) *Human Identity and Identification* (2013). She has published over 40 scholarly articles in books and peer-reviewed journals, including the *American Journal of Physical Anthropology*, *Antiquity*, *Britannia*, and the *International Journal of Osteoarchaeology*.

KYLE HARPER is Senior Vice President and Provost and Professor of Classics and Letters at the University of Oklahoma. He is the author of *Slavery in the Late Roman World, AD 275–425* (2011) and *From Shame to Sin: The Christian Transformation of Sexual Morality in Late Antiquity* (2013). His current work is on the demographic and environmental history of late antiquity: Princeton has just published his third book, *The Fate of Rome: Climate, Disease, and the End of an Empire.*

ROY J. KING is an Associate Professor Emeritus in the Department of Psychiatry and Behavioral Science at Stanford University. He has published widely on the modern DNA archaeogenetics of the Mediterranean Region, with particular interest in both the spread of the Neolithic across Europe and, more recently, models of Late Bronze Age/Archaic Greek colonization of the Eastern and Central Mediterranean.

MICHAEL MACKINNON is Professor of Classics at the University of Winnipeg. As an archaeologist he has worked at over 60 sites throughout the Mediterranean. His particular interests focus upon the role of animals within ancient Greek and Roman societies, as drawn from the integration of zooarchaeological, ancient textual, and iconographical evidence.

MICHAEL MCCORMICK studies the fall of the Roman Empire and the origins of Europe. He is the Francis Goelet Professor of Medieval History at Harvard University, where he chairs the Initiative for the Science of the Human Past (SoHP: https://sohp.fas.harvard.edu/). His books include the prize-winning *Origins of the European Economy* (2002) and *Charlemagne's Survey of the Holy Land* (2011); he recently led the first multiproxy scientific and historical reconstruction of climate under the Roman Empire (*Journal of Interdisciplinary History* 43 (2012)). He edits the free, student-created online *Digital Atlas of Roman and Medieval Civilizations* (http://darmc .harvard.edu/), and is active archaeologically in France and Spain.

TRACY PROWSE is an Associate Professor in the Department of Anthropology at McMaster University. Her research integrates stable isotope and osteological analysis of human remains to investigate weaning, diet, and migration in Roman period populations.

WALTER SCHEIDEL is the Dickason Professor in the Humanities, Professor of Classics and History, and a Kennedy-Grossman Fellow in Human Biology at Stanford University. The author or (co-)editor of seventeen other books, he has published widely on ancient social and economic history, premodern historical demography, and the comparative history of labor regimes, state formation, and inequality.

ALESSANDRA SPERDUTI is a biological anthropologist currently teaching at the University of Naples "L'Orientale." She works on the analysis and interpretation of skeletal samples from Italian archaeological sites, from protohistoric to medieval times. For the Imperial Roman Age, she has been working on Isola Sacra, Velia, Lucus Feroniae, and Herculaneum. She is the author of numerous publications focusing on paleodemography, paleopathology, skeletal occupational markers, methodologies on sex and age-at-death determination, and funerary behaviors. She is also involved in projects focusing on science dissemination and has published articles on "public understanding of science."

NOREEN TUROSS is the Landon T. Clay Professor of Scientific Archaeology at Harvard University.

PETER A. UNDERHILL, PHD, has been a Research Associate in the laboratory of Professor Carlos D. Bustamante, Department of Genetics, Stanford University School of Medicine since 2011. In 1995 he co-invented technology that accelerated the discovery of binary mutations on the Y chromosome. His groundbreaking research set in motion a decades-long period transforming the Y chromosome from a boutique novelty to a mainstream system marking prehistoric demographic events previously beyond the reach of memory. Progress at the genomic level with colleagues in the Bustamante laboratory has rocketed knowledge of Y chromosome diversification into a vivid, time-calibrated albeit male-centric narrative of history valuable for hypothesis testing, interpreting ancient DNA results, and as a guide for researchers modeling the dynamics of population demography by simulation. His Y chromosome research has resulted in numerous peer-reviewed co-authored publications.

MARIJKE VAN DER VEEN is Emeritus Professor of Archaeology at the University of Leicester. Her research focuses on ancient agriculture, the archaeology of food, and the agency of plants. She is author of *Crop Husbandry Regimes* (1992), *Consumption, Trade and Innovation* (2011), and editor of *The Exploitation of Plant Resources in Ancient Africa* (1999), *Luxury Foods* (2003), *Garden Agriculture* (2005), and *Agricultural Innovation* (2010), the latter three issues of *World Archaeology*.

LAUREN WALTHER has recently submitted her PhD in the Department of Archaeology at Durham University. Her research includes the assessment of stature and body proportions from human skeletal remains dating to the Roman and Early Medieval periods in Britain. She will shortly be embarking on a post-doctoral project at Durham University to examine the impact of social change on population health during the development of the kingdom of Northumbria in early medieval Britain.

ACKNOWLEDGMENTS

THE EDITOR OF THIS VOLUME is indebted to Peter Garnsey for suggesting contributors and for his help in coordinating work on Chapter 4; to Peter Shi and several anonymous referees for their comments on the entire manuscript or individual chapters; to Melissa Marturano for her diligence in standardizing the format of the seven main chapters; to Jonathan Weiland for making the maps, which show the sites mentioned in the text; to Rob Tempio for accompanying this project from inception to completion; and to Jay Boggis, David Campbell, Ali Parrington, Matt Rohal, Stephanie Rojas, and Jan Williams for their help in turning our manuscript into a book.

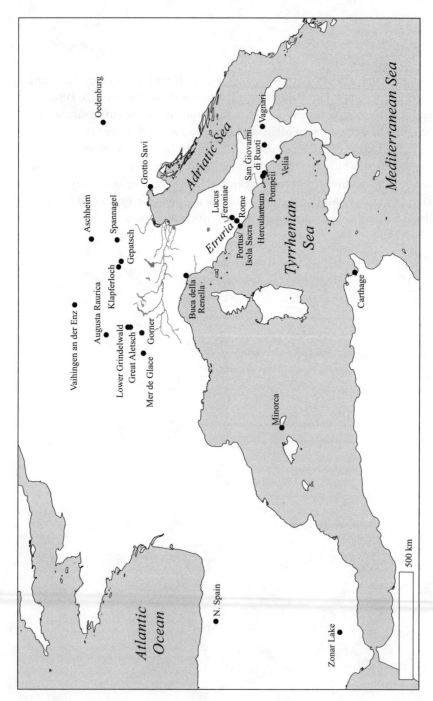

MAP 1. Western Mediterranean.

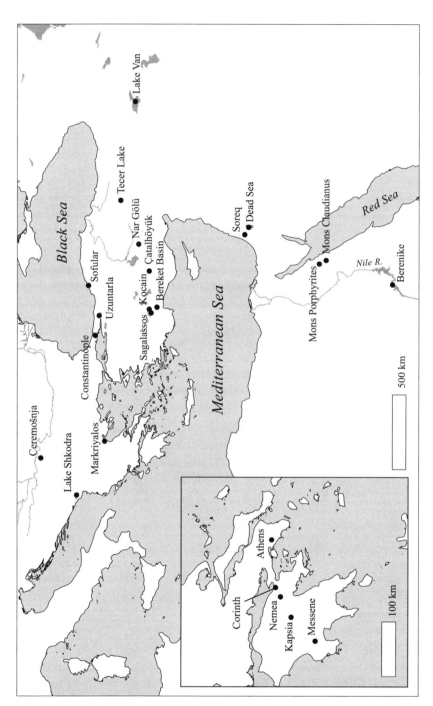

MAP 2. Eastern Mediterranean.

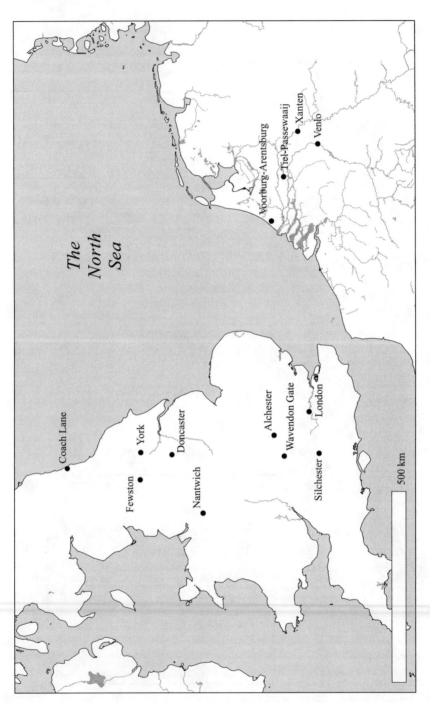

MAP 3. Northwestern Europe.

THE SCIENCE OF ROMAN HISTORY

A HISTORY OF ROMAN HISTORY

Introduction

Walter Scheidel

SCIENCE HAS LONG been making an enormous contribution to our understanding of the ancient past. Archaeology is simply unthinkable without it, and the study of various types of source material from inscriptions and coins to papyri and palimpsests has greatly benefited from scientific analysis. In recent years, the contribution of science has broadened even further as entirely new types of evidence from genetics to climate proxies have been brought to bear on historical inquiries. Thanks to this accelerating expansion, the study of history in general is now approaching a new stage of interdisciplinarity that is firmly grounded in the recognition that human and natural history are intimately and inseparably intertwined.

This book shows that the study of the ancient Roman world is no exception to this trend.[1] Climate is given pride of place (Chapter 1), a powerful influence on the development of agrarian societies that often survived on narrow margins. It remains a formidable challenge to reconstruct meaningful patterns from local data without obscuring local variation. Yet for the first time, we are now able to glimpse the contours of climate change in the long term. Roman power expanded and flourished during a period of favorable conditions— warm, stable, and moist in the right places. Given that Rome's imperial reach turned out to be a unique outlier in the history of western Eurasia, this may well be more than just a coincidence and calls for further inquiry into the interaction of institutions, geopolitics, and environmental factors that produced this outcome.

From the second century CE onwards, growing climatic instability accompanied the fitful decline of Roman power. While a warming trend in the fourth century coincided with temporary imperial recovery in the West, increased precipitation aided development in the East. Prolonged droughts may have been implicated in population movements in the Central Asian steppe in the fourth century and in Arabia two and three centuries later. Even more ominously, the fifth and sixth centuries, a time of upheaval for the Mediterranean, experienced secular cooling coupled with a surge in volcanic activity. The historian's agenda is clear: while the temporal association between trends in macro-social development and climatic conditions is increasingly well documented, the complexity of causal relations remains very much in need of detailed analysis. The history of climate change is also the history of human resilience, and we must ask not only how Roman society was affected by environmental forces but also how it responded to them. Moreover, other ecological factors such as pathogens or deforestation also need to be taken into account.

The study of plant remains is a complementary field of investigation (Chapter 2), closely tied up as it is with that of climatic conditions, even though the connections between them are yet to be explored in depth. Existing research has put emphasis on the spread of cultivable crops under the aegis of Roman rule. The dissemination of naked wheats that were suitable for making bread is one example; charred remains of oil pressings that point to the expansion of oleiculture are another. We can track how particular crops were at first imported, sometimes over long distances, and later incorporated into local farming regimes, and also how widely such crops came to be adopted and consumed. These observations are germane to big questions about the nature of Roman economic development. To what extent were these processes driven by imperial rule as such or were merely the by-product of ongoing long-term growth, just as farming itself had once spread from the Middle East? How "Roman" was the Roman economy, in the sense of being shaped by empire? The food supply of the Roman army is a case in point: how did plant foods found at military sites compare to those present among the local civilian population? Change over time in the sources of food and timber required by the military reveal how state-sponsored demand affected patterns of production. The influence of empire is also visible in the fact that the Roman conquest of Britain closely coincided with the introduction of grain beetles that thrived in large open granaries of the kind set up by the occupiers. Plant remains recovered at Red Sea ports shed light on the dynamics of long-distance trade that would otherwise be irrecoverable, such as the provenance of merchant ships and change over time. Evidence of plant production within urban sites is highly relevant to debates about population size: if the finding that one-sixth of Pompeii's surface area was given over to plant production is anything to go by, Roman cities may not have been as densely inhabited as some would like to think.

But above all, plant remains are a key source of information regarding diet. The discovery of dozens of different plant species at a whole range of Roman-era sites speaks to the scale and scope of economic development: that period's new-found diversity of food consumption was not necessarily restricted to elite settings but was also present in more modest or rural locales. This has considerable ramifications for ongoing debates about Roman well-being and the distribution of gains from growth and commercial integration. The inhabitants of northwestern Europe in particular—a region that felt the transformative power of imperial rule more than many others—enjoyed greatly improved access to and diversity of foodstuffs.

Animal remains offer similar insights (Chapter 3). Patterns of meat consumption have been studied across time and space, linking it to "Romanization" and other processes. Just like crops, animal species spread under Roman rule. Increases in the size of domestic animals in Roman Italy point to productivity gains. A combination of osteometric and genetic investigations helps clarify how much this progress owed to breeding or the introduction of imported varieties. The study of animal remains has enlightened us about various kinds of transfers, from the export of Nile fish to Asia Minor to the migrations of the black rat, which eventually came to be instrumental in the transmission of bubonic plague. Skeletal pathologies, for instance those that document the use of cows alongside oxen for plowing, add to our knowledge of the efficiency of the rural economy. Feeding regimes inferred from dental micro-wear tell us if animals were sustained by pasture or fodder, and variations in heavy metal deposits in goat bones have even been used to track changes in their proximity to human settlements.

Yet however much the remains of ancient plants, livestock, and pests may have to teach us, it is the human body that takes center stage (Chapter 4). It is one thing to observe which crops or animals had spread or were present at a particular site; it is another one entirely to examine how such findings correlate with the physical well-being of people at the time. In the absence of contemporary statistics on food consumption or public health, human bones and teeth are the most important source of information about nutritional status, health, and morbidity in the Roman world. Without them, we cannot hope to observe change over time, both within a given person's life and across generations or centuries. Human skeletal remains form the biggest archive of what it was like, in the most fundamental terms, to "be Roman."

Not all lines of inquiry are equally promising. Longevity is a crucial variable in assessing overall well-being and levels of development, but it is generally poorly or not at all attested outside very narrow settings, most notably the papyrological census record of Roman Egypt. Unfortunately, aggregations of human remains in ancient cemeteries tend to be an unreliable guide to the age structure of past populations. This raises the question whether exceptional cases that have produced demographically plausible patterns are capable of vindicating paleodemographic reconstructions. After all, even a broken clock is sometimes right. But maybe we have been barking up the wrong tree: instead of bemoaning the manifold biases that have shaped (and, from a demographer's perspective, spoiled) the funerary record, these very biases are likely to reflect cultural practices and preferences that are very much worth investigating. In the end, bones may have to tell us more about culture than about demography, a valuable reminder of osteology's ability to shed light on life in the past well beyond the physiological dimension of human existence.

Bones and teeth are of paramount importance in identifying a wide variety of ailments that can often be linked to specific infections, occupational hazards, and cultural norms. It is important to be aware of the limitations of this evidence: the inconclusive debate about the connection between certain types of porotic lesions and malaria stands as a warning against overly confident identifications of Roman pathogen loads. The most common and deadliest diseases of the ancient world, such as gastro-intestinal infections, generally remain hidden from view, and mummified bodies, which allow a wider range of investigations, are confined to just one corner of the Roman world and even there have not fully received the attention they deserve. Even so, considerable progress has been made. The bodies of infants and children hold out particular promise, as dental enamel analysis has begun to shed light on weaning and sanitation practices that would otherwise remain obscure. The early, formative years are in some ways also the ones most worth knowing about, and it should be remembered that children and adolescents would have accounted for over a third of any ancient population. If we ultimately end up with more detailed information about children than adults, this will help offset the general scarcity of information about this critical phase of the ancient life cycle.

Much the same is true for diet. Stable isotope analysis of teeth and bones provides valuable clues about the types of food people used to consume, even though in practice precision remains an elusive goal. Isotopic studies have been at their most successful in ascertaining the relative weight of terrestrial and marine food sources for different groups of people. Given that a sizeable share of the population of the Roman Empire was concentrated in coastal areas where access to seafood was at least an option and that processed marine-based foods were shipped over long distances, this metric is more useful than it might seem in illuminating dietary variations rooted in class and gender as well as geography. However, the biggest question concerns the overall importance of cereals as opposed to animal products in Roman-era diets, and there much work remains to be done.

Last but not least, stable isotope analysis helps us track migration at different stages of the life cycle. Because humans acquire oxygen and strontium isotopic signatures by consuming local food and water—in their dental enamel in childhood and in their bones throughout their lives—comparison of such profiles with local patterns allows inferences about mobility. Complications abound: short-term movement may be hard to track down, imported food and water piped in through aqueducts affect the record, and different regions may exhibit similar isotopic properties. Systematic compilation of local reference data will be the solution to at least some of these problems. Just as previous generations compiled huge editions of inscriptions or papyri, the time has come to create comparable collections of scientific evidence that is relevant to our understanding of life in the past. This applies to isotope signatures just as it does to climate records and genetic information.

The study of body height is yet another branch of osteology (Chapter 5), embedded in a rich tradition of scholarship that seeks to relate stature to various factors such as health and economic development. In the most general terms, height tends to correlate with well-being: however, the fact that the former is the single cumulative outcome of a wide variety of inputs such as genetics, diet and disease greatly complicate causal explanation. In this field, large bodies of data and long-term comparison across space and time are once again of the essence. One key observation that has emerged from the aggregation of local samples is that the Roman period in general was associated with lower body heights than previous or subsequent centuries.[2] The question whether nutrition or pathogen loads played a greater role in this is of fundamental importance to our understanding of the Roman economy. The relationship between imperial rule and physical well-being was bound to be complex, mediated by factors such as economic development, urbanization, connectivity, and inequality that produced conflicting gains and costs in terms of nutrition, health, and thus stature. Once again, as with teeth, the pre-adult record may turn out to be of particular value. The stature evidence points to late menarche and male puberty, in keeping with conditions in current low-income countries and other historical populations. And given enough and sufficiently fine-grained data, class differences in body height—which are well attested for early modern and contemporary societies—may also become apparent. In general, the study of somatic development will greatly benefit from the proper integration of different strands of research, from information about health and diet derived from teeth and bones, about the availability of foodstuffs documented by plant and animal remains, and about geographical and ancestral provenance as documented by stable isotopes and ancient DNA.

The last one of these data sources is derived from most of the other types of ancient remains surveyed so far, from plants to humans and other animals (Chapter 6). Owing to the relatively recent nature of ancient DNA studies and especially the rapid pace of innovation in this field, it has only just begun to contribute to the study of the Roman world. Genetic analysis holds particular promise in identifying the geographical origin of people, livestock, and crops and thus in establishing patterns of human mobility and the transfer of productive resources. Possible genetic discontinuities between ancient Etruscans and more recent Tuscans and connections between Etruscans and the Eastern Mediterranean are of obvious relevance to our assessment of ancient traditions regarding their provenance and to modern models of ethnogenesis. Individual cases of migration over very large distances may catch the eye, but findings of local continuity are equally valuable. Overall, whole-genomic sequencing of larger samples is the best way forward. For antiquity, the most revealing findings made so far concern pathogens rather than humans: the identification of the cause of the sixth-century CE "Plague of Justinian" (as well as the late medieval "Black Death") as *Yersinia pestis* must count as a milestone in the annals of historical epidemiology. The agents behind earlier pandemics such as the second-century CE "Antonine Plague" and the third-century CE "Plague of Cyprian" still await scientific discovery. Among other potent infections, malaria, which is otherwise difficult to infer from skeletal evidence, is also becoming visible, although relative to the likely scale of its spread in the ancient world the existing genetic evidence remains exiguous indeed. All the same, in light of the speed with which this line of research has developed and matured in recent years, it is hard to overestimate its potential for enriching our knowledge of the ancient world.

Analysis of surviving strands of ancient biomolecules is complemented by studies of the genetic makeup of current populations that serves as a giant archive of demographic processes in the past (Chapter 7). Measures of affinity and admixture throw light on the origins of those alive today. In this field, just as with ancient DNA, most existing research has focused on prehistory. A few studies, some of them perhaps already superseded by more recent advances, have identified patterns suggestive of migration from the Levant to North Africa and from the Aegean to Sicily and southern Provence that may be linked to Phoenician and Greek settler activity. Roman history, which lacks similarly distinctive migration events, may prove less fruitful terrain for such studies. One important question that remains to be explored is whether the massive inflow of slaves into select parts of the Italian peninsula has left traces in the genetic record. Both ancient and modern DNA will need to be marshaled to address this problem. Elsewhere, solid evidence of genetic continuity over time could serve as an important antidote to exaggerated notions of population mobility in the Mediterranean environment.

Even this rapid and superficial survey of some of the issues covered in the following chapters should leave no doubt that scientific methods provide insight at all levels of resolution of historical inquiry, from "micro" to "macro." At one end of the spectrum, the individual. Under ideal circumstances, by integrating various approaches, we are now able to tell where someone was from and at what age that person moved to where she died; at which age she was weaned and experienced serious physiological stress; whether she subsisted more on terrestrial or marine foods; and whether she died of the plague. Her somatic data could be compared to those of others at the site and matched with local remains of cultivars, weeds, livestock, and pests, as well as the usual array of inorganic archaeological remains. Never before has it been possible to examine individual Roman lives in such detail.

At the "meso" level, serial analysis of data from a particular locale over the long run and comparison with those from other sites steer us toward broader questions about the impact of empire, of political and economic integration, of urbanization and culture change at the local or regional level and beyond.

Just as the archaeoscience of inanimate objects from ceramics to metals and stone has done for a long time, climatology and bioscience hand the historian additional tools for tackling these questions.

And moving even further to the opposite end of the spectrum, we are now for the first time in a position to try our hand at defensible biohistorical narratives of the Roman Empire as a whole. Kyle Harper's new book meshes climate proxies and scientific data about pathogens with more conventional sources in elucidating the interplay of ecology and human agency over the course of centuries.[3] Much will need to be refined as the scientific evidence expands, but the contours of a truly interdisciplinary history of ancient Rome are now finally in view.

Pursuit of questions about big structures and large processes will require us to think hard about how to integrate conventional evidence with scientific findings. Integration is predicated on the compatibility of observations from different domains of inquiry, a compatibility that arises from *consilience*. Coined in the nineteenth century, this term, to quote Michael McCormick's pithy summary,

> refers to the quality of investigations that draw conclusions from forms of evidence that are epistemologically distinct. The term seems particularly apt for conclusions produced by natural-scientific investigations on one hand and by historical and archaeological studies on the other. Consilience points to areas of underlying unity of humanistic and scientific investigation—a unity arising from that of reality itself.[4]

While this perspective is designed to bridge the gaps between different disciplinary practices and academic precincts of specialized expertise and inquiry, it is worth acknowledging that the underlying premise might also reinforce existing divisions rather than leveling them. Some of our colleagues in the humanities may be skeptical of notions of a "unified reality" or harbor reservations about an encroachment of science. And indeed, the premise of consilient unity leaves little room for the more esoteric varieties of postmodern engagement with the historical record: the very concept is resolutely "modern." To the extent that it will succeed, it may mark a swing of the pendulum towards a more open and, for want of a better word, optimistic perspective on the production of knowledge and our understanding of the world. I believe we ought to welcome such a shift. It is also worth noting that recourse to insights derived from the biosciences readily accommodates historians' concerns about hegemonic discourses and the subaltern: what more immediate way of accessing the history of the "99%" than to study what is actually left of them and the organisms that both sustained and blighted their lives? Archaeobiology gives a powerful boost to history from below, shining a light on those of whom no other record exists.

Nevertheless, biohistorical interdisciplinarity poses genuine challenges. Increasingly sophisticated techniques and falling costs, most dramatically in genetics, keep boosting the contribution of science to historical inquiry. But this progress frequently entails a fair amount of creative destruction. We are faced with perpetual churn in which results made only a few years—never mind decades—ago are called into question or downright superseded by the application of improved methods. This makes for treacherous terrain for the uninitiated. Keeping up to date is an imperfect solution: five or ten years ago, it was perfectly possible for experts to be both up to date and wrong. Caution is the order of the day. Paleodemography and the extrapolation of stature from bone length have long been beset by ongoing confusion about norms and standards. More recently, we have learned that methods and procedures that once seemed state-of-the-art—from trace element analysis in the osteology of ancient nutrition to blood allele studies of modern populations and early work on ancient DNA—cannot bear the weight they had been granted. The enduring lesson is to remain circumspect and resist the ever-present temptation to oversell the latest findings. The very dynamism of scientific research is at once its most attractive feature and a challenge to historians who wish to capitalize on it.

Both the pace of change in the sciences and the professional expertise required to assess and apply its results highlight the need for collaboration across established disciplinary boundaries. Outside archaeology, transdisciplinary research (not to mention teaching) on the ancient world has been rare, and even collaborative work more generally is an exception rather than the norm. Continuing emphasis on individual competence has held back innovation in a variety of areas, from cross-cultural comparative history to Digital Humanities. A biohistorical approach is if anything even more profoundly incompatible with the existing model of training, supporting, and evaluating professional historians as some sort of latter-day master crafts(wo)men. It adds new expectations in terms of what historians ought to know and how they are to cooperate with colleagues from other fields, and draws them deeper into the complex world of grant applications that are the life-blood of their colleagues in the sciences. At the same time, it calls for scientists to partner up with historians in the development of research designs and interpreting the results: transdisciplinarity must not turn into a one-way street that casts historians in the passive role of consumers. Rather, consilient perspectives on the past allow historians to become brokers, by creating ties between discrete communities of scholars that unite them in the pursuit of a richer understanding of the past.

The present volume illustrates only some elements of an engagement with the Roman world that is informed by scientific knowledge. We focus on the human body and on the surrounding biosphere. In so doing, the seven chapters follow an arc from the weather to plants, animals, and humans, and, for humans, from large (skeletons) to small (biomolecules), from phenotype to genotype, and from ancient to modern. For our purposes, the distinction between climatology as part of the Earth Sciences and areas of research that are rooted in biology is merely a formality. Although most climate change in the last few millennia was caused by variations in solar and volcanic activity and the earth's orbit, climate occupies a central position in biohistorical reconstructions because it primarily affected humans indirectly through its impact on flora, fauna, and the water supply.

More could be added. A true "biohistory" of ancient Rome would be broader still, extending into the scientific study of human cognition and behavior, an area that is challenging to access for students of the more distant past and remains outside the scope of this survey. One day, it may be worth pondering how Roman brains and minds were shaped by an environment of endemic slavery and organized violence (from mass conscription to the carnage of the arena), to name just a few prominent features of the historical record.[5]

This volume is meant to offer a guide to different bioscientific approaches and their contribution to the study of Roman history: how they have (or have not) enriched our understanding, and how they might do so in the future.[6] While our focus is on the ancient Roman world broadly defined, the scope of coverage varies from chapter to chapter, and for good reason. Most relevant work on ancient and especially modern DNA deals with earlier periods of human history. Rather than elucidating specific issues of Roman Studies, it gives us a sense of the potential of this research to re-shape our understanding of ancient societies in the coming years. Conversely, the study of bones and teeth presents us with an embarrassment of riches that calls for a degree of selectivity. Chapter 4 therefore concentrates on Roman Italy proper while Chapter 5 privileges stature data from Roman Britain, which has attracted some of the most careful attention. Not every part of the Roman world could be covered in equal measure: evidence from Egypt is particularly rich and would deserve a separate volume, contextualizing Roman finds in the great time depth of Nilotic civilization and making full use of the unique evidence of mummified remains.[7]

One thing is certain. No matter how comprehensive the coverage of a survey of this kind, the rapid progress of scientific research ensures that before long it will seem dated. It cannot be more than a snapshot, capturing a particular moment in the growing entanglement of ancient history and the sciences. We are pushing against the limits of conventional formats of dissemination: the next step may well have to be a continuously updated electronic publication to keep us up to date.

Notes

1. The seven chapters contain over 1,000 references. I therefore largely refrain from adding further bibliography.

2. In addition to the work cited in Chapter 5, this is documented in particular by the dissertation project of Geertje Klein Goldewijk at the University of Groningen, which draws on a larger amount of data than published studies: see Scheidel 2012: 326.

3. Harper 2017. For other times or places, see now especially White 2011; Broodbank 2013; Parker 2013; Brooke 2014; Campbell 2016.

4. McCormick 2011: 257. His article inspires much of what follows in this section.

5. AHR Roundtable 2014 calls on historians to engage with biology more generally. That forum includes contributions on behavior and emotion by Harper 2014, Roth 2014, and Scheidel 2014. See also Harper 2013.

6. Killgrove forthcoming offers a complementary perspective.

7. See Scheidel 2010 for a brief survey of the ancient disease environment.

References

AHR Roundtable 2014. "History meets biology." *American Historical Review* 119: 1492–1629.

Broodbank, C. 2013. *The Making of the Middle Sea: A History of the Mediterranean from the Beginning to the Emergence of the Classical World*. Oxford: Oxford University Press.

Brooke, J. 2014. *Climate Change and the Course of Global History: A Rough Journey*. New York: Cambridge University Press.

Campbell, B. 2016. *The Great Transition: Climate, Disease and Society in the Late-medieval World*. Cambridge: Cambridge University Press.

Harper, K. 2013. "Culture, nature, and history: the case of ancient sexuality." *Comparative Studies in Society and History* 55: 986–1016.

Harper, K. 2014. "The sentimental family: a biohistorical perspective." *American Historical Review* 119: 1547–1562.

Harper, K. 2017. *The Fate of Rome: Climate, Disease, and the End of an Empire*. Princeton: Princeton University Press.

Killgrove, K. forthcoming. *These Old Roman Bones: What Bioarchaeology Tells Us about Life in the Roman Empire*. Baltimore: Johns Hopkins University Press.

McCormick, M. 2011. "History's changing climate: climate science, genomics, and the emerging consilient approach to interdisciplinary history." *Journal of Interdisciplinary History* 42: 251–273.

Parker, G. 2013. *Global Crisis: War, Climate Change and Catastrophe in the Seventeenth Century*. New Haven: Yale University Press.

Roth, R. 2014. "Emotions, facultative adaptations, and the history of homicide." *American Historical Review* 119: 1529–1546.

Scheidel, W. 2010. "Age and health." In *The Oxford Handbook of Roman Egypt*, ed. C. Riggs Oxford: Oxford University Press, 305–316.

Scheidel, W. 2012. "Physical well-being." In *The Cambridge Companion to the Roman Economy*, ed. W. Scheidel. Cambridge: Cambridge University Press, 321–333.

Scheidel, W. 2014. "Evolutionary psychology and the historian." *American Historical Review* 119: 1563–1575.

White S. 2011. *The Climate of Rebellion in the Early Modern Ottoman Empire*. New York: Cambridge University Press.

CHAPTER ONE

Reconstructing the Roman Climate

Kyle Harper & Michael McCormick

Climate and the Science of Antiquity

Environmental history, as a subfield, is now more than a generation old. Traditionally, it has focused on the changing relationship between human societies and the natural world, in both physical and biological dimensions. It has overlapped and connected with related fields such as agrarian history, landscape archaeology, geography, and the study of historical demography and infectious disease. From Braudel to Horden and Purcell, the labors of environmental historians have yielded a much clearer understanding of both the enabling power of the natural world and the constraints it imposes. At the center of the field, it might be suggested, has been an effort to describe how the imperative of extracting energy from the environment has shaped human societies and how, in turn, human societies have exploited and reshaped physical and biological environments in their search for fuel, food, and water.

In the case of Rome, environmental history has built on the traditional study of "the Mediterranean" as a geographical and ecological region.[1] The need to understand the particularities of the zone at the core of the Roman Empire has been primary. From there, study has branched into the exploration of ancient food production, with work spanning from the history of specific crops to the classic work on famine and food shortage by Peter Garnsey.[2] Water systems—from rural irrigation to the monumental urban hydraulics—have often figured prominently in the study of the Roman environment, given the delicacy of water management in the many semiarid regions of the Empire.[3] Forests were once a major theme and are becoming so again, as historians consider how the Romans met their voracious demand for fuel and construction materials.[4] Soils, too, once received attention from historians, although interest has unfortunately abated in recent decades.[5] Human biology has occasionally been placed at the center of environmental history, for instance in the work of Walter Scheidel or Brent Shaw on disease and mortality, or the contributions of Robert Sallares on the history of malaria.[6] In short, ancient environmental history has sought to fulfill the challenge issued by the *Annales* school to write *histoire totale*—to consider human societies in all their material dimensions.[7]

Perhaps the area where the "science of antiquity" is most dramatically changing our understanding of the ancient environment is the study of the paleoclimate.[8] In the last decade or so, climate history has been revolutionized by the discovery and synthesis of new data from unexpected sources. Partly as a by-product of our urgent need to understand anthropogenic climate change, the recovery of paleoclimate records—allowing reconstruction of natural climate variability and change into the deep past—is a boon to the enterprise of environmental history. The global climate system, at some level, frames all the systems and mechanisms that are of concern in environmental history. Where we previously knew next to nothing for ancient history about the backdrop of climate change, recent and ongoing scientific investigations have begun to pierce the veil and illuminate the underlying conditions in which ancient societies developed. The importance of climate in a traditional society is easy to grasp, particularly in societies enmeshed in the favorable but predictably unpredictable precariousness of the Mediterranean.[9] So is the scholarly delicacy of demonstrating precise and rigorous causal connections between environmental conditions and historical change.[10]

In exploring the impact of climate change on ancient societies, and the responses of those societies to the environment, it is essential to state at the outset that both climate change and social impact are complex and multidimensional phenomena that usually cannot be reduced to unilinear cause and effect. Climate change can take many different forms, each with impacts that may differ depending on the circumstances and resilience of the society that experiences them.[11] Changes that have negative consequences in one region may affect other regions more positively. With respect to both temperature and precipitation, it is not only the absolute amount of variation that matters. The timing of these variations could be more or less favorable to particular crops and animals in particular places. Extreme variations could be negative as well as positive: too much wetness can promote blights of crops and animal disease. Speed of change counts as much as timing. In general, slow and gradual climate change is considered less damaging because farmers and pastoralists could adapt to it more easily. The nature of change itself can play a role: unidirectional, or fluctuation back and forth, and fluctuation at different rhythms can modify how climate change affects society. Finally, the clustering of climate events or change can make a big difference. Given the built-in precariousness of the Mediterranean environment, ancient societies developed methods of stocking foodstuffs for the inevitable bad years. But when bad years clustered together, two, three, or more bad years in a row could menace even the most resilient and well-stockpiled of ancient societies. Fluctuation could be as dangerous as unidirectional change.

The resilience of ancient societies likely also changed historically. One of the major "structural" components in Mediterranean resilience was the exceptional agrarian productivity of the Nile valley.[12] Beyond the productivity itself, it was essential that this productivity was determined by factors arising in an Indian Ocean climate system distinct from the prevailing North Atlantic climate that dominates the rest of the Mediterranean basin.[13] Even if general conditions in the Mediterranean were unfavorable for a given harvest, there was a good chance that those conditions would not affect Egypt's harvest. So long as Egypt's food production was integrated into the general Mediterranean economy, and the shipping and distribution systems existed to move that food as needed, much of the Mediterranean was likely resilient against temporary climate-induced shortfalls. Other layers of resilience included the development of interconnected markets around the Mediterranean and the shipping to service them, and the construction of massive granaries, at least for the great cities, to stockpile grain for the inevitable bad years.[14] Such granaries certainly sustained Rome and its ports, as well Constantinople, including Tenedos, on its sea approach.[15] The development of sophisticated water storage and delivery technologies was another important component of resilience in an arid Mediterranean environment. There is a strong case that the aqueduct of Carthage was built in response to a five-year drought that ended in 128 CE; a severe drought explicitly triggered the emperor's decision to rebuild that of Constantinople in 766 CE.[16] Justinianic attention to waterworks in the Holy Land seems similarly to reflect a significant drought.[17]

Until very recently, we had only a few written sources, in combination with archaeological evidence, to reconstruct the history of the ancient climate, and these did not take us very far. With the exception of the Nile floods—which must be reconstructed usually from indirect evidence—antiquity is generally lacking in long series of historical or other reports that would allow us on their own to detect shifts in climate. H. H. Lamb nevertheless did a remarkable job of divining various climate trends from the most scattered and disparate evidence, and his general assessments remain worth reading.[18] But even if the written evidence of antiquity is insufficient in itself to reconstruct ancient climates, judiciously used, it provides a precious check on reconstructions based on other evidence. In a few cases, it is strong enough to extend and deepen climate reconstructions based on paleoclimate proxy evidence: for instance, the great cooling initiated by the "year without sun" in 536 CE.[19] Archaeological evidence will likely play an increasing role in coming years, so long as the chronological and spatial resolution of its information is sufficiently fine to permit precise correlation of cause and effect.

In the absence of direct instrumental records before the most recent generations, climate scientists reconstruct the main features of ancient climate by using proxy data. Proxy data come from tree rings, ice cores, speleothems (stalagtites), lake varves (sedimentary layers), and other natural archives.[20] Comparison of the physical characteristics of proxies with instrumental records of climate over the last century or so has revealed consistent relationships between certain physical characteristics and climate conditions. The assumption that those relationships and the underlying mechanisms have remained constant allows climatologists to extrapolate climate data from similar proxy signals for periods when no instrumental records survive.[21] The rapidly growing precision and detail of the proxy data is mind boggling, and we will offer more details on their testimony below. Historians of the ancient world can be grateful for the proliferating new data and their new insights about the environmental context of ancient economies, societies, and polities.

This chapter tries to frame our current state of knowledge about the physical climate in the period of the Roman Empire's expansion, flourishing, and final fragmentation, ~200 BCE to 600 CE. The emphasis here is on the new evidence of paleoclimate proxy data. We will explore what it is starting to tell historians about the timing and nature of large-scale climate change in the centuries of interest. A final section draws together the disparate sources of evidence into a tentative narrative, highlighting the questions that can be asked about the relationship between climate change and historical change and underscoring the need for more and better data to fill in such a narrative in the future.

Roman Geography and Climate

Both in its scale and its internal diversity, the Roman Empire was an extraordinary geographical entity, a fact obviously relevant to its environment and worth emphasizing in this context. It stretched across three continents surrounding an internal sea, *mare nostrum*. From west to east, it spanned from longitude ca. 9° W to ca. 38° E, excluding its Mesopotamian experiments. Its northernmost reaches were above 56° N, and its southern tip along the Nile was below 24° N. This latter spectrum is particularly exceptional, given that meridional gradients are far more important than zonal ones in determining climate factors. As Walter Scheidel has recently pointed out, "of all the contiguous empires in premodern history, only those of the Mongols, Incas, and Russian czars matched or exceeded the north-south range of Roman rule."[22] The Mediterranean core of the Empire, moreover, sat in one of the globe's more complex climatic regions. The circum-Mediterranean territories are poised at the intersection of the subtropics and mid-latitudes, subject to a diverse array of climate processes; furthermore, the Mediterranean region's highly variegated orography combines with the dynamics of a giant inland sea (ca. 2.5 million km^2) to sharpen microregional differences.[23] In important ways, this diversity acted as an integrated empire's buffer against short-term climatic stress.

The intricate local fabric of sea and mountain shaped the climate of any given point in the Mediterranean, but Mediterranean microclimates were not timeless features of the landscape. Besides the obvious high-frequency variability that drove sharp seasonal changes, there have been more irregular, low-frequency patterns of change on annual, decadal, and longer time-scales. The strongest pattern of large-scale climate variability in the Mediterranean, especially the western and northern Mediterranean, is described by an index known as the North Atlantic Oscillation.[24] The NAO is defined by the atmospheric pressure differences between a persistent zone of high pressure around the Azores (the "Azores high") and a zone of low pressure around Iceland (the "Icelandic low").[25] In a positive mode of the NAO, the strength of the westerlies increases and drives storm tracks to the north, bringing more precipitation to the North Atlantic and Central Europe. The diversion of moisture toward the north, and the blocking properties of the anticyclonic Azores high, mean that a positive NAO mode can be associated with less precipitation, even drought, over the western Mediterranean. And though the links are weaker, a positive NAO can display the opposite effects in the eastern Mediterranean, resulting in more precipitation.[26] In a negative NAO mode, the westerlies track directly across the western Mediterranean, bringing more precipitation to Morocco, Iberia, and Italy.[27]

The controlling mechanisms in the southern and eastern parts of the Mediterranean are more varied. In general, as one moves further south, aridity (caused by the belt of subtropical high pressure) becomes a greater threat, and landscapes turn from predesert into desert. The all-important winter rains are the result of a southward shift in the Intertropical Convergence Zone and the consequent cyclogenesis in low-pressure zones across the region.[28] At a true crossroads of global climate, the eastern territories of the Roman Empire were influenced by a number of large-scale climate mechanisms, including but not limited to those indicated by the North Atlantic Oscillation, the North Sea–Caspian Pattern (NCP), the Asian and African monsoon systems, and even the Pacific's El Niño Southern Oscillation (ENSO).[29] The NCP is an upper atmospheric teleconnection that strongly affects winter weather; in its negative mode, circulation drives air from southwest to northeast across the eastern Mediterranean, while in its positive mode, circulation moves from northeast to southwest.[30] The NCP can be a particularly important determinant of hydroclimatic patterns in the eastern Mediterranean, and unlike the NAO, it can account for patterns that diverge in Greece and Turkey, on the one hand, from the Levant, on the other.[31] The vast and integrated food system of the Roman Empire also depended on the breadbasket of Egypt, where the hydrology of the Nile shaped agriculture; the flow of the Nile was determined by precipitation in central Africa and the Ethiopian highlands, influenced in turn by the

Indian Ocean monsoon systems.[32] This southern arc of the Roman Empire, so important to food production, reacted to climate mechanisms of the southern hemisphere, and thus varied independently of the Atlantic and northern climate mechanisms that affected the rest of the Empire.

The climate of the Roman Empire, then, must be understood as the intersection of local and global factors and as the interplay of stable and dynamic features. Historians of climate, from Ellsworth Huntington to H. H. Lamb and Emmanuel Le Roy Ladurie, have long stressed the importance of climate change for the course of human civilization.[33] During the later Pleistocene, our human ancestors experienced wild swings of climate that brought on ice ages, followed by periods of rapid melt. The Pleistocene climate has been called a "flickering switch," and its huge oscillations were determined principally by celestial mechanics, slight changes in the parameters of the earth's spin and tilt and elliptical orbit around the sun, that entailed massive changes in the amount of energy arriving in the atmosphere. By contrast, the Holocene (ca. 9700 BCE–present) has been characterized by its relative stability and congeniality for human modes of subsistence. But it has become increasingly apparent that even the warm and stable Holocene witnessed continuous and meaningful climate change at various scales.[34] Orbital forcing has continued to operate across the millennia of the Holocene, and within the earth's complex systems, even gradual changes can produced jolting, nonlinear effects. Moreover, on decadal to centennial timescales, several other sources of significant climate forcing are becoming clearer. Changes in the level of solar activity can influence total solar irradiance on earth.[35] Volcanism can affect atmospheric chemistry and emit aerosols that reflect solar radiation back into space.[36] Besides these external forcing mechanisms, the internal dynamics of the ocean-atmosphere system itself can be responsible for climate change on various timescales.

Proxies are natural archives that allow us to extend our knowledge of past climate back in time beyond the beginnings of the instrumental record. Some proxies signal possible forcing mechanisms, such as solar activity or volcanism; others can furnish information about climate parameters like temperature and precipitation. Of critical importance is the scale of resolution. Tree rings, some ice cores, and some lake varves, for instance, allow annual or even subannual resolution and are highly useful for reconstructing climate on timescales of interest to historians of classical antiquity or the middle ages, especially when the years can be absolutely dated, as is generally the case with tree rings; by contrast, proxies like borehole temperatures or marine biota generally reflect only much less finely resolved climate indices, and so may help contextualize late Holocene climate or clarify long-term connections, but furnish little in the way of useful data for historians interested in annual, decadal, or sometimes even centennial timescales.

For almost as long as humans have been writing, our species has recorded observations about environmental phenomena.[37] Such observations

are filtered by the culture of the observer—the assumptions and expectations that implicitly frame his or her view of the world. Often, human observations record anomalous rather than ordinary events. Aside from the Nile
river floods, antiquity has only rarely left records in series and quantities that
invite a robust and independent construction of particular phenomena, like
the wine harvests of medieval France and the light they shed on hot or poor
summers. But, with care, even this imperfect record can be mined for meaningful information and patterns.[38] Indeed, comparison of written and natural
archives validates the idea that *both* preserve significant evidence about past
climates, in ways that are mutually enlightening, challenging, and at times be-
fuddling.[39] Human records are a source of evidence like any other, requiring
cautious interpretation and integration with other streams of data.

The least ambiguous high chronological resolution comes from the dendrodata, the annual rings that mark the growth of trees, in cases where dendro-
ecologists have been able to construct robust, highly replicated, continuous
regional series of overlapping tree rings back to the last centuries before the
turn of the era.[40] While very large sets of carefully vetted and statistically
tested tree ring series can yield satisfactory reconstructions of precipitation
patterns in the seasons of key growth—but not necessarily outside of those
months—seasonal temperature records can also sometimes be deduced from
the special case of trees whose exposed and stressed location makes them
particularly sensitive to temperature shifts.[41] For the ancient world, the most
extensive dendrodata sets come from northwest Europe; comparably scaled
dendrodata sets remain vanishingly rare closer to the Mediterranean heartland, although efforts are underway to improve the situation for the northern,
southern, and eastern rims of the Mediterranean. This limitation is all the
more serious given the highly localized nature of much precipitation.

The chronological accuracy of the dates assigned to annual layers in the
polar ice cores has been subject to more controversy, but recent advances are now
bringing the Greenland cores into a newly tighter alignment with other annual
chronologies before the fourteenth century CE.[42] The foundational evidence
of GISP-2 (Greenland Ice Sheet Project 2) and GRIP (Greenland Ice Core
Project), along with the newer NEEM (North Greenland Eemian Ice Drill-
ing) core, will become even more powerful in reconstructing major features of
the climate history of the northern hemisphere over the last 100,000 years.
However, the Greenland proxies have the limitation of instructing us first and
foremost on conditions in Greenland, from which paleoclimatologists must extrapolate to conditions in western Eurasia and the Mediterranean. An ongoing
project that uses new technology to extract incomparably highly resolved data
from a new ice core drilled on the border of Switzerland and Italy promises to
shed new light on Mediterranean climate signals from the heart of Europe,
and to open new avenues to correlating those signals with the data from Greenland as far back as the first millennium before our era.[43]

Since volcanism and solar activity are the principal forcing mechanisms on the historian's timescale, we should begin exploring the context of Roman climate by focusing on climate forcing in the late Holocene, before turning to the evidence for climate itself in various parts of the Empire. It has long been obvious that changes in orbital parameters, describable simply by celestial mechanics, coincide with significant reorganizations of the Earth's climate system. However, in recent decades, smaller and lower frequency fluctuations in the output of solar irradiance have come to be seen to affect the climate system too.[44] An extreme example is the Maunder Minimum. The virtual absence of sunspot activity from ca. 1645 to 1715 CE, coincides with anomalously low temperatures. A good deal of recent work supports the theory that variable solar activity has profound influence on the climate.[45]

Variability in solar activity is known from almost four decades of instrumental observation; it can be reconstructed for roughly the last four hundred years through records of the direct observation of the number of sunspots. But proxy evidence tracks such variability back much further. As galactic cosmic rays enter the Earth's atmosphere, they produce cosmogenic radionuclides like ^{10}Be and ^{14}C; greater solar activity reduces the energy of galactic cosmic rays as they pass through the heliosphere on their way to Earth, reducing the production of isotopes like ^{10}Be and ^{14}C.[46] There is thus an inverse relationship between the level of solar activity and the production of radionuclides. Since these radionuclides are deposited in physical archives such as ice cores and tree rings that can be dated, they offer a record of solar activity stretching back thousands—and in the case of ice cores, tens of thousands—of years, often with a high degree of resolution.

Efforts to reconstruct total solar irradiance over time have relied on natural archives of ^{10}Be in ice cores and ^{14}C in trees. Each has advantages and disadvantages relative to the other. The production of ^{10}Be is a highly sensitive measure of solar output, and the processes that lead to its deposition in laminated ice sheets are fewer and relatively simpler than those that lead to trees' ^{14}C radiocarbon records.[47] However, ^{10}Be models must rely on data from a handful of deep ice cores. Radiocarbon measurements from trees have the advantage of being more numerous and geographically widespread; they also benefit from the exceptional chronological precision of tree ring data. However, the mediating influence of the cycle by which carbon circulates through the atmosphere, oceans, plants, and so on, requires a greater number of assumptions and a larger number of variables in the model.[48] Fortunately, there is strong agreement about the basic timing of the most important shifts in solar activity for our period. The Roman Empire grew and flourished during a time of relatively high and stable solar activity between two significant grand solar minima in the middle of the fourth century BCE and the seventh century CE.[49] Viewed against the entire Holocene, the Roman Empire expanded and prospered most greatly during a phase of notable stability.

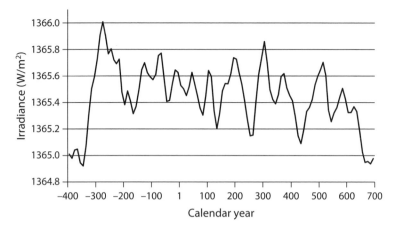

FIGURE 1.1. TSI (Total Solar Irradiance) from [14]C. Source: Data from https://www2.mps.mpg.de/projects/sun-climate/data/tsi_hol.txt.

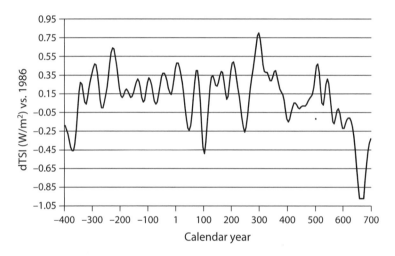

FIGURE 1.2. TSI from [10]Be. Source: Data from ftp://ftp.ncdc.noaa.gov/pub /data/paleo/climate_forcing/solar_variability/steinhilber2009tsi.txt.

It is worth looking at efforts to reconstruct total solar irradiance and zoom in on the Roman period. Figure 1.1. presents a record of solar irradiance from [14]C. A model based on ^{14}C clearly shows the minima of the fourth century BCE and the seventh century CE as well as the elevated and stable levels of solar activity in between.[50] However, it is worth noting possible downturns in solar output in the early second and particularly in the middle of the third and fifth centuries. The fourth century, by contrast appears as a period of high solar activity.

A model based on [10]Be from ice cores, presented in Figure 1.2, shows important similarities in the big picture but also potentially interesting differences within the Roman period.[51] The early second century decline in solar activity is more pronounced; the drop in the middle of the third century is again notable; the fourth century increase is even more dramatic, especially in the first part of the century.

Volcanism is the most clearly understood mechanism of short-term climate forcing in the Holocene.[52] Proxy evidence lets us reconstruct its rhythms into the deep past. Volcanic eruptions cast aerosols sometimes even into the upper atmosphere, where they can block solar radiation, the major source of warmth on earth. Sulfate particles precipitated by the volcanic emissions fall back to earth and become deposited in annually laminated ice sheets. Ice cores can thus preserve highly resolved records of violent volcanic events. Major volcanic eruptions can trigger short-term cooling episodes, sometimes severe enough to incite extreme weather in the following years. Such episodes can be associated with economic and political instability, as has been substantiated for the Carolingian period.[53] Moreover, because of complex feedback effects, it is possible that volcanic events play a role in more profound and systemic climate change. For instance, it has been argued that a period of intense volcanism helped incite the onset of the Little Ice Age.[54] By contrast, most of the Roman period stands out as a phase of moderate to low volcanic activity—with the notable exception of the Late Antique Little Ice Age, whose sudden onset in the spring of AD 536 coincides with a remarkable spate of powerful volcanic eruptions.[55]

The level of volcanic sulfates reconstructed in the GISP2 record from Greenland, presented in Figure 1.3, shows the basic stability of the Roman era, with the exception of a major event in the first century BCE.[56] It bears noting that the relevant segment of the GISP2 core was damaged and offers no evidence for most of the sixth century C.E.

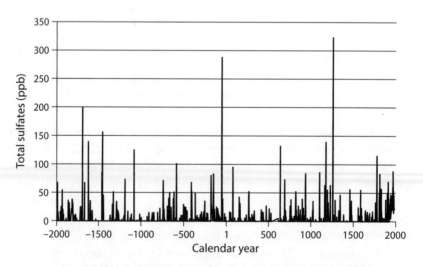

FIGURE 1.3. Volcanic sulfates: GISP2. Source: Data from ftp://ftp.ncdc.noaa .gov/pub/data/paleo/icecore/greenland/summit/gisp2/chem/volcano.txt.

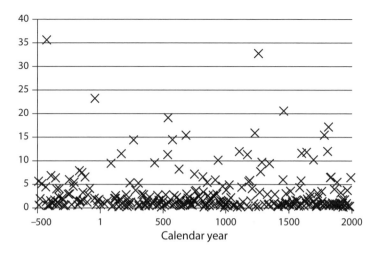

FIGURE 1.4. Estimated global volcanic forcing (negative watts per square meter). Source: Sigl, et al. 2015: Data File, "Volcanic Forcing," col. G.

The GRIP record displays a similar pattern.[57] So too does a more robust reconstruction of global volcanic forcing from multiple calibrated ice-core records recently published by Sigl and colleagues and presented in Figure 1.4.[58]

A study incorporating tree ring evidence with the ice-core record substantiates and enriches the picture of major volcanic events in the past. Using a series of Bristlecone pines from western North America, whose growth is sensitive to the length of the warm season, Salzer and Hughes identified coincidences between volcanic signatures in the ice cores and growth minima in the dendrochronological evidence.[59] Again, the broad absence of major events in the period of the Roman Empire's maximal expansion and prosperity is striking, as shown in Figure 1.5.

The proxies for solar activity and volcanism only tell us about the history of climate forcing mechanisms in the Roman period on a global scale, not the climate itself. One way to begin to approach the paleoclimate itself is through multiproxy temperature reconstructions. There have been many attempts to synthesize proxy data to model the history of global or hemispheric temperature change over the late Holocene. One of the most thorough and impressive is Christiansen and Lungqvist's recent 2000-year reconstruction of Northern Hemisphere temperature, using 91 high-resolution proxies from the extratropical latitudes.[60] Twenty-six of these proxies stretch all the way back to the first century AD. They used only proxies with highly significant correlation to temperature over the calibration period, 1880–1960 CE. Figure 1.6 offers a 50-year smoothed graph of the temperature anomalies (compared against the calibration period). The period of the Roman Empire is notable for its stability and relative warmth. Northern Hemispheric temperatures for the first two and a

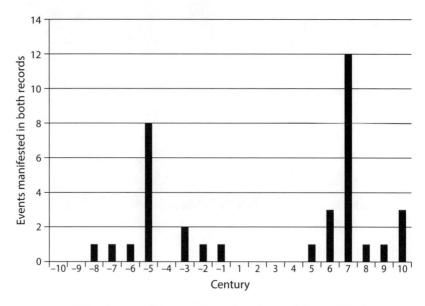

FIGURE 1.5. Volcanic events: ice core and tree rings. Source: Salzer and Hughes 2007.

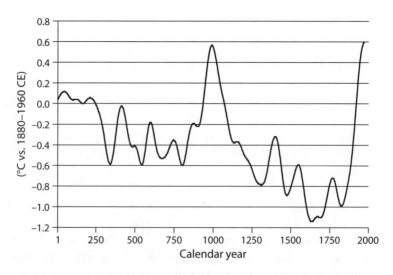

FIGURE 1.6. Temperature anomaly. Data from ftp://ftp.ncdc.noaa.gov/pub/data
/paleo/contributions_by_author/christiansen2012/christiansen2012.txt.

half centuries were at or slightly above those from 1880 to 1960, and strikingly
stable. Around 250 CE begins a phase of instability and general cooling that
lasts until the Medieval Climate Anomaly.

Alpine glaciers furnish another record of climate conditions in regions adjacent to the heart of the Roman Empire.[61] The retreat and advance of glaciers
(in the best cases dated by tree ring series or more frequently by other geophys-

Table 1.1. Physical characteristics of Alpine glaciers

Glacier	Length (km)	Location (N, E)	Terminus (meters asl)	Mean slope (°)	Max thickness	Volume Response Time
Great Aletsch	23.6	46.43, 8.06	1649	14.6	800	Slow (~80 years)
Gorner	13.4	45.95, 7.80	2173	18.7	450	Slow (~60 years)
Mer de Glace	11.6	45.87, 6.92	1531	19.3	380	Fast (~38 years)
Gepatsch	7.5	46.87, 10.75	2140	12.8	210	Fast (~26 years)
Lower Grindelwald	8.0	46.60, 8.06	1376	23.1	230	Fast (~22 years)

ical signatures of trees and moraine debris, particularly the less precise dating by radiocarbon) stem from a complex combination of climate, topographical, and glacier parameters.[62] Winter precipitation and summer temperatures (particularly the latter) are the main controls. Glaciers behave differently depending on their physical characteristics such as slope and depth, so that some glaciers respond more quickly to variation. In recent decades, Alpine glaciers have been extensively studied. The general convergence of the timing of advances and retreats underscores the quality of the data and the broad, regional value of glacier evidence for the study of the paleoclimate. A recent study by Le Roy has helpfully aggregated the physical characteristics of the most intensely studied Alpine glaciers (Table 1.1).[63]

These European glaciers all tell a similar story. A major glacial advance in the archaic Greek period climaxed sometime between 600 and 500 BCE. Thereafter, the ice generally retreated for hundreds of years, reaching a low point under the high Roman Empire. The big, slow Great Aletsch may have reached or shrunk beyond its twentieth-century limits in the early imperial period, pointing to significantly warmer summers. Clear signs mark glacier advance in the second half of the third century CE. The Great Aletsch was expanding by 272 CE, and the newly published data from the Mer de Glace in the Mont Blanc Basin show advance by 287 CE.[64] The fourth-century CE signal is more complex, though it seems clear that at some point in the middle of the century retreat preceded renewed advance in the fifth century CE. The Mer de Glace—fast reacting and well dated—shows a possible peak between 337 CE and the middle of the century, and then *significant* lowering in the space of just a few decades, reaching 1990s levels by 400/402 CE, before expanding rapidly again over the fifth century CE. The forthcoming publication of the Bossons glacier, also in the Mont Blanc Massif, promises to enrich our understanding even further. Sometime in the late sixth or early seventh centuries, Alpine glaciers reached their first millennium maxima, in the newly identified Late Antique Little Ice Age (LALIA).[65]

Reconstructions of forcing mechanisms like solar activity and volcanism from high-latitude ice cores provide generally high chronological resolution but are hemispheric or global in their signal; multiproxy temperature reconstructions are generally superregional and low resolution, but they have the advantage of being robust; glaciers are localized but offer only low chronological resolution. For locally specific *and* chronologically high resolved information, historians of climate turn to tree rings, speleothems, and annually deposited sedimentation stratigraphies like lake varves. Tree rings have traditionally enjoyed pride of place in paleoclimate studies, in the first instance because of the remarkable precision of the chronologies they offer.[66] They have also been crucial in providing a radiocarbon record going back several thousand years. At the same time, because tree biology, local ecology, and environmental effects on tree growth are complex, the quality of the tree-ring data as a proxy for climate parameters depends heavily on sufficiently high replication rates of trees from the same region for each year, and on how well the controls on tree growth are understood.

The robustness, richness, and precision of the tree ring data is the good news; the bad news is that there are, at present, no continuous dendrochronological records from the Mediterranean stretching back to the Roman period. The most relevant data are two long-term tree ring records from the northern arc of the Roman Empire.[67] A series of 7284 oak trees from NE France, NE Germany, and SE Germany allows reconstruction of April-May-June precipitation patterns over the last 2400 years by comparison with a calibration period of 1901–1980. While these trends may say little about precipitation in the Mediterranean, they certainly reflect important environmental conditions in the northwestern provinces of the Empire. The overall picture, presented in Figure 1.7, is one of above-average and broadly stable precipitation levels across the central Roman period (with some possibly meaningful short-term downturns around 40 BCE and 81 CE); from the early third century CE there is a secular decline, with a strong rebound in the fourth century lasting deep into the fifth; the later fifth and sixth centuries CE are marked by relative dryness and sharp instability.

The same study reconstructed summer temperature (June-July-August) from a series of high-elevation trees in the Alps, including 1089 stone pine and 457 European larch.[68] In this case, the controls on growth were reconstructed from correlations with instrumental data stretching from 1864 to 2003. Most important, the tree growth signals correlated strongly with *Mediterranean* temperatures, so that the evidence is of more than local value. This data was powerfully reinforced by a more recent tree ring series of 660 Siberian larch from the Altai mountains, extending from 104 to 2011 CE, scaled against Central Asian summer temperatures from 1900 to 2003. The reconstructed

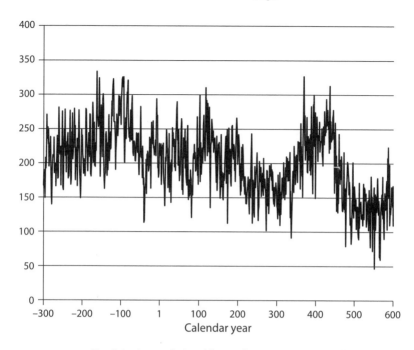

FIGURE 1.7. Precipitation totals (mm) in Northeastern France, Northeastern and Southeastern Germany. Source: Data from Büntgen et al. 2011.

summer temperatures show remarkable agreement with the Alpine data.[69] The Alpine data suggests broadly warmer summers in the period of the late republic, with a cold phase in the 40s BCE. The early Empire is generally warmer; both series show a cooler phase in the late second century that began a century of sharp oscillations. (See below, where we support calling the period 200 BC–150 AD the "Roman Climate Optimum" [RCO]. Sometimes called the "Roman Warm Period" in the literature and inconsistently defined in time, there is a strong case to be made for these three and a half centuries as a period of warm and favorable climate in the circum-Mediterranean). The evidence of tree growth suggests cooling in the first half of the fourth century and, in the Alpine record, gradual warming over the next century and a half before both series signal some of the coldest temperatures in the entire record in the sixth century, especially the 530s and 540s CE. These data are presented in Figure 1.8.

The sharp cooling of the 530s and 540s immediately invokes the much-discussed 536 CE dust veil event.[70] Several credible ancient reports describe in detail an extraordinary and prolonged atmospheric event lasting from the spring of 536 CE to the summer of 537 CE.[71] Procopius (the most important Greek historian of the sixth century) portrays something like an eclipse, Cassiodorus (a western statesman and author of the same period) a sun without brightness and a summer without warmth.[72] In the 1980s, Stothers and

Rampino argued for a volcanic origin of the event, and since then a considerable bibliography has accrued around the subject.[73] The 536 CE event is detectible in the ice-core evidence and a wide range of tree rings around the world; new evidence and a concomitant recalibration of the chronology of the high-latitude ice cores now indicate that it was a northern hemisphere eruption, followed a few years later by a massive volcanic event in the tropics in 540, followed by yet another large-scale event in 547 CE.[74] These events usher in the Late Antique Little Ice Age (LALIA: see below). A connection with the emergence of the first bubonic plague pandemic around the Mediterranean basin in 541–542 CE has seemed an intriguing possibility, although the mechanics of the link have never been satisfactorily explained.[75]

Another valuable and sometimes fairly high-resolution proxy is the stable isotope composition of speleothems. Deposits like stalagmites are formed over time by the accumulation of minerals from drip waters in cave interiors.[76] Stable isotopes are naturally occurring variants of elements in the calcites that form the speleothem. The heavier versions of the element have additional neutrons (such as $\delta^{18}O$, a heavy form of oxygen, or $\delta^{13}C$, a heavy isotope of carbon). The proportion of heavier isotopes in a sample, expressed as parts per mil (‰), is sensitive to the overall climatic conditions under which the water was precipitated as well as the physical properties and processes of deposition at the site. When the local factors are well understood, the ratio of

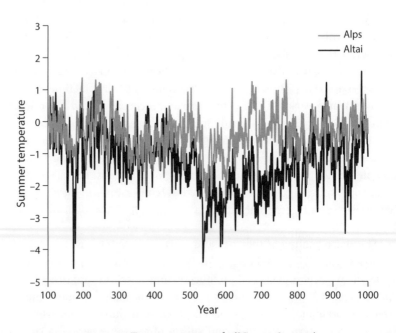

FIGURE 1.8. Temperature anomaly (°C vs. 1961–1990).
Source: Data from (Büntgen et al. 2016).

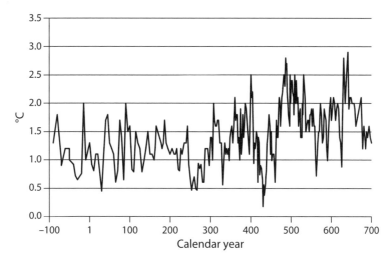

FIGURE 1.9. Temperature reconstruction from Spannagel Cave δ¹⁸O.
Source: Data from ftp://ftp.ncdc.noaa.gov/pub/data/paleo
/speleothem/europe/austria/spannagel2005.txt.

oxygen isotopes can at times furnish information about the overall temperature, source of the precipitated water, season of precipitation, and/or levels of
precipitation. The ratio of carbon isotopes, again when the local environment
and deposition processes are well understood, can furnish information about
the soil and vegetation above the cave, as well as temperature and overall precipitation. When the stable isotope ratios change over time, reflecting changes
in the climate, speleothems form a mineral archive of paleoclimatic data.

The challenge of using speleothems as climate archives lies in securing
a solid understanding of the complex local controls on isotope production.
Great care is required in accounting for the myriad factors that shape and may
distort the signals detected in the deposits in any particular cave. Moreover,
the chronological resolution differs widely, from subannual to centennial (the
table below reflects the average time slice between signals in each record).
But the abundance of karstic topography in the circum-Mediterranean regions
promises a large and growing array of climate data from caves. Compared
to the present paucity of Mediterranean dendrochronologies of comparable scope, at least a dozen speleothem series stem from the territories of the
Roman Empire and stretch back two millennia, as registered in Table 1.2.

The Spannagel cave speleothem from high in the Austrian Alps offers one of
the most important precisely dated and high resolution isotope proxies.[77] A
temperature reconstruction from δ¹⁸O levels, presented in Figure 1.9, shows a
warm and relatively stable climate optimum enduring into the third century,
and sharp variability thereafter. The reconstruction shows a general warming
trend during the fourth century, followed by sharp cooling with the coolest

Table 1.2. Speleothem series

Cave	Location	Resolution	Signal	Source	Paleoclimate Reconstruction
Spannagel (Austria)	47.08N, 11.67E	Near annual	Temp. from $\delta^{18}O$	Mangini, Spötl, and Verdes 2005; Vollweiler et al. 2006	Warm and relative stability until mid-3C CE, then sharp variability; warm ca. 300 CE then cooling; warm ca. 400 CE then sharpest cooling; late 5C CE maximum warmth; 6C CE cooler but still relatively warm
Sofular (Black Sea coast, Turkey)	41.44N, 31.96E	ca. 5.4 years for Holocene, 8.3 years for Roman period	Effective moisture from $\delta^{13}C$	Göktürk 2011; Fleitmann et al. 2009	Holocene moisture peak at 5C BCE thereafter drying; sharply drier ca. 350–450 CE; much moister 450–600 CE
N Spain	~42.79N, 4.26W	Decadal	Temp. from $\delta^{13}C$	Martín-Chivelet et al. 2011	Warm and stable from 200 BCE–200 CE; cold until 450 CE; thereafter warming
Kapsia (Peloponnesus, Greece)	37.62N, 22.35E	~10 years	Precipitation amount from $\delta^{18}O$ and $\delta^{13}C$	Finné et al. 2014	Wet conditions around 160–300 CE; 4C CE wet; dry conditions 1–150 CE and 500 to 700 CE
Klapferloch (Austria)	46.95N, 10.55E	Near annual	Precipitation amount from $\delta^{18}O$ and $\delta^{13}C$	Boch and Spötl 2011	Wet until 4C CE; major oscillations after 350 CE; extremely dry ca. 450–600 CE
Kocain (Southern Anatolia)	37.23N, 30.71E	2.3 years	Winter temperature from $\delta^{13}C$	Göktürk 2011	Cool (vs. whole series mean back to 3700 BCE) 800 BCE to 200 CE; warming 200–500 CE; cooling 500–600 CE; warming afterward

Site (location)	Coordinates	Resolution	Proxy / method	Reference	Findings
Uzuntarla (Thrace, Turkey)	41.59N, 27.94E	2.6 years	Temp. and seasonality of precip. from $\delta^{18}O$; temp. and moisture from $\delta^{13}C$	Göktürk 2011	Roman warm period somewhat warmer; higher $\delta^{18}O$ due to temp. increase "partly compensated by enhanced winter precipitation" Warming 100 BCE–200 CE; cooling from ca. 220 CE to 420 CE; warming to ca. 600 CE; esp. dry late 2C CE and ca. 500 CE
Grotto Savi (Trieste, Italy)	45.61N, 13.88E	18–25 years	Temp. from $\delta^{18}O$	Frisia et al. 2005	RCO peaks 300 BCE–1 CE, similar to present (~2005 CE); cooling in late 2C and late 3C CE; cooling ~450–700 CE; 650 CE coolest until second-millennium "Little Ice Age"
Ceremošnja, Serbia	44.64N, 21.56E	Unclear	Temp. from $\delta^{18}O$	Kacanski et al. 2001	Warm until ca. 50 BCE; cool thereafter
Soreq, Israel (speleothem sample 2–6)	31.45N, 35.03E	Decadal-centennial	Moisture from patterns in $\delta^{18}O$	Schilman et al. 2002; see also Bar-Matthews and Ayalon 2004; Orland et al. 2009	Increased aridity 100–700 CE
Buca della Renella (La Spezia), Apuan Alps, Italy)	44N, 10E	Centennial	Raw $\delta^{18}O$ and $\delta^{13}C$ record	Drysdale et al. 2006	Drier in 1–5C CE, peak early 5C CE

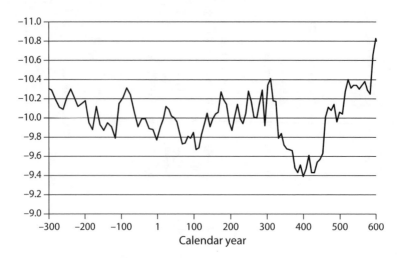

FIGURE 1.10. δ^{13}C from Sofular Cave. Source: Data from ftp://ftp.ncdc
.noaa.gov/pub/data/paleo/speleothem/asia/turkey/sofular2009.txt.

temperatures coming in the early fifth century. Around 500 there was another peak of warmth before cooling again into the sixth century.

The Sofular Cave in northern Turkey offers a valuable relatively highly resolved precipitation record from changing δ^{13}C ratios.[78] The series, presented in Figure 1.10, reflects primarily the climate system of the Black Sea and secondarily Mediterranean circulation. Notably, a precipitation peak around 300 CE turns toward aridity, with a dry peak around 400 CE, and then significantly more precipitation reaching new levels in the sixth century. Although not a direct Mediterranean signal, the Sofular record, with its clear and closely studied pattern, may ultimately provide critical information about climate change in late antiquity.

Finally, we can consider the wide-ranging value of lakes as stores of evidence about the paleoclimate.[79] As is evident from Table 1.3, lacustrine data has the advantage of being relatively abundant and geographically dispersed around the Mediterranean. Lakes also provide multiple kinds of evidence about climate parameters. Lake levels, a register of precipitation runoff or evaporation, can be reconstructed through sedimentation analysis or other archaeological indices. Sedimentation layers can also preserve valuable evidence in the form of stable isotopes and pollen. At the same time, lakes present severe challenges for paleoclimate reconstruction. Understanding the climate controls on the proxy is difficult; especially in the late Holocene, anthropogenic forces can influence hydrological systems or land-use patterns in ways that overwhelm the hydroclimate signal. Moreover, resolution and precision are issues for historians, as lacustrine data series are often too coarse or imprecise to furnish information at annual, decadal, and centennial scales. However, in some important cases, varved deposits provide a high resolution and chronologically precise source of paleoclimate data.

Table 1.3. Lake records

Site	Location	Signal	Source	Paleoclimate Reconstruction
Lake Van (Eastern Turkey)	38.75N, 42.90E	Pollen, isotope data to reconstruct precipitation	Wick, Lemcke, and Sturm 2003	Humid peak ca. 200 BCE then drying to ca. 50 CE; increasing moisture to early 3C CE then gradual aridification to 700 CE.*
Dead Sea	31.60N, 35.47E	Lake levels from sedimentary records	Bookman et al. 2004; cf. Migowski et al. 2006	Wet 200 BCE–200 CE; dry 200–300 CE; wet 300–500 CE; dry 500–750 CE. The dating differs in Migowski 2006.
Bereket Basin (Southwest Turkey)	37.55N, 30.30E	Palynological and sedimentation record of precipitation.	Kaniewski et al. 2007	Dry 40 BCE to 450 CE, driest from 130–350 (+/- 40) CE; wet 450–650 CE.
Nar Gölü (Cappadocia)	38.34N, 34.46E	Moisture from diatom sequence, pollen record, and $\delta^{18}O$.	Dean et al. 2013; Jones et al. 2006; Woodbridge and Roberts 2011	Dry 300–560 CE, drought 440–540 CE; shift to wetter climate after 560 CE; increased snowfall 301–801 CE.

Table 1.3. (*continued*)

Site	Location	Signal	Source	Paleoclimate Reconstruction
Tecer Lake (Central Turkey)	39.43N, 37.08E	Moisture from mineralogical composition of sediment core	Kuzucuoğlu et al. 2011	70 BCE–250 CE warm and moist; 250–400 CE drying increased; 410–500 CE moist again; 500–540 CE high rainfall; 540–630 CE high water discharge to lake; 630–820 CE cool and humid.
Zoñar Lake (Southern Spain)	37.48N, 4.69W	Moisture from varve thickness, geochemical proxies.	Martín-Puertas et al. 2009; for Spain, see also Pérez-Sanz et al. 2013; Corella et al. 2010; Currás et al. 2012	650 BCE–350 CE most humid phase of last 4Kyr; drier 190 BCE–150 CE; wet 150–350 CE; drier after 350 CE.
Lake Shkodra (Eastern Albania)	42.08N, 19.47E	Moisture from pollen and $\delta^{18}O$.	Zanchetta et al. 2012	Wet 550 BCE–50 BCE; wet 150 CE–450 CE; wet 600–700 CE. Significantly drier at 100 CE and 550 CE.
Jura Mountains, French Pre-Alps, Swiss Plateau	~46.63N, ~5.91E	Lake level from sediment and archaeological dates.	Magny 2004; Magny 1993	150 CE–250 CE higher lake levels; thereafter much lower.

* We would like to note that Figure 7b in McCormick et al. 2012 has erroneously reversed the alignment of dry and humid, giving a misleading impression of the direction of change.

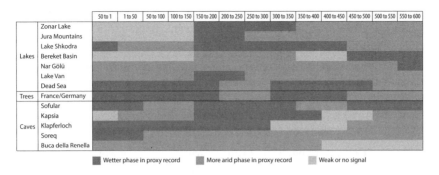

FIGURE 1.11. The complexity of Mediterranean hydrological change (50 BCE–600 CE).

The lake records underscore both the variability of the ancient climate as well as the complexity of Mediterranean hydrology. The most important signals are the generally high levels of moisture in the western and north-central Mediterranean in the early imperial period. The Levant witnessed two significant centennial-scale oscillations, from humid to arid, during the long Roman period; the precise dating is still disputed, but it seems most plausible for now that the third and sixth centuries were phases of drought in the Near East. It is a possibility that the northeastern and southeastern Mediterranean diverged in late antiquity, with parts of Turkey becoming more wet and Palestine, more dry by the sixth century CE.[80]

The regional diversity of Mediterranean hydroclimate response is obvious, particularly when data series from tree rings, speleothems, and lakes are assembled, as in Figure 1.11. Clearly, climate change in the Roman Empire was complex, and reconstructing its contours will require sensitivity to the refraction of large-scale changes through local environments.

A Tentative Reconstruction

Given that most of these palaeoenvironmental records are only beginning to reach the times and places of greatest concern to ancient historians, the following portrait is inevitably partial and destined to undergo considerable revision and improvement. Nevertheless, we know enough to offer some general observations emerging from these early days of palaeoclimate studies of the ancient world, and to begin to present some of the ways that historians could consider environmental change as a factor in historical developments during the centuries of Rome's imperial project.

On a millennial timescale, the Late Holocene (ca. 2250 BCE–present) was a time of cooling. The Sahara and Near East became progressively more arid. The monsoons weakened, there were more El Niños, and the NAO index has tended to decline. In the Mediterranean, seasonal differences became more pronounced. But on shorter timescales, these patterns were variously accentuated, paused, or reversed. The period ~200 BCE to ~150 CE has come to be known as the "Roman Climate Optimum" (RCO), marked by stability, warmth, and precipitation. The phase of Roman expansion took place in the context of a warming climate. Both ^{14}C and ^{10}Be records point to a major solar minimum in the fourth century BCE. From the third century BCE, solar radiation reaching the earth's surface increases and reaches a plateau where it would remain relatively stable for the next thousand years. Volcanism, too, abates in the centuries of the late republic, only showing a meaningful spike around 44 BCE. The stability of the major short-term forcing mechanisms provided the framework for the RCO. These patterns are discernable in a range of written, archaeological, and proxy evidence from around the territories that during this period would become part of a politically and economically integrated empire.[81]

A strong signal for the background trend toward warmth in the Northern Hemisphere mid-latitudes is the prolonged glacier retreat that is evident in the Alps from the middle of the first millennium BCE and continuing until the third century CE.[82] The lengthy tree-ring series constructed for the Alps, which correlates strongly with Mediterranean temperatures, shows a general plateau of warm climate with the exception of a somewhat cooler period from ca. 50 BCE to 30 CE.[83] A temperature reconstruction based on speleothems from northern inland Spain shows warmth from 200 BCE lasting four centuries. The δ^{18}O record from the Spannagel Cave similarly shows warmth and relative stability, although not nearly such warmth as characterized the Medieval Climate Anomaly in the same speleothem. In general, it is apt that the RCO is also known as the Roman Warm Period, which seems accurate with respect to both hemispheric reconstructions as well as the local proxies (predominantly from the western Mediterranean).[84]

The evidence for hydroclimatic conditions during this period is more complex. The tree ring series from France and Germany show an especially moist phase until around 100 BCE, then a gradual drying trend until ca. 300 CE; this signal by no means necessarily reflects Mediterranean precipitation. In the Zoñar Lake record from Spain, the thousand years from approximately 650 BCE to 350 CE is the most humid of the last four millennia; however, the period from ~190 BCE to 150 CE shows a modest drying trend within this longer band of more humid climate.[85] The evidence for frequent flooding of the Tiber River in the late republic and early Empire may be one more indication of more humid conditions in the heartland of the Roman Empire. The overall climate mechanisms that could simultaneously produce more humid conditions in Spain, Italy, and north-central Europe remain to be illuminated, and the possibility of anthropogenic causes for some of the signals carefully sorted out.[86]

In the east the record is even more divergent and at present uncertain. It seems best to separate the evidence from the regions located in modern Turkey and Israel. In Israel, the two principal sources of data are the Dead Sea and the Soreq Cave. The Dead Sea shows wetter conditions from ca. 200 BCE to 200 CE, with a reasonable degree of certainty.[87] Reconstructions of the spe-leothem record from the Soreq Cave differ wildly. The analysis of Schilman and colleagues is probably most consistent with the evidence of Dead Sea lev-els; based on both the $\delta^{18}O$ record from the cave and from a quite different marine sample of planktonic foraminifera, it indicates an arid peak around 100 BCE and thereafter greater humidity.[88] In Turkey, the trend is broadly op-posite. The Sofular Cave record from northwest Turkey shows a gradual trend toward aridification ca. 300 BCE–100 CE, with a wet interval however around 100 BCE.[89] The Bereket Basin record from southwestern Turkey shows arid-ity from ~40 BCE for the next five centuries.[90] At Lake Van, well to the east, aridity peaks earlier, around 150 BCE, after which there is greater humidity.[91]

From about 150 CE, the climate appears less stable than in the preceding centuries in both parts of the Empire, and multiproxy indicators are some-times less consistent. Broader climate conditions seem to have impacted dif-ferently the eastern and western parts of the Empire. Multiple indicators point to cooling in the northwestern provinces of the Empire in the third century: solar activity indicates a cooling episode about 260 CE, while sea ice slowly increased off Greenland until ca. 290 CE, which is consistent with central Greenland temperatures.[92] A major volcanic event in 266 was the largest of the violent eruptions recently documented and redated between 169 and 304; all of them could have contributed to cooling.[93] Such rapid short-term changes would have had a great capacity to disrupt food production in the most diffi-cult decades the Roman Empire had faced so far, for the political, military, and monetary crisis peaked between ca. 250 and 280–290.[94]

The cooling certainly reached the western provinces of the Empire for it is unmistakable in the Alpine glaciers: the centuries-long retreat of the large and slow-reacting Great Aletsch glacier came to an end, and it was growing around 272 CE when it reached an extent comparable to ca. 1982 CE.[95] In the eastern Alps, the dendrodata indicate cooling setting in around 200; after a few warmer years from 221 to 231 and sharp cooling from 243 to 253, gradual cooling pre-vailed until it stabilized around 315 and shifted to warming around 365. This pattern is not inconsistent with the Spannagel speleothem record, although Spannagel suggests the cooling ended a bit earlier. Early summer precipitation in northeastern France and central Europe becomes exceptionally variable around 250 and would continue thus until about 650. The variability was initially ac-companied by a marked trend toward dry conditions that peaked around 300.

Multiple proxies agree that some warming occurred in the fourth century in the northwestern part of the Empire. The isotope values show that cen-tral Greenland experienced stable temperatures or gentle warming through around 375, punctuated by cool spells ca. 305 and 335.[96] The Austrian Alpine dendrodata is interpreted as indicating gradual cooling, but renewed warm-ing starts around 365.[97] The dendrodating of *in situ* trees proves that the Gepatschferner Glacier advanced around 335, reaching an extent similar to ca. 1930 CE. But glacier growth stops before 400; by around 400, the Lower Grindelwald was definitely in retreat, which appears to fit the warming signal from the Austrian dendrodata and speleothem.[98] In Britain, the warmth-loving nettlebug reappears outside its main twentieth-century range in fourth-century deposits.[99] Northeastern French and central European dendrodata indicate that dry conditions began yielding to relatively wet early summers after 300 and that wet summers persisted from 350 until about 450, when conditions were moister than under the early Roman Empire.[100] The written sources shed relatively little light overall, but they too suggest more frequent flooding in the second half of the fourth century.[101] On balance, the proxy data points to a fairly stable fourth century that warmed in the second half, at least in the northwestern provinces of the Empire.

In the eastern provinces, the Talmud mentions droughts in Palestine most certainly ca. 210–220, and, less compellingly, ca. 220–240 CE and 255–270; multiple historical records document a general drought that lasted from ca. 311 to 313.[102] The Dead Sea levels show a sharp drop in precipitation ca. 200 CE followed by a steep rise in precipitation lasting as much as 200 years, although radiocarbon dates for the return of wetter weather disagree. One study dates renewed moist conditions around 300 CE, which would fit archaeological features that have been attributed to the fourth century and show a substantial rise in the Dead Sea shoreline; springs in Roman Palestine may also have been more productive then.[103] Another investigation places the return of wetter weather around 400 CE, which would be consistent with the Greenland Cl- values that reflect longer summers and therefore less precipitation in the Middle East until about 400.[104] Although the numbers are small, the proportion of historical records of drought to reports of high precipitation tips toward precipitation between 375 and 475.[105] The Sofular cave also indicates a drying and/or cooling period in Asia Minor followed by a very long wetter and/or warmer era, although the dates assigned to both periods run a little later than the other proxies.[106] Whether wetter conditions returned to the eastern Roman Empire closer to 300 or to 400 CE, the eastern Roman Empire's prosperous fifth century coincided with increased moisture, precisely the most critical variable in its relatively arid environment.

We have observed elsewhere two major climate developments originating beyond Rome's frontiers that may have contributed to the environmental and political conditions that the Empire faced in late antiquity. The productivity of the Nile, whose annual floods reflect precipitation in Central Africa and especially Ethiopia, seems to have undergone a subtle but significant change in the second century CE. Down to 299, when the data become sparse, the overall proportion of better-than-normal floods appears nearly identical on either side of 155 CE.[107] However, the early Roman Empire enjoyed substantially more abundant and therefore agriculturally excellent floods than the later Roman Empire. Before 156 CE, 19% of floods were of the two most favorable categories, while only 7% fit that description after 155. Conversely, the three most serious categories of deficient floods occurred more frequently in the later period (31%) as opposed to the earlier Empire (21%). In other words, as the Empire reached its historical peak, and the great grain fleets sailed north every year to feed the capital and swell the cereal resources of the Empire, the resilience in Mediterranean food supplies owed to Egypt's productive farms was enhanced. After ca. 155 CE, the best harvests became substantially more infrequent, and worse harvests, more common, as the Empire struggled to face mounting political, military, and economic challenges even as its capacity for climate resilience was diminished. The written records suggest that unusually favorable climate conditions for Egyptian food production prevailed over the first two centuries of the Roman Empire, while the conditions underpinning food production appear to have been consistently less good from 156 to 299 CE.

The second climate development beyond the Roman frontiers that may have contributed to imperial difficulties in the fourth century concerns the succession of wetter and drier periods in Central Asia that could have affected nomadic groups whose expansion impinged on Eurasia's sedentary empires. The pastoral component of their economy likely made them sensitive to fluctuating patterns of precipitation, just as the Romans explicitly observed of Arab pastoralists in the Persian Empire.[108] Recent dendrodata indicate that the first half of the third century was wetter than average, and the second was marked by drought conditions from about 242 to 293, with intermittent returns to more normal conditions. But a severe drought lasted nearly forty years in the fourth century, one of the worst in 2000 years. Documented by the Dulan-Wulan tree ring chronology, prevailing drought conditions began in 338 CE and continued until 377, when wetter conditions returned. The extent of this drought in time and space has suggested that it could have played a critical role in driving the mobile pastoral federation that coalesced around the name of "Huns" somewhere east of the Don River to seek pastures and predation farther to the west and south; in this case, the dendrodata would confirm speculation about an environmental factor in the Hunnic invasion that goes back at least a century.[109] Historical sources indicate that the Huns had reached the Don River by the 370s, and crossed it around 375, driving the Goths settled in the area north of the Black Sea to flee into the Roman Empire and ultimately to attack it, destroying the Roman army in 378 at Adrianople (mod. Idirne, Turkey).[110]

Multiple indicators point to fifth- and especially sixth-century cooling. In fact, this cooling now appears strong and extensive enough to constitute a Late Antique Little Ice Age.[111] Solar activity indicates a cooling episode around the middle of the fifth century. Greenland sea ice also indicates cooling that peaked around 540 CE. In the western provinces, after the signs of Alpine glacier retreat—and therefore warming—already noted around 400 CE, fifth-century signals are less clear. Around 430, the slow-reacting Great Aletsch was still advancing.[112] Alpine tree rings suggest that summer temperature was neither especially warm nor cool in the fifth century, although it may have increased slightly over the century. The Austrian speleothem indicates cooling in the first half of the century, followed there too by warming in the second half of the fifth century. The fifth century now looks volcanically active as well, implying considerable potential for disruptive volcanic winters regardless of the overall temperature trend.[113]

Written sources across the entire empire document the seriousness of the veiling of solar radiation in 536 and 537 which caused crop failures in different areas.[114] Many northern European tree rings display a stress signal in these years, and the solar veiling and cooling have now been connected with three powerful volcanic eruptions separated by just a few years.[115] However, most revealing, the sharp drop in summer temperatures starting in Europe in 536 reconstructed from Alpine tree rings has now been documented by a robust new tree ring record from central Asia. By this evidence Alpine summer temperatures cooled in the 540s by an estimated *average* of −1.9° C; in the Altai mountains, summer temperatures dropped simultaneously by a reconstructed average of −3.2° C. Consonant with the solar veiling and the onset of the Late Antique Little Ice Age, the Greenland sea ice evidence for cooling peaks around 540. What is more, there was glacier advance in both the French and the Swiss Alps.[116] The Lower Grindelwald Glacier (Swiss Alps) advanced around the mid-sixth century.[117] It now appears that glacier advance began in the fifth century and reached its first-millennium maximum in the seventh century. The sharp cooling across western Eurasia would prevail at least until ca. 660, and provides new insight into the turmoil that arose in this era as Asian and northern peoples impinged on the sedentary empires of Rome, Persia, and China.[118]

Against the backdrop of overall conditions that around 450 had shifted to drier in northeastern France and Central Europe, historical records document serious floods indicating a relatively wet final quarter of the sixth century in Italy and Gaul.[119] In seeming contrast to broader dry conditions in France, the increase in eastern precipitation and generally humid conditions in the eastern Roman Empire resumed or continued. The longer winters over Greenland in the fifth and sixth centuries are consistent with increased precipitation in the Levant, and we have seen that Dead Sea levels testify to a steep increase in precipitation that lasted about two centuries and started either around 300 or 400. In the eastern written records from ca. 375 to 475, precipitation events also outpace drought reports. That changes dramatically around 500. In the early sixth century, in addition to reports of sharp cooling and crop failures at the time of the 536 event, mentions of eastern droughts and unusual, especially unseasonable heat events exceed precipitation reports, notably the long drought that affected Palestine from 523 to 538, when even normally reliable springs dried up. Contemporaries explicitly connect a 536 drought with the migration into the empire of Arab pastoralists.[120] As we noted earlier, Roman water works in Palestine appear more frequent in this period.[121] Summer water shortages that affected Constantinople in the 520s could also reflect a decline in precipitation compared to the fourth and fifth centuries during which the capital's great aqueducts had been constructed.[122] The two centuries of favorably wet Levantine conditions documented by Dead Sea levels came to an end at some point in the sixth century, and arid conditions seem to have persisted through the eighth century.[123] Further south and east, however, it is possible that the cooler summers of the LALIA worked different effects. The Altai record seems to correlate positively with summer temperatures in parts of the Arabian peninsula. If this new correlation should be confirmed and more fully understood, one would expect cooler summers in Arabia to correspond with possible increases in precipitation, and surely with decreased evapotranspiration, with a resulting increased hydrological balance. That in turn might imply increased scrub vegetation, precisely the main fodder for the camel herds that would play such a critical role in the sudden expansion of the new civilization launched from the Arabian peninsula.[124]

Conclusions

Rapidly accumulating new proxy evidence, deepening understanding of climate mechanisms and teleconnections, and renewed scrutiny of the historical and archaeological evidence mean that the picture we have just sketched is bound to undergo considerable refinement and improvement, particularly with respect to the highest-resolution proxy evidence for the presently underrepresented areas of southern Europe, the Levant, and northern Africa. Nevertheless enough is clear to make indubitable that, in antiquity, climate changed, sometimes swiftly and strongly, and that such changes likely elicited multiple and diverse human responses.

But we are only in the first stages of establishing the patterns of complex environmental changes, much less their impact on ancient society. The first task will be to expand the amount and precision of the paleoclimate data. The second, and even more challenging one, will be to seek rigorous and compelling understanding of the complex interactions between climate changes and ancient society's responses. A broad *association* between stable and generally favorable conditions and the period of the ancient economy's growth, and Roman imperial expansion, seems undeniable, as does an association of increasing climate instability, at least in some parts of the Roman world, with mounting challenges to the Roman imperial system. But the devil is in the details, details which must be worked out from an ever richer natural archive of paleoclimate records, a very large and growing but somewhat intractable set of archaeological evidence, and a nearly static set of written records.

This will require some sophisticated thinking about exactly how climate can affect an ancient economy. Food production is the first, obvious avenue of approach: Aggregate amounts of food production, certainly, but also changing organizational aspects of agrarian economy related to food production, such as changing environmental niches of food production.

Climate change likely affected the overall ecological community of living organisms. For instance, did drier periods contribute to contracting environments favorable to insect parasites such as mosquitoes, and therefore influence the rate of malarial infection in some zones? Or the converse? To take a specific example, how exactly could the drier springs and colder weather we think we can see in third-century Gaul connect to political, military, and economic trends? How might they have affected the domestic animals that powered the local economy? Did they change the expectations for housing construction—more central heating for the rich, more thatch for the poor—or for clothing norms?[125] Were precipitation changes strong enough to affect water tables, and thereby change which zones were most suitable for human habitation and agrarian economies as they were experienced by ancient cultural norms? Culture does indeed shape human water consumption even as biology determines the minimal threshold. Did drier climates challenge the profligate use of water (and wood) required by Roman bath culture? Surely there are grounds for seeing specific Roman and Byzantine aqueduct projects as responses to increasing aridity, underscoring the resilience of which ancient societies were sometimes capable.[126] Can climate scientists find ways of translating their data into measures that ancient historians and archaeologists can make direct use of? For example, regionally specific estimates of changing lengths, onsets, and ends of growing seasons provide readier materials for historical analysis than average temperature changes.

In order to accomplish the ambitious task of integrating the new environmental evidence into a broader explanation of ancient history, we will need greater critical understanding of ancient society's changing resilience to such challenges. It has been observed that the ancient Mediterranean's economic integration reduced the threat of famine, so long as it was possible to move food and, as it were, the virtual water that fostered the food that grew outside of areas stricken by drought.

In sum, in the last few years our knowledge of ancient environments and their potential impact on history has expanded near exponentially. But the next years will make clear just how big the challenges are to integrate the new evidence with our knowledge of ancient history to produce new, more rigorous understanding of the unfolding of the rise and fall of the ancient world in all its glorious complexities.

Notes

1. Grove and Rackham 2001; Horden and Purcell 2000; Hughes 1994; Sallares 1991.

2. E.g., Garnsey 1988, 1998, 1999.

3. E.g., Leone 2012; Shaw 1995.

4. Harris 2013 and 2011; Hughes 2011; Meiggs 1982.

5. E.g., Huntington 1917. See Williamson 2013, focused on Anglo-Saxon England, for the potential today.

6. Scheidel 2001 and 1996; Shaw 1996; Sallares 2002.

7. E.g., Febvre and Bataillon 1966 (orig. 1922); Le Roy Ladurie 1966.

8. For an overview, McCormick 2011.

9. For Mediterranean climate, see, in general, Lionello 2012; Xoplaki et al. 2004; Xoplaki 2002.

10. Harper 2015b; Brooke 2014; Butzer 2012; Caseldine and Turney 2010; McAnany and Yoffee 2010; Folke 2006; Chew 2002; DeMenocal 2001.

11. On types of climate change, see Bradley 2015, 14–16.

12. Erdkamp 2005, 225–237, on Egyptian grain in Rome.

13. Marriner et al. 2012; Eltahir 1996.

14. Garnsey 1988.

15. Rickman 1980; Mango 1985, 40 and 54–55; Keay 2012.

16. Leveau 2014; Theophanes, *Chronographia*, A.M. 6258, ed. C. De Boor (Leipzig, Teubner, 1883–5), 1.440.14–24 and Nicephorus I, *Short history*, 85, ed. and trans. C. Mango (Dumbarton Oaks Texts, 10, Washington, D.C., Dumbarton Oaks Research Library and Collection), p. 160.1–12.

17. McCormick et al. 2012; C. P. Jones 2007.

18. Lamb 1982, 156–169.

19. See below.

20. McCormick et al. 2012; Luterbacher et al. 2012.

21. For a case study of Irish chronicle sources and tree ring data, Ludlow et al. 2013.

22. Scheidel 2014.

23. As Horden and Purcell 2000 emphasize. See further Broodbank 2013.

24. P. D. Jones, Jonsson, and Wheeler 1997; Xoplaki et al. 2004.

25. Olsen, Anderson, and Knudsen 2012.

26. Xoplaki 2002, 163–164; see Black 2012, for some correlation between positive winter NAO and high precipitation in the Levant, but the lack of correlation for negative phases of NAO and climate in Israel/Jordan.

27. Trouet et al. 2009.

28. Finné et al. 2011, 3154.

29. Xoplaki 2002.

30. Black 2012, 1655; Kutiel and Benaroch 2002.

31. The lower-level East Atlantic/West Russia pattern may also account for such differences: Dermody et al. 2012; Krichak and Alpert 2005.

32. Marriner et al. 2012; Eltahir 1996.

33. Huntington 1907; Lamb 1982; Le Roy Ladurie 2004; a more recent synthesis, Brooke 2014.

34. Wanner et al. 2015; Finné et al. 2011; Wanner et al. 2008; Mayewski et al. 2004.

35. Gray et al. 2010; see below.

36. Sigl et al. 2015; Gao, Robock, and Ammann 2008.

37. Telelēs 2004 for the Byzantine world; for the Roman and late antique, see the dataset the authors have published as Historical Evidence on Roman and Post-Roman Climate, 100 BCE to 800 CE, at darmc.harvard.edu.

38. E.g., Ludlow et al. 2013; McCormick et al. 2007.

39. Haldon et al. 2014, on the coherence of written and proxy evidence in medieval Anatolia.

40. Manning 2013, 136.

41. As in Büntgen et al. 2011.

42. Sigl et al. 2015.

43. See https://sohp.fas.harvard.edu/historical-ice-core-heart-europe for updates on this team project, co-led by Paul Mayewski and Michael McCormick.

44. Beer, Vonmoos, and Muscheler 2007.

45. On solar forcing, e.g., Gray et al. 2010; Beer, Mende, and Stellmacher 2000; Bond et al. 2001; Shindell et al. 2003; Shindell 2001.

46. Steinhilber et al. 2012; Beer, Vonmoos, and Muscheler 2007.

47. Vonmoos, Beer, and Muscheler 2006.

48. Manning 2013, 122; Usoskin and Kromer 2005.

49. Steinhilber et al. 2012.

50. Vieira et al. 2011.

51. Steinhilber, Beer, and Fröhlich 2009.

52. Sigl et al. 2015.

53. McCormick et al. 2007.

54. Miller et al. 2012; see also Crowley et al. 2008; Büntgen et al. 2016.

55. Rossignol 2012; Rossignol and Durost 2010.

56. Data from ftp://ftp.ncdc.noaa.gov/pub/data/paleo/icecore/greenland/summit/gisp2 /chem/volcano.txt. The 44 BCE eruption has not received its due: see now Sigl et al. 2015.

57. Data from ftp://ftp.ncdc.noaa.gov/pub/data/paleo/icecore/greenland/summit/grip /chem/gripacid.txt.

58. Sigl, et al. 2015.

59. Salzer and Hughes 2007.

60. Christiansen and Ljungqvist 2012; see also Ljungqvist 2009.

61. Holzhauser, Magny, and Zumbühl 2005.

62. Six and Vincent 2014; Hoelzle et al. 2003; Haeberli et al. 1999.

63. Le Roy et al. 2015.

64. Le Roy et al. 2015.

65. Holzhauser, Magny, and Zumbühl 2005, esp. Fig. 6; Le Roy et al. 2015, esp. Fig. 7 and p. 14. See on the LALIA p. 37f., below.

66. Hughes 2002.

67. Büntgen et al. 2011.

68. Büntgen et al. 2011.

69. Büntgen et al. 2016.

70. An event with an extensive bibliography. For a recent discussion, claiming both volcanic and cometary origins of the event, see Abbott et al. 2014. More generally, Gunn 2000.

71. On the written evidence, Arjava 2005. See now Harper 2017, Chapter 6.

72. Procopius, *De bellis*, 4.14.5–6; Cassiodorus, *Variae*, 12.25.

73. Stothers and Rampino 1983.

74. See now esp. Sigl et al. 2015. For tree evidence, see Salzer and Hughes 2007; Eronen et al. 2002; D'Arrigo et al. 2001. For ice-core evidence, Larsen et al. 2008.

75. Another small conundrum lies in a tree-ring series constructed for the years 398– 610 CE from archaeological wood recovered in the excavation of the Yenikapi harbor in Istanbul. The oaks of unknown origin used in sixth-century harbor construction at Constantinople show no clear signs of the 536 event and might seem to challenge the impact of the 536 event despite eyewitness reports and unambiguous evidence from other proxy archives. But our present ignorance of the place of origin and therefore of the growth conditions of these still provisionally dated trees and the similar absence of strong responses in contemporary precipitation-driven tree rings from central Europe (Büntgen et al. 2011, the Supplementary Data For Fig. S4, Oak Extremes) caution against drawing stark conclusions from the data. In any event, a new and robust tree ring record from the Altai mountains in Central Asia confirms and deepens the testimony of the Alpine temperature series, especially, as we will see, with respect to the onset of the LALIA across western Eurasia (at least) in 536. On possible links with plague, see now Harper 2017, Chapter 5.

76. Göktürk 2011; F. McDermott et al. 2011; McDermott 2004.

77. Vollweiler et al. 2006; Mangini, Spötl, and Verdes 2005.

78. Göktürk 2011; Fleitman et al. 2009.

79. Luterbacher et al. 2012, 112–119.

80. See above, on the EA-WR and NCPI climate patterns, as well as Finné et al. 2014, 224; Tudryn et al. 2013; M. D. Jones et al. 2006.

81. Hin 2013. See now Harper 2017, Chapter 1.

82. Holzhauser, Magny, and Zumbühl 2005; Le Roy et al. 2015.

83. Büntgen et al. 2011.

84. Christiansen and Ljungqvist 2012.

85. Martín-Puertas et al. 2009; Martín-Puertas et al. 2010; see also Corella et al. 2010; Pérez-Sanz et al. 2013; Currás et al. 2012.

86. Indeed, this pattern of humidity in the western Mediterranean complicates any interpretation that foregrounds a positive-NAO as the dominant regime of the Roman Climate Optimum, as Brooke 2014.

87. Bookman et al. 2004.

88. Schilman et al. 2002.

89. Fleitmann et al. 2009.

90. Kaniewski et al. 2007.

91. Wick, Lemcke, and Sturm 2003.

92. Büntgen et al. 2011; Mayewski et al. 2004.

93. Sigl et al. 2015 according to whose calculations of Global Volcanic Forcing (Supplementary Date 6, Sheet 1, col. G) the reconstructed impact of the eruptions was, in ascending order, those now dated 206, 281, 304, 236, 169 and 266 CE.

94. Witschel 1999; compounded by, and possibly related to, the outbreak of a major pandemic: Harper 2015a.

95. Holzhauser, Magny, and Zumbühl 2005.

96. Mayewski et al. 2004; McCormick et al. 2012.

97. Büntgen et al. 2011.

98. Nicolussi and Patzelt 2001. However, trees germinating around 400 CE in the forefields of the Lower Grindelwald Glacier and the Suldenferner indicate that this advance ended in the second half of the fourth century: see Figure 2 in Holzhauser, Magny, and Zumbühl 2005; and Nicolussi et al. 2006.

99. Kenward 2004, Table 1, where of eleven Roman sites, four are dated to the fourth century, one to the "?3rd, 4th" and another to fill dated between the late third and late fourth.

100. Büntgen et al. 2011, 578.

101. Perhaps reflecting, in part, the survival of the historian Ammianus Marcellinus' work for these years: Res gestae 14.10.2 (354), 17.12.4 (358), 27.5.5. (368), 29.6.17 (374–375); Claudian, De bello Gildonico 1.40–42 and Symmachus Ep. 6.7.1 (397–398); further references in McCormick Harper, More and Gibson 2012.

102. For references to the written sources, see McCormick, Harper, More and Gibson 2012 under these years.

103. Bookman et al. 2004; Hirschfeld 2004.

104. Migowski et al. 2006, Fig. 3.

105. Among the most securely attested events, there is one report each of flooding and drought from 300 to 350. From 375 to 475, one or two droughts and four precipitation/flooding events are mentioned in the eastern empire: see McCormick Harper, More and Gibson 2012 for details.

106. Drying and/or cooling from ca. 300–450 CE, followed by wetter/warmer from ca. 450–625 CE.

107. McCormick et al. 2012.

108. Marcellinus Comes, *Chronicon, Auctarium* a. 536, ed. T. Mommsen, Monumenta Germaniae historica, Auctores antiquissimi, 11 (Berlin, 1894), p. 105.

109. Huntington 1907, 329–385, argued for a drought-induced migration chiefly from what he took to be archaeological evidence of the fluctuation of Caspian sea levels; cf. Lamb 1982, 159–60. On the Hunnic expansion in the fourth century, see de la Vaissière 2012, 144–146.

110. Maenchen-Helfen 1973, 1–36; A. H. M. Jones 1964, 1,152–154.

111. See now Büntgen et al. 2016 and Boch and Spötl 2011 for an additional isotopic signal from an Austrian cave flow stone, showing a "Dark Age Cold Period."

112. Hanspeter Holzhauser, Magny, and Zumbühl 2005.

113. Sigl et al 2015, Extended Data Figure 2 and Supplementary data, Nature 14565-s4. xls, Sheet4.

114. Harper 2017, Chapter 6; Arjava 2005; Gunn 2000.

115. Sigl et al. 2015.

116. Holzhauser, Magny, and Zumbühl 2005 suggest that the Aletsch glacier at least advanced to nearly Little Ice Age maximum extensions around 600 CE. The evidence for the French Alps supports that picture: Le Roy et al. 2015. However, this suggestion is not supported by recent results from the eastern Alps. Detrital wood from the forefield of the Gepatschferner glacier covers the whole time period between ca. 510 and 809 CE, suggesting that this glacier's sixth-century advance did not exceed glacier limits of about 1920/1940 CE.

117. Terminal tree rings between ca. 546 and 579 CE now clarify the chronology of previously published materials in Holzhauser, Magny, and Zumbühl 2005; cf. H. Holzhauser and Zumbühl 1996.

118. Büntgen et al. 2016.

119. See McCormick et al. 2012, as well as McCormick, Harper, More and Gibson 2012.

120. Written sources document reliably fourteen drought and heat events in the eastern Roman Empire between 500 and 599 CE: in Palestine, ca. 500 (destroying valuable vineyards in the Negev); a fifteen-year drought from 523 to 538 that dried up the Siloah spring; winter heat events occurred in Cilicia in 550 and, in 600–601, in Palestine and Syria; droughts struck Syria in 525–526 and 568–569, and the Persian Empire in 536, provoking the Arab migration; droughts befell Constantinople in autumn 530, in 556, and 562, not to mention an exceptional late heat spell ca. September-November 551; heat events occurred in Mesopotamia in 501, 502, and 595; in Cilicia in 550 a spring heat event melted snow and caused flooding noted below. See McCormick, Harper, More and Gibson 2012, for detailed records. Eight excessive precipitation events and/or floods are recorded with similar levels of certainty in Mesopotamia in 501–502, 525 (including Syria), in 553–554, in 580, and ca. 582–602, and in 597–598; and in Cilicia in 537 and 550. See McCormick, Harper, More and Gibson 2012, under these years for further details, as well as the next note.

121. Five of seven Roman inscriptions documenting the construction, maintenance and protection of public water works in Palestine cluster in the sixth century: Di Segni 2002; Avigad 1977; C. P. Jones 2007.

122. Procopius, *De aedificiis*, 1, 11, 10–14, 42.26–43.23; Malalas, *Chronographia*, 18, 17, ed. J. Thurn, Corpus fontium historiae byzantinae, 35 (Berlin, 2000), 364.39–41; Theophanes, *Chronographia*, A.M. 6020, ed. C. De Boor, 1 (Leipzig, 1883), 176.26–7.

123. Bookman et al. 2004; Migowski et al. 2006. The discrepancy in the proposed starting date of the wetter conditions (and therefore of the subsequent dry conditions) at ca. 300 or ca 400 CE lies well within the two-hundred-year range of the samples' radiocarbon accuracy at two standard deviations (see especially Migowski et al. 2006 Appendix A), but both studies concur that drying set in sometime between ca. 500 and 600 CE.

124. Büntgen et al., 2016.
125. McCormick 2013, here 84–87.
126. Leveau 2014.

References

Abbott, D. H., et al. 2014. "What Caused Terrestrial Dust Loading and Climate Downturns between A.D. 533 and 540?" *Geological Society of America Special Papers* 505 (September): 421–438. DOI: 10.1130/2014.2505(23).

Arjava, A. 2005. "The Mystery Cloud of 536 CE in the Mediterranean Sources." *Dumbarton Oaks Papers* 59: 73–94. DOI: 10.2307/4128751.

Avigad, N. 1977. "A Building Inscription of the Emperor Justinian and the Nea in Jerusalem (Preliminary Note)." *Israel Exploration Journal* 27 (2/3): 145–151.

Bar-Matthews, M., and A. Ayalon. 2004. "Speleothems as Palaeoclimate Indicators, a Case Study from Soreq Cave Located in the Eastern Mediterranean Region, Israel." In *Past Climate Variability through Europe and Africa*. Vol. 6. eds. W. Battarbee, F. Gasse, and C. E. Stickley. Netherlands: Springer, 363–391. DOI: 10.1007/978-1-4020-2121-3_18.

Beer, J., W. Mende, and R. Stellmacher. 2000. "The Role of the Sun in Climate Forcing." *Quaternary Science Reviews* 19 (1–5): 403–415. DOI: 10.1016/S0277-3791(99)00072-4.

Beer, J., M. Vonmoos, and R. Muscheler. 2007. "Solar Variability Over the Past Several Millennia." *Space Science Reviews* 125 (1–4): 67–79. DOI: 10.1007/s11214-006-9047-4.

Black, E. 2012. "The Influence of the North Atlantic Oscillation and European Circulation Regimes on the Daily to Interannual Variability of Winter Precipitation in Israel." *International Journal of Climatology* 32 (11): 1654–1664. DOI: 10.1002/joc.2383.

Boch, R., and C. Spötl. 2011. "Reconstructing Palaeoprecipitation from an Active Cave Flowstone." *Journal of Quaternary Science* 26 (7): 675–687. DOI: 10.1002/jqs.1490.

Bond, G., et al. 2001. "Persistent Solar Influence on North Atlantic Climate During the Holocene." *Science* 294 (5549): 2130–2136. DOI: 10.1126/science.1065680.

Bookman, R., et al. 2004. "Late Holocene Lake Levels of the Dead Sea." *Geological Society of America Bulletin* 116 (5–6): 555–571. DOI: 10.1130/B25286.1.

Bradley, R. S. 2015. *Paleoclimatology: Reconstructing Climates of the Quaternary*. Amsterdam: Elsevier.

Broodbank, C. 2013. *The Making of the Middle Sea: A History of the Mediterranean from the Beginning to the Emergence of the Classical World*. Oxford; New York: Oxford University Press.

Brooke, J. 2014. *Climate Change and the Course of Global History: A Rough Journey*. New York: Cambridge University Press.

Büntgen, U., et al. 2011. "2500 Years of European Climate Variability and Human Susceptibility." *Science* 331 (6017): 578–582. DOI: 10.1126/science.1197175.

Büntgen, U. et al. 2016. "Cooling and Societal Change during the Late Antique Little Ice Age from 536 to around 660 AD." *Nature Geoscience* 9: 231–236. DOI: 10.1038/ngeo2652.

Butzer, K. W. 2012. "Collapse, Environment, and Society." *Proceedings of the National Academy of Sciences of the United States of America* 109 (10): 3632–3639. DOI: 10.1073/pnas.1114845109.

Caseldine, C. J., and C. Turney. 2010. "The Bigger Picture: Towards Integrating Palaeoclimate and Environmental Data with a History of Societal Change." *Journal of Quaternary Science* 25 (1): 88–93. DOI: 10.1002/jqs.1337.

Chew, S. C. 2002. "Globalisation, Ecological Crisis, and Dark Ages." *Global Society* 16 (4): 333–356. DOI: 10.1080/0953732022000016081.

Christiansen, B., and F. C. Ljungqvist. 2012. "The Extra-Tropical Northern Hemisphere Temperature in the Last Two Millennia: Reconstructions of Low-Frequency Variability." *Climate of the Past* 8 (2): 765–786. DOI: 10.5194/cp-8-765-2012.

Corella, J. P., et al. 2010. "Climate and Human Impact on a Meromictic Lake during the Last 6,000 Years (Montcortès Lake, Central Pyrenees, Spain)." *Journal of Paleolimnology* 46 (3): 351–367. DOI: 10.1007/s10933-010-9443-3.

Crowley, T. J., et al. 2008. "Volcanism and the Little Ice Age." *PAGES News* 16 (2): 22–23.

Currás, A., et al. 2012. "Climate Change and Human Impact in Central Spain during Roman Times: High-Resolution Multi-Proxy Analysis of a Tufa Lake Record (Somolinos, 1280 M Asl)." *CATENA* 89 (1): 31–53. DOI: 10.1016/j.catena.2011.09.009.

D'Arrigo, R., et al. 2001. "Spatial Response to Major Volcanic Events in or about AD 536, 934 and 1258: Frost Rings and Other Dendrochronological Evidence from Mongolia and Northern Siberia: Comment on R. B. Stothers, 'Volcanic Dry Fogs, Climate Cooling, and Plague Pandemics in Europe and the Middle East' (Climatic Change, 42, 1999)." *Climatic Change* 49 (1–2): 239–246. DOI: 10.1023/A:1010727122905.

Dean, J. R., et al. 2013. "Palaeo-Seasonality of the Last Two Millennia Reconstructed from the Oxygen Isotope Composition of Carbonates and Diatom Silica from Nar Gölü, Central Turkey." *Quaternary Science Reviews* 66 (April): 35–44. DOI: 10.1016/j.quascirev.2012.07.014.

De la Vaissière, É. 2012. "Central Asia and the Silk Road." In *Oxford Handbook of Late Antiquity*, ed. S. F. Johnson. New York: Oxford University Press, 142–169.

DeMenocal, P. B. 2001. "Cultural Responses to Climate Change during the Late Holocene." *Science* 292 (5517): 667–673. DOI: 10.1126/science.1059827.

Dermody, B. J., et al. 2012. "A Seesaw in Mediterranean Precipitation during the Roman Period Linked to Millennial-Scale Changes in the North Atlantic." *Climate of the Past* 8 (2): 637–651. DOI: 10.5194/cp-8-637-2012.

Di Segni, L. 2002. "The Water Supply of Palestine in Literary and Epigraphical Sources." In *The Aqueducts of Israel*, eds. D. Amit and J. Patrich. Portsmouth, RI: Journal of Roman Archaeology, 36–67.

Drysdale, R., et al. 2006. "Late Holocene Drought Responsible for the Collapse of Old World Civilizations Is Recorded in an Italian Cave Flowstone." *Geology* 34 (2): 101–104. DOI: 10.1130/G22103.1.

Eltahir, E. 1996. "El Niño and the Natural Variability in the Flow of the Nile River." *Water Resources Research* 32 (1): 131–137.

Erdkamp, P. 2005. *The Grain Market in the Roman Empire: A Social, Political and Economic Study*. Cambridge, UK; New York: Cambridge University Press.

Eronen, M., et al. 2002. "The Supra-Long Scots Pine Tree-Ring Record for Finnish Lapland: Part 1, Chronology Construction and Initial Inferences." *The Holocene* 12 (6): 673–680. DOI: 10.1191/0959683602hl580rp.

Febvre, L., and L. Bataillon. 1966, *A Geographical Introduction to History*, London: Routledge & Kegan Paul.

Finné, M., et al. 2011. "Climate in the Eastern Mediterranean, and Adjacent Regions, during the Past 6000 Years—A Review." *Journal of Archaeological Science* 38 (12): 3153–3173. DOI: 10.1016/j.jas.2011.05.007.

Finné, M., et al. 2014. "Speleothem Evidence for Late Holocene Climate Variability and Floods in Southern Greece." *Quaternary Research* 81 (2): 213–227. DOI: 10.1016/j.yqres.2013.12.009.

Fleitmann, D., et al. 2009. "Timing and Climatic Impact of Greenland Interstadials Recorded in Stalagmites from Northern Turkey." *Geophysical Research Letters* 36 (19): L19797. DOI: 10.1029/2009GL040050.

Folke, C. 2006. "Resilience: The Emergence of a Perspective for Social–Ecological Systems Analyses." *Global Environnemental Change* 16 (3): 253–267. DOI: 10.1016/j.gloenvcha.2006.04.002.

Frisia, S., et al. 2005. "Climate Variability in the SE Alps of Italy over the Past 17, 000 Years Reconstructed from a Stalagmite Record." *Boreas* 34 (4): 445–455. DOI: 10.1080/03009480500231336.

Gao, C., A. Robock, and C. Ammann. 2008. "Volcanic Forcing of Climate over the Past 1500 Years: An Improved Ice Core-Based Index for Climate Models." *Journal of Geophysical Research: Atmospheres* 113 (D23): D23111. DOI: 10.1029/2008JD010239.

Garnsey, P. 1988. *Famine and Food Supply in the Graeco-Roman World: Responses to Risk and Crisis.* Cambridge; New York: Cambridge University Press.

Garnsey, P. 1998. *Cities, Peasants, and Food in Classical Antiquity: Essays in Social and Economic History,* Cambridge: Cambridge University Press.

Garnsey, P. 1999. *Food and Society in Classical Antiquity,* Cambridge: Cambridge University Press.

Göktürk, O. M. 2011. "Climate in the Eastern Mediterranean through the Holocene Inferred from Turkish Stalagmites." PhD dissertation, Universität Bern, Switzerland.

Goosse, H., et al. 2012. "The Role of Forcing and Internal Dynamics in Explaining the 'Medieval Climate Anomaly.'" *Climate Dynamics* 39 (12): 2847–2866. DOI: 10.1007/s00382-012-1297-0.

Graham, N., et al. 2011. "Support for Global Climate Reorganization during the 'Medieval Climate Anomaly.'" *Climate Dynamics* 37 (5): 1217. DOI: 10.1007/s00382-010-0914-z.

Gray, L. J., et al. 2010. "Solar Influences on Climate." *Reviews of Geophysics* 48 (4): 1–53. DOI: 10.1029/2009RG000282.

Grove, A. T., and O. Rackham. 2001. *The Nature of Mediterranean Europe: An Ecological History.* New Haven: Yale University Press.

Gunn, J. 2000. *The Years without Summer: Tracing A.D. 536 and Its Aftermath.* Oxford: Archaeopress.

Haeberli, W., et al. 1999. "On Rates and Acceleration Trends of Global Glacier Mass Changes." *Geografiska Annaler: Series A, Physical Geography* 81 (4): 585–591. DOI: 10.1111/1468-0459.00086.

Haldon, J., et al. 2014. "The Climate and Environment of Byzantine Anatolia: Integrating Science, History, and Archaeology." *Journal of Interdisciplinary History* 45 (2): 113–161. DOI: 10.1162/JINH_a_00682.

Harper, K. 2015a. "Rethinking the Plague of Cyprian: Pandemics and Passages to Late Antiquity." *Journal of Roman Archaeology* 28: 223–260. DOI: 10.1017/S1047759415002470.

Harper, K. 2015b. "Civilization, Climate, and Malthus: The Rough Course of Global History." *Journal of Interdisciplinary History* 45 (4): 549–566. DOI: 10.1162/JINH_a_00758.

Harper, K. 2017, *The Fate of Rome: Climate, Disease, and the End of an Empire.* Princeton: Princeton University Press.

Harris, W. V. 2011, "Bois et déboisement dans la Méditerranée antique," *Annales. Histoire, sciences sociales* 66: 105–140.

Harris, W. V. 2013, "Defining and Detecting Mediterranean Deforestation, 800 BCE to 700 CE," in W. V. Harris, ed., *The Ancient Mediterranean Environment between Science and History,* Leiden: 173–194.

Hin, S. 2013. *The Demography of Roman Italy: Population Dynamics in an Ancient Conquest Society (201 BCE–14 CE).* Oxford; New York: Oxford University Press.

Hirschfeld, Y. 2004. "A Climatic Change in the Early Byzantine Period? Some Archaeological Evidence." *Palestine Exploration Quarterly* 136 (2): 133–149. DOI: 10.1179/003103204225014184.

Hoelzle, M., et al. 2003. "Secular Glacier Mass Balances Derived from Cumulative Glacier Length Changes." *Global and Planetary Change* 36 (4): 295–306. DOI: 10.1016/S0921-8181(02)00223-0.

Holzhauser, H., M. J. Magny, and H. J. Zumbühl. 2005. "Glacier and Lake-Level Variations in West-Central Europe over the Last 3500 Years." *The Holocene* 15 (6): 789–801. DOI: 10.1191/0959683605hl853ra.

Holzhauser, H., and H. J. Zumbühl. 1996. "To the History of the Lower Grindelwald Glacier during the Last 2800 Years—Paleosols, Fossil Wood and Historical Pictorial Records—New Results." *Zeitschrift für Geomorphologie Supplementband* 104: 95–127.

Horden, P., and N. Purcell. 2000. *The Corrupting Sea: A Study of Mediterranean History*. Oxford: Blackwell Publishers.

Hughes, D. J. 1994. *Pan's Travail: Environmental Problems of the Ancient Greeks and Romans*, Baltimore: The Johns Hopkins University Press.

Hughes, D. J. 2011. "Ancient Deforestation Revisited." *Journal of the History of Biology* 44: 43–57.

Hughes, M. K. 2002. "Dendrochronology in Climatology—The State of the Art." *Dendrochronologia* 20 (1–2): 95–116. DOI: 10.1078/1125-7865-00011.

Huntington, E. 1907. *The Pulse of Asia: A Journey in Central Asia Illustrating the Geographic Basis of History*. Boston; New York: Houghton, Mifflin and Co.

Huntington, E. 1917. "Climatic Change and Agricultural Exhaustion as Elements in the Fall of Rome." *Quarterly Journal of Economics* 31: 173–208.

Jones, A.H.M. 1964. *The Later Roman Empire, 284–602: a Social Economic and Administrative Survey*. Oxford: Blackwell.

Jones, C. P. 2007. "Procopius of Gaza and the Water of the Holy City." *Greek, Roman, and Byzantine Studies* 47 (4): 455–467.

Jones, M. D., et al. 2006. "A High-Resolution Late Holocene Lake Isotope Record from Turkey and Links to North Atlantic and Monsoon Climate." *Geology* 34 (5): 361–364. DOI: 10.1130/G22407.1.

Jones, P. D., T. Jonsson, and D. Wheeler. 1997. "Extension to the North Atlantic Oscillation Using Early Instrumental Pressure Observations from Gibraltar and South-West Iceland." *International Journal of Climatology* 17 (13): 1433–1450. DOI: 10.1002/(SICI)1097-0088(19971115)17:13<1433::AID-JOC203>3.0.CO;2-P.

Kacanski, A., et al. 2001. "Late Holocene Climatic Change in the Balkans; Speleothem Isotopic Data from Serbia." *Radiocarbon* 43 (2B): 647–658. DOI: 10.2458/azu_js_rc.43.3896.

Kaniewski, D., et al. 2007. "A High-Resolution Late Holocene Landscape Ecological History Inferred from an Intramontane Basin in the Western Taurus Mountains, Turkey." *Quaternary Science Reviews* 26 (17–18): 2201–2218. DOI: 10.1016/j.quascirev.2007.04.015.

Keay, S. J. (ed.). 2012. *Rome, Portus and the Mediterranean*. London: British School at Rome.

Kenward, H. 2004. "Do Insect Remains from Historic-Period Archaeological Occupation Sites Track Climate Change in Northern England?" *Environmental Archaeology* 9 (1): 47–59. DOI: 10.1179/env.2004.9.1.47.

Krichak, S. O., and P. Alpert. 2005. "Decadal Trends in the East Atlantic–West Russia Pattern and Mediterranean Precipitation." *International Journal of Climatology* 25 (2): 183–192. DOI: 10.1002/joc.1124.

Kutiel, H., and Y. Benaroch. 2002. "North Sea–Caspian Pattern (NCP)—An Upper Level Atmospheric Teleconnection Affecting the Eastern Mediterranean: Identification and Definition." *Theoretical and Applied Climatology* 71 (1–2): 17–28. DOI: 10.1007/s704-002-8205-x.

Kuzucuoğlu, C., et al. 2011. "Mid- to Late-Holocene Climate Change in Central Turkey: The Tecer Lake Record." *The Holocene* 21 (1): 173–188. DOI: 10.1016/j.quascirev.2007.04.015.

Lamb, H. H. 1982. *Climate, History, and the Modern World*. London; New York: Methuen.

Larsen, L. B., et al. 2008. "New Ice Core Evidence for a Volcanic Cause of the A.D. 536 Dust Veil." *Geophysical Research Letters* 35 (4): L04708. DOI: 10.1029/2007GL032450.

Le Roy, M., et al. 2015. "Calendar-Dated Glacier Variations in the Western European Alps during the Neoglacial: The Mer de Glace Record, Mont Blanc Massif." *Quaternary Science Reviews* 108 (January): 1–22. DOI: 10.1016/j.quascirev.2014.10.033.

Le Roy Ladurie, E. 2004. *Histoire humaine et comparée du climat*. Paris: Fayard.

Le Roy Ladurie, E. 1966, *Les paysans de Languedoc*. Paris: Mouton.

Leone, A. 2012, "Water Management in Late Antique North Africa: Agricultural Irrigation." *Water History* 4: 119–33.

Leveau, P. 2014. "Évolution climatique et construction des ouvrages hydrauliques en Afrique romaine." In *Regards croisés d'Orient et d'Occident: Les barrages dans l'antiquité tardive*, eds. F. Baratte, C. J. Robin, and E. Rocca. Paris: Fondation Simone et Cino del Duca, 125–138.

Lionello, P. (ed.). 2012. *The Climate of the Mediterranean Region: From the Past to the Future*. 1st ed. Elsevier Insights. London; Waltham, MA: Elsevier.

Ljungqvist, F. C. 2009. "Temperature Proxy Records Covering the Last Two Millennia: A Tabular and Visual Overview." *Geografiska Annaler: Series A, Physical Geography* 91 (1): 11–29. DOI: 10.1111/j.1468-0459.2009.00350.x.

Ludlow, F., et al. 2013. "Medieval Irish Chronicles Reveal Persistent Volcanic Forcing of Severe Winter Cold Events, 431–1649 CE." *Environmental Research Letters* 8 (2): 024035. DOI: 10.1088/1748-9326/8/2/024035.

Luterbacher, J., et al. 2012. "A Review of 2000 Years of Paleoclimatic Evidence in the Mediterranean." In *The Climate of the Mediterranean Region from the Past to the Future*, ed. P. Lionello. Netherlands: Elsevier, 87–185. DOI: 10.1016/B978-0-12-416042-2.00002-1.

Maenchen-Helfen, O. 1973. *The World of the Huns: Studies in Their History and Culture*. Berkeley: University of California Press.

Magny, M. 1993. "Holocene Fluctuations of Lake Levels in the French Jura and Sub-Alpine Ranges, and Their Implications for Past General Circulation Patterns." *The Holocene* 3 (4): 306–313. DOI: 10.1023/A:1008195401085.

Magny, M. 2004. "Holocene Climate Variability as Reflected by Mid-European Lake-Level Fluctuations and Its Probable Impact on Prehistoric Human Settlements." *Quaternary International* 113 (1): 65–79. DOI: 10.1016/S1040-6182(03)00080-6.

Mangini, A., C. Spötl, and P. Verdes. 2005. "Reconstruction of Temperature in the Central Alps during the Past 2000 Yr from a δ18O Stalagmite Record." *Earth and Planetary Science Letters* 235 (3–4): 741–751. DOI: 10.1016/j.epsl.2005.05.010.

Mango, C. A. 1985. *Le développement urbain de Constantinople, IVe–VIIe siècles*. Travaux et mémoires du Centre de recherche d'histoire et civilisation de Byzance. Monographies, 2. Paris: de Boccard.

Mango, C. A. 1993. *Studies on Constantinople*. Aldershot, Hampshire, UK; Brookfield, VT: Variorum.

Manning, S. W. 2013. "The Roman World and Climate: Context, Relevance of Climate Change, and Some Issues." In *The Ancient Mediterranean Environment between Science and History*, ed. W. V. Harris. Leiden: Brill, 103–170.

Marriner, N., et al. 2012. "ITCZ and ENSO-like Pacing of Nile Delta Hydro-Geomorphology during the Holocene." *Quaternary Science Reviews* 45: 73–84. DOI: 10.1016/j.quascirev.2012.04.022.

Martín-Chivelet, J., et al. 2011. "Land Surface Temperature Changes in Northern Iberia since 4000yr BP, Based on δ13C of Speleothems." *Global and Planetary Change* 77 (1–2): 1–12. DOI: 10.1016/j.gloplacha.2011.02.002.

Martín-Puertas, C., et al. 2009. "The Iberian–Roman Humid Period (2600–1600 Cal Yr BP) in the Zoñar Lake Varve Record (Andalucía, Southern Spain)." *Quaternary Research* 71 (2): 108–120. DOI: 10.1016/j.yqres.2008.10.004.

Martín-Puertas, C., et al. 2010. "Late Holocene Climate Variability in the Southwestern Mediterranean Region: An Integrated Marine and Terrestrial Geochemical Approach." *Climate of the Past Discussions* 6 (5): 1655–1683. DOI: 10.5194/cpd-6-1655-2010.

Mayewski, P. A., et al. 2004. "Holocene Climate Variability." *Quaternary Research* 62 (3): 243–255. DOI: 10.1016/j.yqres.2004.07.001.

McAnany, P. A., and N. Yoffee. 2010. *Questioning Collapse: Human Resilience, Ecological Vulnerability, and the Aftermath of Empire.* Cambridge: Cambridge University Press.

McCormick, M. 2011. "History's Changing Climate: Climate Science, Genomics, and the Emerging Consilient Approach to Interdisciplinary History." *Journal of Interdisciplinary History* 42 (2): 251–273. DOI: 10.1162/JINH_a_00214.

McCormick, M. 2013. "What Climate Science, Ausonius, Nile Floods, Rye, and Thatch Tell Us about the Environmental History of the Roman Empire." In *The Ancient Mediterranean Environment between Science and History,* ed. W.V. Harris. Leiden; New York: Brill, 61–88.

McCormick, M., P. E. Dutton, P.A. Mayewski, and N. Patterson. 2007. "Volcanoes and the Climate Forcing of Carolingian Europe, A.D. 750–950." *Speculum* 82 (4): 865–895. DOI: 10.1017/S0038713400011325.

McCormick, M., K. Harper, A.F.M. More, and K. Gibson. 2012. "Geodatabase of Historical Evidence on Roman and Post-Roman Climate." DARMC Scholarly Data Series, Data Contribution Series # 2012-1. DARMC, Center for Geographic Analysis. Harvard University, Cambridge, MA 02138: https://darmc.harvard.edu/data-availability.

McCormick, M., et al. 2012. "Climate Change during and after the Roman Empire: Reconstructing the Past from Scientific and Historical Evidence." *Journal of Interdisciplinary History* 43 (2): 169–220. DOI: 10.1162/JINH_a_00379.

McDermott, F. 2004. "Palaeo-Climate Reconstruction from Stable Isotope Variations in Speleothems: A Review." *Quaternary Science Reviews* 23 (7–8): 901–918. DOI: 10.1016/j.quascirev.2003.06.021.

McDermott, F., et al. 2011. "A First Evaluation of the Spatial Gradients in $\delta18O$ Recorded by European Holocene Speleothems." *Global and Planetary Change* 79 (3–4): 275–287. DOI: 10.1016/j.gloplacha.2011.01.005.

Meiggs, R. 1982, *Trees and Timber in the Ancient Mediterranean World.* Oxford: Clarendon Press.

Migowski, C., et al. 2006. "Holocene Climate Variability and Cultural Evolution in the Near East from the Dead Sea Sedimentary Record." *Quaternary Research* 66 (3): 421–431. DOI: 10.1016/j.yqres.2006.06.010.

Miller, G., et al. 2012. "Abrupt Onset of the Little Ice Age Triggered by Volcanism and Sustained by Sea-Ice/ocean Feedbacks." *Geophysical Research Letters* 39 (2): L02708. DOI: 10.1029/2011GL050168.

Nicolussi, K., and G. Patzelt. 2001. "Untersuchungen zur Holozänen Gletscherentwicklung von Pasterze und Gepatschferner (Ostalpen)." *Zeitschrift für Gletscherkunde und Glazialgeologie* 36: 1–87.

Nicolussi, K., et al. 2006. "Precisely Dated Glacier Fluctuations in the Alps over the Last Four Millennia." In *Global Change in Mountain Regions,* ed. M. F. Price. London; New York: Parthenon Publishing Group, 59–60.

Olsen, J., N. J. Anderson, and M. F. Knudsen. 2012. "Variability of the North Atlantic Oscillation over the Past 5,200 Years." *Nature Geoscience* 5 (11): 808–812. DOI: 10.1038/ngeo1589.

Orland, I. J., et al. 2009. "Climate Deterioration in the Eastern Mediterranean as Revealed by Ion Microprobe Analysis of a Speleothem That Grew from 2.2 to 0.9 Ka in Soreq Cave, Israel." *Quaternary Research* 71 (1): 27–35. DOI: 10.1016/j.yqres.2008.08.005.

Pérez-Sanz, A., et al. 2013. "Holocene Climate Variability, Vegetation Dynamics and Fire Regime in the Central Pyrenees: The Basa de La Mora Sequence (NE Spain)." *Quaternary Science Reviews* 73: 149–169. DOI: 10.1016/j.quascirev.2013.05.010.

Rickman, G. 1980. *The Corn Supply of Ancient Rome.* Oxford; New York: Clarendon Press; Oxford University Press.

Rossignol, B. 2012. "Le climat, les famines et la guerre: éléments du contexte de la peste antonine." In *L'impatto della "peste antonina,"* ed. E. L. Cascio. Bari: Edipuglia, 87–122.

Rossignol, B., and S. Durost. 2010. "Volcanisme global et variations climatiques de courte durée dans l'histoire romaine (Ier s. av. J.-C.-IVème s. ap. J.-C.): Leçons d'une archive glaciaire (GISP2)." *Jahrbuch des Römisch-Germanischen Zentralmuseums Mainz* 54 (2): 395–438. DOI: 10.11588/jrgzm.2007.2.30692.

Sallares, R. 1991, *The Ecology of the Ancient Greek World.* Ithaca, NY: Cornell University Press.

Sallares, R. 2002, *Malaria and Rome: A History of Malaria in Ancient Italy.* Oxford: Oxford University Press.

Salzer, M. W., and M. K. Hughes. 2007. "Bristlecone Pine Tree Rings and Volcanic Eruptions over the Last 5000 Yr." *Quaternary Research* 67 (1): 57–68. DOI: 10.1016/j.yqres.2006.07.004.

Scheidel, W. 2014. "The Shape of the Roman World: Modelling Imperial Connectivity." *Journal of Roman Archaeology* 27: 7–32. DOI: 10.1017/S1047759414001147.

Scheidel, W. 1996. *Measuring Sex, Age and Death in the Roman Empire.* Roman Archaeology Supplementary Series. Ann Arbor: Journal of Roman Archaeology.

Scheidel, W. 2001. *Death on the Nile: Disease and the Demography of Roman Egypt.* Leiden: Brill.

Schilman, B., et al. 2002. "Sea-Land Paleoclimate Correlation in the Eastern Mediterranean Region during the Late Holocene." *Israel Journal of Earth Sciences* 51 (3): 181–190. DOI: 10.1560/504G-007U-5NKY-GUN1.

Shaw, B. D. 1995, *Environment and Society in Roman North Africa.* Aldershot.

Shaw, B. D. 1996, "Seasons of Death: Aspects of Mortality in Imperial Rome." *Journal of Roman Studies* 86: 100–138.

Shindell, D. T. 2001. "Solar Forcing of Regional Climate Change During the Maunder Minimum." *Science* 294 (5549): 2149–2152. DOI: 10.1126/science.1064363.

Shindell, D. T., et al. 2003. "Volcanic and Solar Forcing of Climate Change during the Preindustrial Era." *Journal of Climate* 16 (24): 4094–4107.

Sigl, M., et al. 2015. "Timing and Climate Forcing of Volcanic Eruptions for the Past 2,500 Years." *Nature* 523 (7562): 543–549. DOI: 10.1038/nature14565.

Six, D., and C. Vincent. 2014. "Sensitivity of Mass Balance and Equilibrium-Line Altitude to Climate Change in the French Alps." *Journal of Glaciology* 60 (223): 867–878. DOI: 10.3189/2014JoG14J014.

Steinhilber, F., and J. Beer. 2011. "Solar Activity—The Past 1200 Years." *PAGES News* 19 (1): 5–6.

Steinhilber, F., J. Beer, and C. Fröhlich. 2009. "Total Solar Irradiance during the Holocene." *Geophysical Research Letters* 36 (19): L19704. DOI: 10.1029/2009GL040142.

Steinhilber, F., et al. 2012. "9,400 Years of Cosmic Radiation and Solar Activity from Ice Cores and Tree Rings." *Proceedings of the National Academy of Sciences of the United States of America* 109 (16): 5967–5971. DOI: 10.1073/pnas.1118965109.

Stothers, R. B., and M. R. Rampino. 1983. "Volcanic Eruptions in the Mediterranean before A.D. 630 from Written and Archaeological Sources." *Journal of Geophysical Research: Solid Earth* 88 (B8): 6357–6371. DOI: 10.1029/JB088iB08p06357.

Telelēs, I. 2004. *Meteōrologika phainomena kai klima sto Byzantio: symboles stēn ereuna tēs hellēnikēs kai latinikēs grammateias 1.1.* Athens: Akadēmia Athēnōn.

Trouet, V., et al. 2009. "Persistent Positive North Atlantic Oscillation Mode Dominated the Medieval Climate Anomaly." *Science* 324 (5923): 78–80. DOI: 10.1126/science .1166349.

Tudryn, A., et al. 2013. "A 2300-Year Record of Environmental Change from SW Anatolia, Lake Burdur, Turkey." *Journal of Paleolimnology* 49 (4): 647–662. DOI: 10.1007 /s10933-013-9682-1.

Usoskin, I. G., and B. Kromer. 2005. "Reconstruction of the ^{14}C Production Rate from Measured Relative Abundance." *Radiocarbon* 47 (1): 31–37. DOI: 10.2458/azu_js_rc.47.2798.

Vieira, L.E.A., et al. 2011. "Evolution of the Solar Irradiance during the Holocene." *Astronomy & Astrophysics* 531: A6. DOI: 10.1051/0004-6361/201015843.

Vollweiler, N., et al. 2006. "A Precisely Dated Climate Record for the Last 9 Kyr from Three High Alpine Stalagmites, Spannagel Cave, Austria." *Geophysical Research Letters* 33 (20): L20703. DOI: 10.1029/2006GL027662.

Vonmoos, M., J. Beer, and R. Muscheler. 2006. "Large Variations in Holocene Solar Activity: Constraints from ^{10}Be in the Greenland Ice Core Project Ice Core." *Journal of Geophysical Research* 111: A10105. DOI: 10.1029/2005JA011500.

Wanner, H., et al. 2008. "Mid- to Late Holocene Climate Change: An Overview." *Quaternary Science Reviews* 27 (19–20): 1791–1828. DOI: 10.1016/j.quascirev.2008.06.013.

Wanner, H., et al. 2015. "Holocene Climate Variability and Change; a Data-Based Review." *Journal of the Geological Society* 172 (2): 254–63. DOI: 10.1144/jgs2013-101.

Wick, L., G. Lemcke, and M. Sturm. 2003. "Evidence of Lateglacial and Holocene Climatic Change and Human Impact in Eastern Anatolia: High-Resolution Pollen, Charcoal, Isotopic and Geochemical Records from the Laminated Sediments of Lake Van, Turkey." *The Holocene* 13 (5): 665–675. DOI: 10.1191/0959683603hl653rp.

Williamson, T. 2013. *Environment, Society and Landscape in Early Medieval England: Time and Topography.* Rochester, NY: The Boydell Press.

Witschel, C. 1999. *Krise, Rezession, Stagnation?: Der Westen des römischen Reiches im 3. Jahrhundert n. Chr.* Frankfurt am Main: Marthe Clauss.

Woodbridge, J., and N. Roberts. 2011. "Late Holocene Climate of the Eastern Mediterranean Inferred from Diatom Analysis of Annually-Laminated Lake Sediments." *Quaternary Science Reviews* 30 (23–24): 3381–3392. DOI: 10.1016/j.quascirev.2011.08.013.

Xoplaki, E. 2002. "Climate Variability over the Mediterranean." PhD dissertation, University of Bern, Switzerland.

Xoplaki, E., et al. 2004. "Wet Season Mediterranean Precipitation Variability: Influence of Large-Scale Dynamics and Trends." *Climate Dynamics* 23 (1): 63–78. DOI: 10.1007 /s00382-004-0422-0.

Zanchetta, G., et al. 2012. "Multiproxy Record for the Last 4500 Years from Lake Shkodra (Albania/Montenegro)." *Journal of Quaternary Science* 27 (8): 780–789. DOI: 10.1002 /jqs.2563.

CHAPTER TWO

Archaeobotany

THE ARCHAEOLOGY OF HUMAN-PLANT INTERACTIONS

Marijke van der Veen

Introduction

Plants are essential to human and animal life on earth: they create the oxygen we breathe and the food we consume. Additionally, plants provide the fibres for our clothes, the building materials for our shelter, the fuel that keeps us warm, the ingredients for our medicines, and the flowers that give us beauty. Importantly, plants are also the 'materials' with which we create and maintain group identity, social relations and a sense of community (food sharing) or social distinction (luxury foods), and individual identity (clothes, colour (plant dyes) and smells (perfumes, plant resins). Plants thus engage with our everyday lives in a variety of different ways, affecting our nutrition and health, our social practices, our emotions and our work. The cultivation, distribution, selection, preparation and consumption of foodstuffs and the use of plants in many other day-to-day activities, are practices deeply embedded in our cultural norms. Importantly, this routine engagement with plants, enacting the same set of actions over and over again, day after day, year after year, makes us who we are. Archaeobotany, the study of the plant remains recovered from archaeological excavations, can thus provide insights into our different modes of being, as well as trace past social and cultural behavior and continuity and change therein. While some of these activities and choices are recorded in surviving texts from the period, many are not, either because they concern individuals and social groups that did not use texts and were not written about, or because they concerned activities nobody perceived worthy of recording. Together with zooarchaeology, human bone and stable isotope analyses (see Chapters 3 and 4) archaeobotany can offer a significant contribution to our

understanding of daily life in the past. While it forms part of the archaeological sciences, and uses a variety of scientific methods, its focus is firmly on
human-plant interactions.

Here the contribution of archaeobotany to our understanding of life in
the Greco-Roman world is reviewed. This chapter does not offer a synthesis
of the current archaeobotanical evidence (regional and temporal variability
across the region are both too great to allow a synthesis in one chapter); instead, this chapter aims to highlight what can be achieved through archaeobotany by focusing on one aspect: food. It is divided into five main sections, each
concentrating on one of the five phases of food, as first described by Goody:
food production, the realm of the farm and the landscape; food distribution
and trade, the realm of the granary, the market, and long-distance transport;
food preparation, the realm of the kitchen; food consumption, the realm of
the meal and, in many instances also, the realm of the table; and finally food
disposal, the realm of the dustbin or refuse deposit and, par excellence, the
realm of archaeology.[1] Other human-plant interactions (body treatment in life
and in death; ideological role of plants and trees; selection of wood for fuel,
artifacts, and building materials; impact of and on local vegetation and environment, fodder crops, utilization of wild plants, etc.) are mentioned in pass-
ing, but for reasons of space cannot be treated in any detail here. The chapter
will conclude with a brief reflection on how these interactions helped create
many different modes of being, how daily life in antiquity varied across time
and space. Finally, it is worth emphasizing here that I regard plants recovered
from archaeological excavations as a form of material culture, shaped by and
shaping their interactions with people, to be studied in a similar fashion to
and alongside other lines of evidence, including faunal remains, human remains, isotopes, ceramics, tools, buildings, and texts. Each dataset has its own
strengths and weaknesses, and only by combining all the available evidence
are we likely to get nearer to the many and varied realities of the past.

Agriculture: How Was the Food Produced?

Farming was the principal occupation of many in antiquity, with most farmers
engaged in small-scale agrarian technologies, rather than in capital-intensive
estates.[2] The period under study here saw many changes in agricultural practice, such as greater divergences in the scale of cultivation, the development
of new technologies to improve water management and soil maintenance, the
rise of arboriculture, the greater mobilization of agricultural produce over
long distances (e.g., supply of the Roman armies, supporting growing urbanization, the trade in spices), and the introduction of new crops, but also many
elements of continuity. Archaeobotany can help identify these, and many different approaches are available.

Such studies usually start by establishing which crops were cultivated and when this changed. For example, naked wheats had been part of European agriculture from the Neolithic onwards, but their rise to prominence is a relatively late phenomenon. Across large parts of the Mediterranean and northern and central Europe, we see the hulled wheats (einkorn, emmer, and spelt) replaced by naked, or free-threshing, wheats (bread, durum, and rivet wheat) during the later first millennium BCE and the early first millennium CE.[3] This transformation is not synchronous across the region; for example, in France the shift to naked wheat started in the south and occurred progressively later in the north.[4] The hulled wheats tend to be associated with smaller-scale, subsistence production, and the naked wheats with production for a surplus and market exchange. Additionally, one of the naked wheats, bread wheat, has superior bread-making qualities. The growing reliance on naked wheats during the first millennium CE is often linked to an increase in the need for grain to support the Roman conquest, the rise of towns, and economic expansion more widely. That other factors play a significant role too is clear from the fact that in certain areas hulled wheat, in this case spelt, maintains its position, for example in parts of southwest Germany and northern Switzerland, where ecological factors (spelt's ability to tolerate high altitudes) and agronomic ones (the continued use of the three-field system due to a lack of fertilizers), combined with a strong cultural preference for spelt, clearly outweighed any economic disadvantages.[5] Bread-making quality is another factor, and in places where a decline in soil fertility affected the successful cultivation of bread wheat, as was the case in second-century CE northern France, we see a switch back to spelt wheat, a species of hulled wheat less demanding on soil type than bread wheat, but, like bread wheat, with good bread-making properties.[6] Bread is not just a source of nutrition, of course; it is also an artifact, a cultural object, and the increased usage of bread wheat, a type of wheat that can produce a leavened, white loaf, is also linked to the rise of Christianity in the Mediterranean and northwest Europe.[7]

The ecological and agronomic requirements of these different crops leads us to the identification of cultivation methods, such as sowing, tillage, maintaining soil fertility (manuring, fallowing or crop rotation), weeding, and irrigation. These practices are linked to crop yields, crop reliability, land ownership, labor costs, integration with animal husbandry, and intensity of cultivation, as well as with the location of the fields. Did ancient farming start as a form of low-intensity, shifting cultivation, and then progress to more labor-intensive continuous cropping, or are these types of husbandry regimes related to specific local circumstances? Traditionally, we have used indirect methods to infer cultivation regimes, by studying the ecology of the arable weeds associated with the crops. Their life form (annual/perennial) and ecology (preference for nutrient rich or poor, acid, or neutral and wet or dry soils) help to identify the conditions in the arable fields, from which cultivation techniques and scales of production can be inferred. For example, an autecological analysis of the cereal crops and their associated weed floras at six Iron Age sites in northeast England has revealed two distinct crop husbandry regimes, one representing intensive, small-scale subsistence agriculture, while the other was indicative of a more extensive regime, suggesting arable expansion.[8] Similarly, monitoring variations in weed species' tolerance for soil pH, as well as their ability to recover from soil disturbance through tillage and weeding, has helped identify marked differences between cultivation plots at a Neolithic Linearbandkeramik site in southwest Germany, with some characterized by high disturbance and high pH, others by lower levels of disturbance and ambiguous pH, and others intermediate between these two. Remarkably, these different plots and practices were linked to specific groups of houses within the settlement and maintained over several generations.[9] Furthermore, weed seed dormancy has been used to reveal a shift in plough technology and agrarian practice (from ard to mouldboard plough) in first millennium CE Britain.[10]

Increasingly, weed ecology is studied using FIBS (Functional Interpretation of Botanical Surveys). This method measures functional attributes of arable weeds (e.g., leaf area, canopy size, rooting depth, size and number of stomata, date of flowering onset, length of flowering) in modern nonmechanized farming practices and uses these as indicators of the potential of species to cope within a particular (manmade) environment. It moves away from formal analogies and, as such, avoids the problems associated with the previously used approaches of phytosociology and autecology.[11] By establishing the ecological significance of each attribute, FIBS enables us to identify which aspect of husbandry is indicated by the weeds, thus facilitating the recognition of cultivation practices, including ones no longer in existence. To date, this method has succeeded in recognizing present-day irrigated versus dry-farmed fields in Jordan, intensively manured and weeded plots in Greece, crop rotation regimes in Jordan, and sowing time in central Europe.[12]

Recently, a further method has become available, stable isotope analysis, which studies the chemical signatures in the crops themselves. To date, the focus has been on nitrogen and carbon. For example, manuring the cereal fields will raise the nitrogen values ($\partial^{15}N$) in the grains, while water availability and irrigation can be inferred from stable carbon values ($\Delta^{13}C$).[13] This work has, in turn, important implications for our reconstruction of human diet. Stable nitrogen isotope ratios ($\partial^{15}N$) from human bone collagen have been used to infer the relative importance of animal versus plant foods in the diet, as enrichment of $\partial^{15}N$ occurs higher up the food chain (see Chapter 4). If, as has now been demonstrated, manuring can significantly raise $\partial^{15}N$ in cereal grain and chaff, human diets containing a major component of manured cereal grain might, erroneously, be interpreted as indicating a high animal-based component in the diet.[14] This highlights the importance of studying the isotopic values of human skeletal material together with those on faunal *and* plant remains at each location, to avoid problems of equifinality and to integrate our understanding of foodways with that of crop management and food production.[15] Another application of this technique concerns the relationship between climate and agriculture.[16] (See also Chapter 1). For example, Riehl has linked a reduction in drought-susceptible crops in the Early Bronze Age Near East with an increase in aridity after 4000 BP.[17] Additionally, recent experimental work is now supporting the hypothesis that the atmospheric conditions during the last glaciation would have restricted the productivity of potential crop progenitors, meaning that the rise in atmospheric CO_2 concentration in the immediate postglacial period might have been beneficial to their domestication.[18] It goes without saying that there remain many methodological challenges to be resolved, but recent large-scale charring experiments have shown that the effects of charring on stable isotope values in cereal grain and pulse seeds are small and predictable.[19]

Alongside the cereals and pulses, fruit trees, and in particular grape vines and olive trees, were and are of significant economic and cultural importance in the Mediterranean region, and this is reflected in the wide range of studies concerning grape and olive cultivation, as well as fruits of other trees, such as the *Prunus* genus (cherries, plums).[20] These include various attempts to distinguish between the seeds of wild versus domestic fruit trees, establishing time and geographical location of domestication, and identifying the earliest evidence for wine and olive oil production. Initially, such studies relied primarily on seed dimensions, ratios and surface sculpture descriptions to differentiate shape types. While often successful, not all archaeological specimens could be allocated to species or type, partly because surface sculpturing and hilum did not always survive on older specimens, and partly because centuries of cultivation and hybridization have caused size overlap between species and varieties.[21] More recently, these methods have been supplemented with geometric morphometrics (Elliptic Fourier Transform method), which includes measurement and capturing of the overall three-dimensional shape of each seed, combined with statistical analyses to evaluate the diversity within and between populations. Achievements to date include the recognition of a relationship between seed shape and domestication, thus improving our ability to detect the start of domestication, and, tentatively, degrees of biodiversity and regional variability; see, for example, recent studies on olive, grape, cherry, and date.[22]

The analysis of ancient DNA (aDNA) is now an additional, if not yet mainstream, part of archaeobotanical research. Along with the tried and tested polymerase chain reaction (PCR) method, which has limitations due to the small amounts of fragmentary aDNA that survive in ancient seeds, the new "next generation" sequencing (NGS) method is offering many new possibilities.[23] One key issue in all these studies is the survival of biomolecules in ancient plant material. The survival of aDNA in desiccated plant material is remarkably good, and has already revealed unusual genetic features in desiccated barley grains from Egypt, which may reflect adaptation to the local, dry environment, as well as contributed to our understanding of the evolutionary processes underlying domestication in cotton.[24] There are, of course, few locations in the world where plant materials will survive in desiccated form, but where they do survive, their preservation is exceptional, and its full analysis thus all the more important; see, for example, the remains from Berenike, Qasr Ibrim, and Quseir al-Qadim—all in Egypt—and from Xinjiang, China, Gran Canaria, Spain, and from historic buildings in Central Europe and Britain.[25] aDNA also survives in many, though not all, plants preserved in waterlogged, anoxic, conditions, as demonstrated in grape seeds, plum stones, and wheat grains.[26] Survival in charred plant material is much more problematic, however, and is heavily dependent on charring regime, but the NGS method may ultimately prove successful here too.[27] This is important, as most plant material from archaeological sites is preserved by charring, and an exclusive reliance on desiccated and waterlogged remains would exclude large parts of the world.

Areas of research currently addressed by archaeogenetics include the identification of plant material where conventional methods fall short (e.g., in the naked wheats where chaff fragments are absent), the number of domestication events for each crop, the trajectory of the spread of agriculture, the identification of landraces and biodiversity, and the adaptive evolution of crops after domestication, especially once they move into regions outside of their natural environments (flowering behavior and day-length responsiveness, nutritional value, tolerance to drought or waterlogging). Phenotypic characterization and genome sequences may, of course, be difficult to achieve, considering the complexity of the genetic basis to many phenotypes.[28] In fact, some evolutionary questions may more easily be extracted from extant landraces, considering the relatively short evolutionary history of many of the crop plants—in the case of vegetatively propagated fruit trees and vines, this may concern just a few generations—and this has recently been done for barley.[29] In all this work, the use of specialized laboratories and a strict protocol are, of course, essential prerequisites.[30]

Monitoring changes in the scale of agricultural production partially relies on the identification of changes in the density of remains deposited in the archaeological record. Here it is important to appreciate that the archaeobotanical record, and in particular the deposition of charred remains, is created by both routine activities and occasional accidents and/or deliberate conflagrations, and that great care is needed to distinguish between the two.[31] The charring of plant material during routine, day-to-day household-based activities such as grain dehusking, cleaning, drying, and food preparation immediately prior to consumption will result in low-density deposition of remains, especially by-products such as chaff and weed seeds, rather than grain. In contrast, the accidental or deliberate burning of produce (storerooms catching fire, acts of violence) will lead to high-density deposition of plant material (grain, pulses, other foodstuffs). These latter events will occur more frequently in places where produce is handled and stored in bulk, which tends to be at large producer sites rather than in small domestic settings. Thus, an increase in the predominance of grain-rich samples is likely to be an indicator of an increase in the scale of production and consumption. This approach has been used to interpret the increase of grain-rich samples at selected Iron Age sites in Britain as evidence for the production of surpluses consumed during feasting.[32]

An increase in the visibility of large quantities and high densities of agricultural by-products used deliberately as fuel (e.g., chaff, olive pressings) is another marker of the increase in agricultural production. Pomace, the pressings of olive oil, burns at a high and constant temperature and produces little smoke, making it an ideal fuel indoors, but also for industrial production.[33] Across the Mediterranean charred remains of olive pressings have been found, but an expansion in its use is visible during the Roman period, highlighting a marked increase in olive oil production and thus the availability of large quantities of pomace as fuel for the growing urban population, in urban bakeries and in the growing pottery industry (e.g., Herculaneum and Pompeii). Olive oil production may have reached up to one billion litres each year during the height of the Roman Empire, which would translate into 1 million tons of pomace and 2–4.5 billion hours of heat.[34] Similarly, in Roman Britain we see a proliferation of samples rich in chaff at rural sites, often, though not exclusively, associated with so-called corn-driers, together with a rise in large barns, mills, and other agricultural structures, all pointing to an expansion of agriculture in response to greater demand after the Roman conquest of the region.[35] At the same time, the disappearance of agricultural by-products at certain sites, such as the disappearance of chaff and weed seeds from proto-urban settlements such as Pompeii and Silchester, has been taken to mean that these now became more fully urban in character.[36] Research addressing similar issues is currently ongoing in Rome.[37]

Here it is worth emphasizing that food was produced not just in the countryside, but in the towns as well. For example, the suburbs of Rome and many other towns were surrounded by market gardens and orchards, and many townhouses had gardens too, used for decorative purposes and food. Their abundance and importance became clear during excavations at Pompeii, Herculaneum, and nearby villas, all destroyed by 79 CE eruption of Vesuvius. Root cavities, charred seeds and fruits, pollen, planting trenches, and plant pots were found in many garden plots, both large and small; even entire orchards, vineyards, and market gardens were present within the city walls of Pompeii. Food plants recovered from these gardens include almonds, beans, citrus, figs, grapes, hazelnuts, pears, and herbs such as dill, rosemary, and thyme. The importance of garden production is reflected in the fact that an estimated 17 percent of the excavated area of Pompeii was allocated to gardens and the growing of plants.[38]

Further indications of agricultural change, other than those mentioned above, include the expansion of agriculture into (or contraction out of) regions less suited to agriculture combined with the adaptation to local climatic and edaphic conditions and the development of suitable crop management practices (drainage, irrigation, heavy plough).[39] As a final point, both palynology and charcoal analysis make a significant contribution to our understanding of vegetation change and the human impact on the local environment, including the expansion of arable land and the sometimes devastating effect of deforestation on the landscape, but these studies lie outside the scope of this chapter.[40]

Combined, the evidence reveals huge and complex variations in the type and scale of agricultural practices, meaning that the ancient agricultural texts and plant treatises (e.g., those by Cato, Columella, Pliny, Theophrastus, Varro), valuable though they are, should be read in their temporal, cultural, and regional contexts, rather than as reliable guides to agriculture across the entire Greco-Roman world. Agricultural practices develop through interactions with many different variables, including cultural (e.g., scale of land use, form of land tenure and degree of market involvement) and natural ones (e.g., climate, altitude, soils, hydrology, physical requirements of plants), and are thus historically contingent and in continuous flux.[41]

Distribution and Trade— Where Did the Food Come From?

The storage, distribution, and exchange of agricultural produce are part of every farming regime, but when these practices move away from household or domestic settings to larger communal or empire-wide requirements, significant structural and organizational changes are needed. The Roman period in particular saw increased mobilization of resources over vast areas, including the feeding of Rome with grain from North Africa, the supply of the Roman army at the frontiers of the Empire, and the trade in exotic luxuries such as spices from the Indian Ocean to satisfy the growing demand from the elite. Archaeobotany can contribute to our understanding of each of these processes.

Storage of grain and other foodstuffs beyond the domestic scale is visible in the archaeological record through the appearance of large storage and processing facilities, such as granaries, storage pits, corn-driers, mills, and barns, through an increased occurrence of deposits full of charred grain or other stored food crops, and through evidence for inadequate storage in the form of batches of germinated grain or crop seeds spoiled by insect damage. For example, we now have convincing evidence that inadequate storage became a serious problem in Roman Britain. Grain pests (*Coleoptera*) that thrive in poorly ventilated storage buildings and in grain that is not fully dry when put into storage, make their first appearance in Britain during this time.[42] These grain beetles have not been recorded on Iron Age or earlier sites and are not thought to be native to Britain. They appear from the very start of the Roman Conquest, probably as adventitious inclusions in grain brought into Britain by the Roman army during its early campaigns. Examples include the first century CE finds of grain weevils (*Sitophilus granarius*) at Alchester (here together with other imports, such as millet and coriander), London, and York.[43] The sudden appearance of these grain pests can be linked to the increased use of large, open grain stores containing bulk quantities of grain (in contrast to domestic-scale household storage previously), which created environments in which these grain pests could thrive. Additionally, the large-scale trade and movement of grain—both across the Channel and within Britain—facilitated their rapid spread.[44] Examples of stored grain that had sprouted due to poor storage conditions were found in Roman York and London; the latter assemblages comprised between 23% and 44% of sprouted grain.[45]

Evidence for such medium to long-distance trade can be detected through the presence of 'exotic arable weeds' within stored produce. For instance, the presence of seeds of *Orlaya platycarpa* in a shipment of wheat and in a batch of spelt chaff found adjacent to granaries, both in the Netherlands, points to imported grain. *Orlaya* is a sub-mediterranean species, not native to the region, and its presence thus suggests that the grain was brought in from Belgium or further south.[46] Similarly, fruits of *Myagrum perfoliatum*, a species of southern European and Near Eastern origin that will not grow successfully north of the Loire, found in bread wheat and spelt wheat at several Roman sites in northern France also points to grain transport to the northern parts of the Roman Empire.[47] In the same way, the presence of a few grains of einkorn, as well as seeds of lentils and bitter vetch amongst a deposit of spelt grain in first-century Roman London identified this batch of grain as originating from either the Mediterranean or the Near East.[48]

This raises questions about the supply of the Roman army when settled along the frontiers. Should we envisage centralised long-distance supply routes, local compulsory requisition, temporal and regional adaptation to local circumstances, or a combination of these at various times? What about the ability of local landscapes and agricultural populations to sustain the additional burden? Did the military presence create unsustainable local pressure, destabilizing local production, or, instead, generate stimulus and agricultural growth? The evidence of grain shipments reaching northern France and the Netherlands from further south suggests the need for medium- to long-distance supplies, but the modeling of data derived from landscape reconstruction, archaeozoology, archaeobotany, and wood analysis in the Lower Rhine Delta shows a more nuanced and complex pattern, with the region initially likely able to sustain the food and wood requirements of the army, but with increasing pressure on resources from the second century CE onwards.[49] Some local provisioning was in evidence throughout, but supplemented with extraregional resources. That the increase in demand put pressure on local farming is apparent in parts of northern France, where we see a switch from bread wheat, a crop that had been on the rise since the late Iron Age, back to spelt wheat during the second century CE, probably due to soil exhaustion—bread wheat is a more demanding crop than spelt wheat.[50]

Grain was not the only product needed at the northern frontiers—timber was another—and the application of dendrochronology combined with the identification of the wood used in river barges, in the construction of a harbor quay and in road building, again point to the movement of resources across considerable distances, as well as offering exact dates for specific construction events. For example, the oak piles used in the construction of the harbor quay at Voorburg-Arentsburg, the Netherlands, in ca. 160 CE originated from southeast Netherlands and southern Germany, while the rebuilding of the quay shortly after 205 CE used oak from the Mosel region.[51] Similar techniques established that two Roman river barges and a Roman punt from Utrecht, The Netherlands, must have been constructed in the Lower-Scheldt region and thus points to inland navigation between this region and the Rhine-based limes, while wood used in the construction of a road joining the limes in the Lower Rhine region in 124–125 CE, perhaps related to the visit of the emperor Hadrian to the region, was all derived from a single source, probably that between Xanten and Venlo, and transported some 100 kilometres over water using barges.[52]

At the opposite end of the Roman Empire, wood analysis of timbers, artifacts, and charcoal also reveal long-distance contacts, with ship timbers and ship-related artifacts made of Indian teak wood (*Tectona grandis*) at the ports of Berenike and Quseir al-Qadim, both located on the Red Sea coast of Egypt, underlining the role of these ports in the Indian Ocean spice trade. Temporal changes in the range of exotic versus native woods used for everyday artifacts and ship timbers at Quseir al-Qadim point to changes in shipping practice, with ships built according to the Mediterranean tradition as well as Indian Ocean vessels frequenting the harbors during the Roman period, in contrast to the Islamic period when Indian Ocean vessels tended to terminate their journeys at Aden, leaving Egyptian or Yemeni vessels to carry the goods up the Red Sea.[53]

Recent excavations at both Berenike and Quseir al-Qadim (Myos Hormos as it was known as in antiquity) have also provided a rich new archive of archaeobotanical evidence for the spice trade. Both sites represent key transport-hubs in the Indian Ocean trade, and the hyperarid climate at the Red Sea coast of Egypt has resulted in the spectacular preservation of botanical remains of spices and other food remains.[54] Here, temporal change in the number of imported species and their numerical frequency in the Roman and medieval Islamic deposits has helped us identify how the spice trade differed in both nature and scale between these two time periods, with black pepper the most abundant spice in both periods, but with many other spices too rare and precious to be accessible to those working in the Roman port, and, in fact, to most living elsewhere in the Empire. By the medieval Islamic period, this had changed, with a wider range of spices, including ginger and cardamom, now consumed in the port, and by a wider, if still elite, group across the Mediterranean and beyond.[55]

Work at other harbors is augmenting our understanding of the role and importance of these long-distance networks, and this increasingly also includes studies of the actual harbor environments and changes in the vegetation and landscape of their immediate surroundings, through geoarchaeological and pollen analyses.[56]

Questions concerning the logistics of supplying food, timber, and fuel are not restricted to the Roman army of course; the provisioning of the growing urban population as well as specialist workforces operating at mines and quarries needs further study. An example of the latter comes from two Roman quarry settlements, Mons Claudianus and Mons Porphyrites. Both are marble quarries that were subject to imperial monopoly with the stone used for imperial projects, such as the Pantheon in Rome (grey granodiorite columns in the portico) and for statuary made of purple porphyry. The distance from civilization—the quarries are located in a remote part of the Eastern Desert of Egypt, some seven days travel from the Nile valley—was clearly no obstacle to a rich and varied diet, as the archaeobotanical assemblages produced not just staples such as cereals, pulses, dates, and onions, but also luxuries including black pepper, artichoke, pomegranate, persea, various nuts, as well as many herbs and condiments. Moreover, seeds of plants normally eaten as "greens," such as leaf or spinach beet, lettuce, endive/chicory, cabbage, mint, basil, and rue, suggest that the soldiers or quarry workers were able to supplement these foods with fresh greens grown in small vegetable plots in the desert.[57] Additional pollen analysis and charcoal identifications brought to light that the working animals were fed barley grain, chaff, and straw, all brought in from the Nile valley, and that fuel consisted of chaff and straw as well as desert shrubs and trees, with charcoal of two acacia species brought in to be used in the smithies. Furthermore, the ceramic evidence points to the ample supply of wine and olive oil from across the Empire.[58] When we compare the botanical evidence for foods with those listed in the ostraca, there is good agreement between the two for cereals, pulses, and vegetables.[59] The texts also mention processed foods (e.g., bread, cakes, malt, wine, olive oil, vinegar), but, remarkably, are almost silent on the many herbs, fruits, and nuts that feature so prominently in the botanical assemblage, which demonstrates why it is so critically important to always use all lines of evidence when reconstructing food and agriculture. Combined, the evidence highlights that these quarry sites were not malnourished or undersupplied desert-stations, but settlements which had access to most foods that were available in the Nile valley. The importance of the stone as symbols of imperial prestige meant that these quarries were embedded in a complex logistical network linking the Eastern Desert with Rome, the eastern and western Mediterranean, India, the Red Sea coast, and with the Nile valley.

Long-distance trade in foodstuffs is, of course, not a Roman phenomenon, though current evidence suggests that this period in particular saw a major growth in the translocation of foodstuffs. For example, some 50 new food plants were brought to Britain and other parts of northwest Europe as part of the Roman conquest of this region, initially as supplies for the Roman armies, but subsequently to meet demand of soldiers and civilians more widely. Some of these foods were widely imported from the start (e.g., fig), others became more abundant in the middle Roman period (e.g., coriander), while others still only gradually increased in popularity (e.g., plum). In this context, it is important to note that many of these plants became part of local agriculture, thus switching status from imported foods to introduced crops (e.g., apple, pear, plum, cherry, walnut, cabbage, leaf beet), which had a significant impact on local agricultural practices (see above under "agriculture"), and also resulted in a major widening of dietary breadth and nutrient availability for large sections of the population (see below under "consumption").[60]

This long-distance exchange of foodstuffs in northwest Europe started when the cultural contact between this region and the Mediterranean increased. This is visible through the presence of wine amphorae, as well as the remains of olive, celery, coriander, and dill, in mid to late Iron Age sites across the region. Current evidence suggests that these foods go primarily if not exclusively to elite locations, such as the *oppida*, as part of the wider phenomenon of Roman-style products being desired and acquired by local elites.[61] This changes in the early Roman period when both the range and scale of such imports increased and such foods became available to more sections of society (see above).

The analysis of plant DNA is offering crucial additional data to our understanding of such translocations of crops. For example, many of the newly introduced food plants concern species that are exotic to northwest Europe, such as pear, plum, walnut, coriander, leek, onion, cucumber, and lettuce, but others are natives, that is, wild forms do grow in the region, such as celery and apple. For this latter group it raises the question whether the Romans brought actual cultivars of these crops with them, or, instead, introduced the concept of their cultivation and encouraged the cultivation of local species. Here DNA analysis is proving invaluable. For example, the DNA of modern apple cultivars (*Malus domestica*) indicates that the wild progenitor of our domestic apple is *Malus sieversii*, a native of the mountain region of Kyrgyzstan and northwest

China, rather than the European crab apple, *Malus sylvestris*.[62] This reveals that in the case of apple the Romans brought the cultivated apple to northwest Europe, and did not use the local wild variety, although some subsequent hybridization between the two is likely. The introduction of cultivated fruit trees into northwest Europe would thus have required the import of budded stems (scions), which could be grafted on to local rootstock (wild crab apple, sloe, etc.) that was specially developed for the purpose, or by also bringing in the rootstock, that is, the live plant. Evidence for the transportation of live plants is available in the form of *ollae perforatae*, purpose-made pots used to plant and transport trees, vines, and shrubs, which are found across the Roman Empire, including Britain, and are dated to the late first century BCE to the mid second century CE.[63] That these fruits soon became widely available is clear from the hundreds of apple pips found at several British sites, including second-century Doncaster, London, and Late Roman Silchester.[64]

The strength of DNA analyses is also evident in a recent study of historical landraces of barley. This study identified the presence of three separate groups of barley in Europe, revealing that barley was introduced into Europe more than once, each originating from a different part of southwest Asia. The strain of barley that can cope with long growing seasons and wet summers, originally domesticated in Iran, was introduced later than the others and is found predominantly in northwest Europe.[65]

Finally, and just briefly, the cargoes of shipwrecks provide further and very direct evidence of these often long-distance food transports. Finds include shipments of wheat in a sunken river barge in The Netherlands, of pomegranates in a shipwreck off the Turkish coast, an amphora full of olives found in the Thames estuary, as well as cotton seeds, coffee beans, and spices in a shipwreck in the Red Sea.[66]

Preparation—How Was the Food Prepared and Consumed?

The preparation of food includes a wide variety of processes, all designed to improve absorption and digestion of the plant nutrients, remove toxins, increase palatability, change the physical form of a food, or convert raw ingredients into storable foodstuffs. Such processes include pounding, milling, boiling, roasting, steaming, parboiling, baking, and fermenting.[67] Thus, cereal grains can be converted to porridge, bread, bulgur, and beer, grapes to raisins or wine and olives to olive oil. Studies to determine these processes from archaeological remains of food are a growing area of research in archaeobotany. Several approaches are used, often in combination. Apart from establishing which parts of the plants are preserved, breakage patterns are studied, using charring experiments and Scanning Electron Microscopy (SEM), and combined with ethnographic observations.[68] For example, Valamoti demonstrates that the charring of fragmented grain causes the endosperm to ooze out, generating a characteristic bulging appearance, while the breakage of grain *after* charring shows surfaces that are porous and irregular in appearance.[69] Shiny glassy surface textures are more typical of grain that had been soaked in boiling water, broken, and then charred. This experimental work led her to conclude that grain fragments from Bronze Age sites in Greece represented bulgur (i.e., boiled and then ground cereal grain). Such preprocessing of grain for later consumption is important, as it converts seasonally available produce into nutritious and storable foodstuffs for consumption at a later date.

Similar techniques were used by Samuel to study preserved fragments of bread and residues of beer from ancient Egypt.[70] Using SEM, she was able to identify yeast cells, bacteria, and starch granules, the latter heavily pitted, indicating that enzymes had started to break down the starch, as part of the malting process. Together with experimental work, the archaeological evidence for ovens, milling tools, ceramic vats, as well as the rich artistic record from Egypt and documentary evidence, the many processes and ingredients involved in the baking and brewing traditions of ancient Egypt could be reconstructed. High magnification tissue analysis has also helped determine the type of cereal represented in the so-called amorphous charred objects, now generally assumed to represent cereal-based products, found at many archaeological sites. Likewise, a remarkably well-preserved charred flat bread (galette) from a first-century Roman cemetery in France was identified as prepared from finely ground flour of barley mixed with some einkorn or emmer, and without leavening.[71]

Beer was produced throughout prehistory but on a household scale, using ordinary vessels and ovens, and thus not easily detectable in the archaeological record, though when large deposits of germinated grain are discovered, malting and beer brewing may be in evidence.[72] In some regions and periods, we see the appearance of specialized structures, indicative of cereal processing and beer brewing on an "industrial" scale. In Roman Britain, for example, beer may have represented a cash crop, where a surplus of grain could be turned into a product that had added value and thus could be sold at a profit.[73] Here, germinated grain and detached sprouts or coleoptiles (part of the malting process) are regularly found associated with so-called corn-driers. The archaeobotanical evidence suggests they were multifunctional structures, with the more intensely heated ovens thought to have been used to dry spelt grain and the more moderately heated ones to germinate grain and produce malt.[74] Archaeobotanical evidence for beer flavorings such as sweet gale (*Myrica gale*) and hop (*Humulus lupulus*) becomes prominent from ca. 500 CE in northwest Europe.[75]

The processes involved in the extraction of olive oil or the production of wine have seen comparable studies combining archaeobotany, ethnography, scanning electron microscopy, and experimentation. Residues of these processes, including fragmented olive stones and pressed fruit flesh of grapes and olives, can and have been identified in the archaeological record, though distinguishing between whole grapes and raisins remains problematic.[76] Fats and liquids such as oil, wine, and beer may also be studied through chemical analysis of organic residues, and include the identification of an early wine through the presence of tartaric acid in a pottery jar from a prehistoric site in Iran, and the differentiation between installations for oil and wine production.[77]

Other food types have seen less work to date, but methods are now being developed to determine which part of a plant was consumed or whether the fruit was consumed fresh or dried. For example, a study of the breakage pattern of seeds of watermelon from Roman and Islamic period sites in Egypt revealed that the consumption of the seeds, rather than the just the fruit flesh, on current evidence appears to be an Islamic-period introduction.[78] The preparation of pulses by soaking these prior to boiling speeds up the cooking process, and, importantly, in certain pulses also removes harmful toxins (e.g., grass pea and bitter vetch).[79]

Consumption—Who Ate What?

Daily food intake and adequate nutrition levels are day-to-day concerns for most people, with the lack of sufficient food a concern for many, and ample availability a pleasure for some. Apart from the need to meet basic nutritional requirements, food is used in the construction and maintenance of social relations, power relations, and many other cultural, ethnic, and religious identities. Being able to determine what was eaten, how the diet changed over time or differed between social groups is thus an important aspect of archaeobotanical research. At a basic level archaeobotany can establish which plant foods were available to the inhabitants of a site and region, but in several parts of the Greco-Roman world the database is now substantial enough to allow identification of different consumer groups and temporal changes in these.

A survey of sites with excellent preservation of botanical remains across the region suggests that by the Roman period the range of food plants available to many of its inhabitants is considerable, and far beyond mere subsistence. For example, the number of food plants recovered at two Roman quarry sites in the Eastern Desert of Egypt was 50+ (desiccated remains), at the Roman port of Myos Hormos 50+ (desiccated remains), at Roman Carthage 20+ (charred and waterlogged remains), at Pompeii 40+ (mineralized and charred remains), at Herculaneum 30+ (mineralized and charred remains), at Roman London 40+ (waterlogged, mineralized, and charred remains), at the civilian settlement Oedenburg (France) 50+ (waterlogged remains), at the minor rural settlement Wavendon Gate (England) 12+ (waterlogged and charred), and the village of Nantwich (England) 10+ (waterlogged).[80] While many of these sites have an elite presence (military or civilian), which might partially explain this rich array of foods, this is not the case at Herculaneum, Wavendon Gate, and Nantwich. At Herculaneum a sewer servicing a number of shops as well as domestic, non-elite, accommodation, produced a wide variety of foodstuffs, while Nantwich is a village and Wavendon Gate a small rural settlement, indicating that this diversity of plant foods was not restricted to elite sections of the population. What is more, this diversity of food plants does not just include local plants, but exotics and/or newly introduced foods as well, such as black pepper and date at Herculaneum, coriander, leek, fig, dill, and celery at Nantwich, and coriander, plum, cherry, celery, and summer savory at Wavendon Gate. To explore this differential social access to food plants further, the analysis of plant assemblages from non-elite sites with good preservation might usefully form a future research priority.

The diet of specific individuals is usually beyond the reach of archaeobotany, except where mummies, bog bodies, or coprolites are preserved. Here stable isotope analysis can offer great insights (see below, Chapter 4), and where possible, archaeobotanical results should thus be combined with those from zooarchaeology and stable isotope analysis.[81] While the advantage of stable isotope analysis is that it can study individuals, its disadvantage is that it can only identify very broad dietary variation (terrestrial versus marine foods, C3 versus C4 crops),[82] and then only on sites where human remains are preserved. The strength of archaeobotany lies in the fact that it can identify individual plant species and that plant remains are recovered from all settlement sites (in contrast to predominantly mortuary contexts for human remains), thus offering the potential for large-scale regional and chronological surveys.

A significant increase in availability of nutrients and flavorings has been demonstrated for the Roman period in northwest Europe. In this region the plant-based diet of the entire population throughout prehistory consisted of cereals and pulses, a limited range of wild fruits, nuts, and berries, and several wild plants used as greens, flavorings, and in medicinal recipes. Any social differentiation in diet was expressed primarily in the quantities of these foods consumed, including that of meat and better cuts of meats. This changed very rapidly with the incorporation of the region into the Roman Empire; though this process started during the later Iron Age (see "Distribution" section). A large range of fruits, nuts, vegetables, herbs, spices, and oil-rich seeds was introduced into northwest Europe at this time, initially forming part of army supplies, but soon accessible to a wider range of people.[83] In Britain for example, some 50 new plant foods were introduced. Most remained very rare, but fig, for example, is found at 40% of sites in the Early Roman period, dropping to 25% by the Late Roman period, while coriander starts at 28% and increases to just over 40% by the Mid-Roman period. While many of these foods disappear again with the withdrawal of the Roman army (e.g., olive), others stay, having become—or starting to be—integrated into British agriculture (e.g., apple, plum, cherry, walnut, cabbage, leaf beet, dill) and thus available to a wider section of society.

Where the database is substantial enough, it is possible to identify the development of different consumer groups. For example, in Roman Britain the

major towns, especially London, the military sites and the rural sites form separate consumer groups, with London sites having access to the largest range
of imports, fruits, and nuts, the military sites showing a larger-than-average
emphasis on herbs, while rural sites show a greater reliance on vegetables and
wild foods. Marked regional differences are visible too. Remarkably, the villas
(elite rural sites) do not stand out as a separate group; some show similarities
with the military sites (many imported foods), but others look no different
from non-elite rural sites. In fact, some minor rural sites (hamlets) have a
range of foodstuffs similar to that of certain villas and military sites. Thus,
here the plant remains highlight the presence of considerable within-group
variation, which appears linked to economic opportunity (proximity to major
road and river transport, markets, presence of a shrine, economic prosperity
of the region) as well as social aspirations.[84]

The interaction between food, identity and geopolitics is also in evidence
at the opposite end of the Roman Empire, at Quseir al-Qadim, located on the
Red Sea coast of Egypt. During the Roman period the diet of those working
and living in the port reflects a strong connection with the Roman world. By
the Islamic period, the residents of the port had adopted foodways more characteristic of parts of the Middle East; the port had become part of the Islamic
world. These changes in diet are part of the geopolitical realignment of the
Red Sea and its ports at that time, and they are an integral part of making
those transformations and identities real. In other words, geopolitics does not
concern only high-level political transformations, it also changes the way people live their day-to-day lives; it is through the daily routines of food procurement and consumption that these transformations become real.[85] Findings
like this make archaeobotany such a rewarding discipline.

The selection of foods used in offerings and burials offers further insights
into social and cultural choices and mortuary practices. In the past the basic
concerns of everyday life—food availability and the continuity of the agricultural cycle—were often ritualized through the provision of offerings (agricultural produce, foodstuffs), and archaeological evidence for these has
been found at many public and domestic altars, temple sites, and sacrificial
pits, as well as in a range of funerary contexts.[86] The types of food recovered
from such sites include charred bread and cake or pastries, cereal grains and
pulses, a variety of fruits, nuts, and wild plants. Some of these may have been
chosen because of their association with a particular deity, others for their
scent or ornamentation or as kindling material. For example, at the classical
necropolis at Thasos, northern Greece, foods such as pomegranates, garlic,
grapes, and bread were found to have religious significance, while at the third-
century BCE sanctuary at Messene in the Peloponnese the selection concerned
cones and seeds of stone pine, olives, grapes, almonds, and chestnuts.[87] The
state of the foods when placed on the fires—offerings of complete fruits or
breads, as against leftovers from funerary meals—can be determined using

similar techniques to those described in the section on food preparation above (scanning electron microscopy, charring experiments, fragmentation studies). Additionally, the application of combined gas chromatography–mass spectrometry has identified plant exudates (gums, resins) in late Roman burials in Britain, including resins from European pine trees and mastic/terebinth from Mediterranean *Pistacia* trees, as well as, remarkably, frankincense from Southern Arabia or eastern Africa, the latter at both Dorchester and York, Britain.[88]

Detecting patterning in these datasets is hampered by the fact that the sampling of botanical remains at burial sites, temples, and altars was often unsystematic during early excavations and, consequently, by the lack of adequate numbers of case studies from across the region. Nevertheless, the range of plants found in burials and at shrines or temples is usually very similar to that on domestic sites in the same region and period, in line with the notion that these offerings are reflections of everyday concerns surrounding food. Thus, the link between status, degree of Roman influence, and availability of newly introduced foods is seen not just at settlement sites, but in funerary contexts too.[89] Associating certain foodstuffs with particular deities is, to date, largely done through reliance on classical sources and the surviving artistic record, with the association between pine cones and the Isis cult the one most commonly referred to.[90] A recent review of the Roman period evidence for dates (*Phoenix dactylifera*) in northwest Europe suggests that this imported fruit was primarily associated with ceremonial contexts; it rarely occurs in settlement sites. It is thought to be linked to particular cults, making it more a symbolic object than a food.[91]

A special case is that of the "Lady of the Sarcophagus," the burial of a young woman discovered in an undisturbed sarcophagus in Milan, dated to the third century CE.[92] Not only could food and drink offerings be identified in the deposits associated with the sarcophagus, but microexcavation and laboratory analyses of the sarcophagus' interior also proved very informative. These identified her dress, and the possessions, gifts, or offerings placed inside the burial, including a bunch of grapes, garlands of flowers, nuts, and fruits, the latter suggesting an autumn burial. Additionally, pollen, botanical, and chemical analyses highlighted the use of resins, aromatic herbs, and unguents, such as terebinth and mastic.

Plants and plant substances were, of course, also consumed for their medicinal, aromatic, psychoactive, and decorative properties. These comprise both cultivated and wild plant species. It can be difficult to determine whether certain wild plants were used for any of these purposes, as they are often part of the local vegetation or weed flora, meaning that there are several possible mechanisms by which they arrived on site. Where such plants are found in pure and dense concentrations, as is the case for terebinth, poppy, and *Lallemantia* at several prehistoric sites in northern Greece—in quantities of 50 seeds or more—the evidence that they were used for specific purposes is convincing, even if we cannot be certain what that purpose was.[93] As with all archaeobotanical evidence, density of the remains combined with contextual information is crucial here.

Disposal—What Is Left for Us to Find?

Of all stages, this is the one most critical for archaeologists. After all, food is eaten and thus disappears. As a result, archaeobotanists have to reconstruct what was consumed, who ate what, how it was produced, distributed, or prepared through the leftovers and the waste discarded at each settlement. This means that with rare exceptions (mummies, bog bodies, time capsules such as Pompeii), we are dealing with a partial and fragmentary dataset, and one that is reduced further by the fact that dead plant tissues on or in the ground normally decay after a number of years, meaning that plants survive in the archaeological record only in certain specific circumstances. Consequently, a whole host of methodological procedures needs to be adhered to, to ensure that the data are collected and interpreted correctly. Fortunately, research has shown that archaeobotanical data are structured in a very consistent way, thus facilitating cross-cultural and temporal comparisons.[94]

The four most common modes of preservation encountered are charring or carbonization, waterlogging, desiccation, and mineralization (mineral replacement). The actual mode of preservation matters greatly, because each type of preservation preserves a slightly different range of plant types. For example, cereals and pulses are typically found in charred form, while remains of fruits, vegetables, herbs, and spices are more commonly recovered in waterlogged or mineralized form. Nuts, oil-rich seeds, and fiber plants such as flax, take an intermediate position; they are commonly found in both carbonized and waterlogged state.[95] Desiccated plant material can include all categories of crops, including vegetative parts of these crops, often in a remarkable state of preservation, but they are rare. Thus, the reconstruction of agricultural practices and consumption of staple foods (cereals and pulses) is best carried out using charred remains, which, fortunately, are found on virtually all settlement sites. In contrast, questions concerning food consumption patterns of other types of food (esp. fruits, nuts, herbs, and spices) may be better addressed using assemblages of waterlogged, desiccated, or mineralized material, the latter primarily found in sewers, latrines, or cess pits.

The strengths and weaknesses of these different modes of preservation have been highlighted by some regional assessments. For instance, a comparison between charred and waterlogged remains of wild food plants from central European Neolithic sites has indicated that charred assemblages possess on aggregate about 35% of the range of edible wild plants documented in waterlogged samples.[96] Similarly, at Roman North African sites with charred and desiccated preservation, the charred component of assemblages comprises just

20% of the total number of identifications, while the desiccated material comprises, on average, twice as many food and other economic plant taxa than the charred component. Finally, a sewer at Roman Herculaneum, containing primarily mineralized plant remains, produced very few cereals, even though these would have been a significant component of the diet.[97] This leads to two important observations. Firstly, in instances where excellent preservation is expected (sites with a high potential for waterlogged, desiccated, or mineralized preservation) sites should be sampled in great detail, to provide full evidence for activities that are not or only partially traceable at other sites. Secondly, the fact that charred remains are found at virtually all settlement sites and many ceremonial sites, combined with the fact that these assemblages show remarkable consistency in the range of plant materials they comprise (grain, pulses, cereal chaff, arable weeds and occasional nut shells and fruit stones) makes these very suited to reconstructions of agricultural practices and regional and chronological comparisons of these.

The remains of food and other plants are generally not visible with the naked eye and thus not routinely recovered during excavation; a carefully designed sampling strategy should, therefore, be part of each excavation project, aiming to collect material from the full range of activities that occurred on site. As total sampling (i.e., collecting samples from all excavated deposits) is not always practical on large-scale excavations, a sampling strategy that combines random and judgment sampling is likely to be the most successful.[98] Sample size should be adjusted to ensure retrieval of at least 100+, but preferably 300+, identifications per sample.[99] In many cases this will mean a sample size of 60 litres from deposits with charred remains and up to 10 litres where waterlogged, desiccated, or mineralized remains are present.[100] Sieving should be appropriate to the type of deposit and mode of preservation, with water flotation or wet sieving over an 0.5mm mesh practiced as standard today, though with an 0.3mm mesh used where waterlogged deposits are encountered. It goes without saying that partial sampling, small sample sizes, the use of too wide a mesh, or not sieving at all, will produce assemblages not representative of the target population and thus of little value.

Establishing the formation processes of each sample and the route of entry into the archaeological deposit for each species and plant component is a critical aspect of the interpretation of each sample. Understanding these processes has relied heavily on ethnographic studies of traditional farming and the sequence of crop processing activities taking place after the harvest, as well as on charring and digestion experiments, to establish the direction of loss.[101] Such studies rely on calculating ratios for the main crop components, densities of remains per liter of sieved deposit, frequency of each species in the samples, diversity indexes, and identification of spatial patterning of remains across sites and regions, as well as correspondence analysis and other multivariate analyses to identify correlations and associations between samples, taxa, occupation phases, and types of site. To ensure that such calculations are reliable, samples should have a sufficient number of identified remains. Ideally each sample contains at least 300 identifications, though those with 100+ can be used for less demanding analyses. As in all quantitative analyses, it is critical to think carefully about what data go into each analysis, to determine the formation process of each sample before deciding to include samples in any analysis, to ensure that each compares like with like.[102] The acronym GIGO (Garbage In, Garbage Out) is a helpful mnemonic here.

Critically important too is the dating evidence for all samples, and direct dating of individual plant specimens is advisable where archaeological dating is imprecise, where residuality in a deposit is suspected, where the result seems to be unusual for the time or region concerned, or where the introduction of new crops is monitored.[103]

It goes without saying that the archaeobotanical data need to be compared and integrated with the results of other bodies of evidence from the same sites, regions, and periods. Here the formulation of research questions is beneficial. While each project will have research questions that are specific to each line of evidence, it will be advantageous to create a number of shared research questions, where each dataset addresses the same set of questions, to identify whether similarities in the direction of change are present in all datasets. This way, the data within each line of evidence can be studied and quantified according to agreed-upon practices and methodologies within each subdiscipline, and the *answers* to each of these questions by each dataset, rather than the *data* of each line of evidence, can be integrated into a wider interpretative framework of the transformations seen at that time and place.

Daily Lives—Can We Identify Different Modes of Being?

While archaeobotany inevitably is much concerned with methodologies, the true aim of the discipline is to contribute to our understanding of the mutual interactions between humans and plants and the roles of these interactions in the cultural process. Previously, there has been a tendency towards materialism and environmental determinism, seeing production and consumption as key foci and economic and environmental factors as key drivers in changing practices. This, in turn, was replaced by a greater emphasis on human agency, an approach that recognizes and emphasizes the key role played by human action and human choices, thus moving away from notions of human actions as determined by external forces (climate change, demographic pressure, ecological stress). In this approach social factors are regarded as the key drivers in people's behavior, and people are viewed as agents that choose to use plants in order to achieve or maintain a certain outcome, such as a certain social status or a specific identity. Within this approach, however, plants are viewed merely as passive objects. Today, there is a growing understanding that both humans *and* plants have agency, and that both affect one another, that daily lives were and are shaped by the day-to-day interactions or "relationalities" between people and plants.[104] Plants were, and are, an integral part of our lives, our nutrition and health, our work, our body image, and our social relations. The properties that plants possess and display, both as growing organisms and as harvested resources, influence what we can do with them and how we can relate to them—not only in practical terms, but also in terms of the social and cultural meanings and values that they carry. Plants were, and are, used every day *and* discarded every day. Archaeobotany is thus ideally placed to identify these routine practices, to distinguish between routine and more unusual events, between group practice and individuality, and can, consequently, contribute to our understanding of past daily lives. Here a few of these interactions are briefly considered.

The routine, day-to-day engagement with food plants, in the sense of gathering, tending, cultivating, pruning, weeding, harvesting, and processing plants creates daily, monthly, and yearly rhythms, which, in the case of farmers and plant collectors, are tied to the life cycle of the crops they grow or gather. This process also includes the engagement with particular types of tools and the movements made with those tools (spade, plough, traction animal, scythe, pruning hook, threshing stick or sledge, sieve, basket), enacting the same set of actions over and over again, year after year, and all these engagements together make farmers who and what they are. These embodied routines condition how farmers see and interact with the world, the landscape, and the plants and animals, as well as other humans; they are their life.[105] By doing it they become farmers, a particular mode of being, but it is an ontology that is rooted in particular historically arisen relationships, relationships that are in a continuous process of transformation and becoming, through their interaction with both natural and cultural factors. A simple dichotomy between farming and nonfarming lifestyles is unhelpful. Each type of crop will bring its own rhythms and each environment, each social and each historical context its own set of possibilities and constraints. Plants are affected too, of course, as is clear from the fact that only some were domesticated, others became extinct, some (including weeds) spread across the globe, others did not, and so forth.

One of the best examples of these mutualistic human-plant relationships is the transition to farming and the associated emergence of sedentism, ownership, and wealth accumulation, in that this transition brought about fundamental changes in plants, animals, society, vegetation, and the material world. Complex interactions between natural factors and human agency played an important role at different stages of this transition.[106] Other examples of mutualistic human-plant relationships include the spice trade, where the potency and desirability of tropical spices combined with the social aspiration for luxury foods resulted in long-distance trade, new ports of trade, shipping and navigation innovations, and, ultimately, the rise of globalization; the introduction of so-called summer crops into the Middle East, where the potential of certain tropical and subtropical crops (e.g., sugar, cotton) combined with their physical requirements (irrigation) impacted on agriculture and labor relations in the Middle East and North Africa; the Columbian Exchange in which the European demand for cheaper produce of sugar, tobacco, and cotton, combined with the suitability of these crops to plantation cultivation, led to their introduction into the Americas and the need for a cheap labor force, which brought about the triangular slave trade; the role and attraction of sugar (and tea) in sustaining the workers during long working hours in the industrialization process; the current obesity crisis; and last, but not least, the constant battle between farmers and weeds associated with the evolution of cultivation techniques and the parallel response in seed dormancy mechanisms.[107]

On a more local scale, archaeobotany can contribute to our understanding of the daily realities of people living side by side in the same village and engaged in the same agrarian activities (provided large-scale excavation and intensive sampling were carried out). For example, at the Neolithic site of Vaihingen an der Enz, southwest Germany, a study of the crops and associated weed floras, combined with artifact assemblages at each of the houses, identified several different but contemporary house groups, each cultivating the same crops, but, according to the weed evidence, in plots at different distances from the settlement, suggesting that land was owned by "clans."[108] This differential location of plots per house group was long-lived (continuing over several generations), but not ecologically "neutral"; the best land was not equally shared between the house groups. The areas closest to the village, located on the loess soils and with high pH, could benefit from greater levels of soil disturbance and manuring and thus had higher yield potential, but these were preferentially cultivated by people from one particular house group. Other house groups cultivated lands at great distances away, on thinner loess soils, with ambiguous pH levels, less soil disturbance, and, consequently, likely lower yields. Thus, some households/ groups had an advantage over others, and notably, these differences were also expressed in the spatial patterning of the households/groups within the settlement and, as mentioned, continued over generations.[109]

The degree of social cohesion in a community may also be studied through storage practices. Here we need to acknowledge the different potential for storage between plants and animals. While plants can be consumed piecemeal and can be stored in individual households, animals, especially larger animals such as cattle, cannot; these need to be shared between households.[110] At the Neolithic site of Çatalhöyük, Central Anatolia, where families lived side by side in conjoined dwellings, plant foods (grain, fruit, nuts, condiments) were often stored in special bins in relatively inaccessible and invisible parts of the house, a potentially divisive practice. In contrast, the "storage" of animal protein was not at the household level, but through social sharing of meat, during feasts, with evidence of these communal activities that enhance social cohesion commemorated by the display of the heads and horns of aurochs near the entrances of the house.[111] This highlights how social practices are therefore not simply the imposition of arbitrary human practices on a passive world of plants and animals, but, instead, emerge, in particular historical contexts, from dynamic relationships between people, plants, animals, and things, all of which are active participants in these relations.

At a more individual or personal level, the physical ingestion of plants into the body is another arena in which plants affect our daily lives. The impact of plant substances on our physical and mental state are well known, but not yet widely studied in archaeology. Here cultural norms and belief systems govern what is regarded as edible or acceptable to eat, and research into this cultural context of food has included the identification and role of communal and elite feasting, the use of foods, including the avoidance of specific foodstuffs, in the construction of ethnic or religious identities, social relations and positions of power, as mentioned above. The "you are what you eat" view has also been used in stable isotope studies in terms of both the chemical signatures left in the bones and the nutritional deficiencies visible in the skeletal remains of individuals (see below, Chapter 4). The material properties of plants, such as their sweetness, bitterness, proteins, carbohydrates, vitamins, minerals, toxicity, and psychoactive substances, not only affect our enjoyment of and emotional reaction to foods, they are also implicated in certain addictions, sought after for their stimulant or mind-altering properties, and affect our physical well-being in other ways (overconsumption, especially of sugary and fatty foods—a current concern). Tooth decay and its associated discomfort and pain may serve as an example. High rates of caries tend to be associated with sedentary, agricultural communities as they rely heavily on cariogenic foods (foods producing or promoting the development of tooth decay), and analyses of bacterial DNA from ancient dental calculus deposits confirm that oral microbiota implicated in the development of caries become more prevalent after the transition to farming.[112] Archaeobotanical evidence for poor oral health comes from Gran Canaria, Spain, where fig seeds were found embedded in the pulpar cavities of pre-Hispanic human remains.[113] Evidence that it is the foods rather than the sedentary lifestyle that matter here can be seen in a Pleistocene community of hunter-gatherers in Morocco, where an unusually high prevalence of caries was linked to a reliance on highly cariogenic wild plant foods, such as the sweet acorns of the Holm oak (*Quercus ilex*).[114] The role of psychoactive substances in human culture and social life has so far primarily focused on the role of alcohol as a social lubricant and as a political tool.[115]

Plants have the ability to raise strong emotional reactions, and these embrace all aspects of life, including the significance of certain food taboos in religious beliefs, the association of certain foods with a foreign culture or foreign power, moral objections to luxury foods, and the role of foods in celebrations and other social occasions. Nonfood plants affect our emotions and our being too, as can be seen in the placement of flowers and garlands with the dead, in the use of ointments, in the construction of gardens, in the planting of sacred pine and elm groves at cemeteries, and of palm groves at places of recreation.[116] Finally, the place of body treatment and the use of plants, dyes, and resins deserve further investigation.[117] Combined gas chromatography–mass spectrometry (GC-MS) can now be used to identify archaeological plant resins, opening up new avenues for research. For example, these substances were used in mortuary practices to disguise the odor of decomposition, to aid soft-tissue preservation, to signify the social status of the deceased, and, most importantly, to facilitate the transition to the next world.[118]

Conclusion

Archaeobotany has contributed greatly to our understanding of daily life in the Greco-Roman world (and in past daily life more generally). It informs about mundane activities rarely discussed in surviving texts, about the annual routine of producing food, the daily chore of preparing food and disposing of the leftovers, about the daily social encounters over a meal, about nutrition and health, about social status and identity, about the ideological role of plants in personal lives, about different ontologies. It speaks about those not represented in the written record and adds extra information about those that are. The apparent "vocality" of texts[119] has meant that the contribution of archaeobotany has been less prominent in the core regions of classical archaeology than elsewhere. We must hope that this brief survey and this volume highlight and convince that the application of multiple lines of evidence will enhance our understanding of the past and will illuminate more clearly the great complexity and diversity of practice and being.

Acknowledgements

I would like to thank Walter Scheidel for inviting me to contribute to this volume, and I am grateful to Amy Bogaard, Terry Hopkinson, Valerie Maxfield, and Jacob Morales for helpful comments on earlier drafts of this paper.

Notes

1. Goody 1982.
2. Margaritis and Jones 2009.
3. E.g., Campbell 2008; M. Jones 2007, 260–266; Nesbitt and Samuel 1996; Pelling 2008; Van der Veen 1995; Van der Veen 2014b; Zech-Matterne et al. 2014.
4. Zech-Matterne et al. 2014.
5. Rösch et al. 1992.
6. Zech-Matterne et al. 2014.
7. M. Jones 2007, 260–269.
8. Van der Veen 1992.

9. Bogaard et al. 2011; see also Section 7 below.

10. M. Jones 1988, 2009.

11. Autecology focuses on the relationship between an individual plant species and its environment, while phytosociology deals with plant communities and the relationships between the species within them.

12. Bogaard et al. 1999; Charles et al. 2003; G. Jones 2002; G. Jones et al. 1999, 2000, 2010.

13. Bogaard, et al. 2013; Fraser, et al. 2011; Wallace, et al. 2013.

14. Bogaard et al. 2007.

15. Bogaard and Outram, 2013; Fiorentino et al. 2015; Fraser et al. 2013.

16. Araus et al. 1997; Riehl et al. 2014.

17. Riehl 2009.

18. Cunniff et al. 2010.

19. Fiorentino et al. 2015; Nitsch et al. 2015.

20. E.g., Marvelli et al. 2013; Rodríguez-Ariza and Moya 2005.

21. Burger et al. 2011.

22. Newton et al. 2014; Terral et al. 2004 (olive); Orrù et al. 2013; Pagnoux et al. 2015; Terral et al. 2010 (grape); Burger et al. 2011 (cherry); Gros-Balthzard et al. 2016; Terral et al. 2012 (date).

23. Allaby et al. 2014; Brown et al. 2015.

24. Allaby et al. 2014; Li, et al 2011; Palmer et al. 2009; Palmer et al. 2012.

25. Cappers 2006; Rowley-Conwy 1994; Van der Veen 2011; Van der Veen and Morales 2015 (Egypt); Jiang et al. 2015 (China); Morales et al. 2014; Oliveira et al. 2012 (Spain); Ernst and Jacomet 2005; Letts 1999 (buildings Europe).

26. Manen et al. 2003; Pollmann et al. 2005; Schlumbaum and Edwards 2013.

27. Bunning et al. 2012; Fernández et al. 2013.

28. Brown et al. 2015.

29. G. Jones et al. 2012; G. Jones et al. 2013.

30. Brown et al. 2015; Cooper and Poinar 2000.

31. Fuller and Stevens 2009; Fuller et al. 2014; Hillman 1981; Van der Veen 1992, 2007; Van der Veen and Jones 2006.

32. Hillman 1984; Van der Veen and Jones 2006, 2007.

33. Neef 1990; W. Smith 1998; Rowan 2015.

34. Rowan 2015.

35. Van der Veen 2014b.

36. Ciaraldi 2007; Robinson 1999; Robinson 2012.

37. Motta, 2002.

38. Giesecke 2013; Jashemski 1979; Kron 2013.

39. E.g., Bouchaud 2011; Bouchaud et al. 2011; M. Jones 1981; Van der Veen et al. 1996.

40. See Bottema et al. 1990; Cheddadi et al. 2015; Foxhall et al. 2007; Harris 2013; Mercuri et al. 2015; Roberts et al. 2004; Sadori and Giardini 2007; Veal 2012, 2013, 2014.

41. Halstead 2014, chapter 7; see also Van der Veen 2010.

42. Smith and Kenward 2011.

43. Booth et al. 2007, 24, 281; Kenward and Williams 1979; Smith and Kenward 2011.

44. Smith and Kenward 2011.

45. Kenward and Williams 1979; Straker 1984. See also Kislev and Simchoni 2007 for an example from Masada, Israel.

46. Pals et al. 1989; Pals and Hakbijl 1992.

47. Zech-Matterne et al. 2014.

48. Straker 1984.

49. Kooistra et al. 2013; Van Dinter et al. 2014.

50. Zech-Matterne et al. 2014.

51. Domínguez-Delmás et al. 2014.

52. Jansma et al. 2014; Visser 2015.

53. Vermeeren 1999; Van der Veen and Gale 2011.

54. Cappers 2006; Van der Veen 2011.

55. Livarda 2011; Van der Veen 2011; Van der Veen and Morales 2015.

56. Bouby et al. 2011; Sadori et al. 2010; Sadori et al. 2014; Van Zeist et al. 2001; Vittori et al. 2015.

57. Van der Veen 1998a and b, 2001; Van der Veen and Tabinor 2007.

58. Tomber 1996.

59. Bülow-Jacobsen 1997, 2003; Cuvigny 1996, 2000.

60. Bakels and Jacomet 2003; Livarda 2011; Livarda and Van der Veen 2008; Van der Veen 2008; Van der Veen et al. 2008.

61. Kreuz 2004; Lodwick 2014; Zech-Matterne et al. 2009.

62. Harris et al. 2002.

63. Macauley-Lewis 2006.

64. Buckland and Magilton 1986, 198; Robinson et al. 2006.

65. G. Jones et al. 2012, 2013.

66. Pals and Hakbijl 1992; Sealey and Tyres 1989; Ward 2001, 2003.

67. Rowan 2014; Stahl 1989.

68. al-Azm 2009; Samuel 2000; Valamoti et al. 2008.

69. Valamoti 2011.

70. Samuel 2000.

71. Heiss 2014; Heiss et al. 2015.

72. Bouby et al. 2011; Stika 2011.

73. M. Jones 1981.

74. Campbell 2008; Cunliffe 2009; Van der Veen 1989.

75. Behre 1992.

76. Margaritis and Jones 2006, 2008a and b; Marinova et al. 2011; Miller 2008; Valamoti et al. 2007.

77. Evershed 2008; McGovern et al. 1995, 1996; Pecci et al. 2013.

78. Cox and Van der Veen 2008. For other examples, see Van der Veen 2011, chapter 4.4.

79. Valamoti et al. 2011

80. Ciaraldi 2007; Davis 2011; Pearson and Letts 1996; Rowan 2014; Robinson and Rowan 2015; Tomlinson 1987; Van der Veen 2001, 2011; Van der Veen and Tabinor 2007; Vandorpe and Jacomet 2011a; Van Zeist et al. 2001.

81. E.g., Papathanasiou et al. 2013.

82. Plants do not all have the same photosynthesis pathway; they do not all fixate carbon in the same way. Most food plants in temperate environments (for example, wheat, barley, rice, apples, carrots, spinach, sugar beet, trees) have a so-called C_3 pathway, while many tropical food plants (for example, maize, sorghum, millet, sugar cane) have a so-called C_4 pathway. The introduction of tropical, C_4, plants into a temperate environment can thus be identified in the carbon isotope ratios of those that consumed those plants.

83. Bakels and Jacomet 2003; Jacomet et al. 2002; Kreuz 2004; Livarda and Van der Veen 2008; Van der Veen et al. 2008. See also Van der Veen 2003.

84. Van der Veen 2008; Van der Veen et al. 2008.

85. Van der Veen 2011; Van der Veen and Morales 2017.

86. E.g., Bouby and Marinval 2004; Heiss 2014; Kohler-Schneider et al. 2015; Kučan 1995; Megaloudi 2005; Megaloudi et al. 2007; Petrucci-Bavaud and Jacomet 1997; Robinson 2002; Rottoli and Castiglioni 2011; Vandorpe and Jacomet 2011b; Zach 2002.

87. Megaloudi 2005; Megaloudi et al. 2007.

88. Brettell et al. 2015.

89. Bouby and Marinval 2004; Robinson 2002; Rottoli and Castiglioni 2011.

90. E.g., Kislev 1988.

91. Livarda 2013.

92. Rossignani et al. 2005.

93. Valamoti 2012/13.

94. Fuller et al. 2014; Hillman 1981; Jacomet 2013; Van der Veen 2007; Van der Veen and G. Jones 2006.

95. Van der Veen et al. 2013, Fig. 7.

96. Colledge and Conolly 2014.

97. Van der Veen 2007 (North Africa); Rowan 2014 (Herculaenum).

98. M. Jones 1991; Van der Veen 1984.

99. Van der Veen and Fieller 1982.

100. E.g., Campbell, Moffett and Straker 2011.

101. E.g., Boardman and Jones 1990; Braadbaart 2008; Charles 1998; Hillman 1981, 1984; Jacomet 2013; G. Jones 1984, 1998; Kreuz 1990; Miller and Smart 1984; Van der Veen and Jones 2006; Van der Veen 2007; Wallace and Charles 2013.

102. G. Jones 1991; Van der Veen and Fieller 1982.

103. Pelling et al. 2015.

104. E.g., Ingold 1993, 1996; Van der Veen 2014a.

105. Ingold 1993, 1996; Halstead 2014.

106. E.g., Cunniff et al. 2010; Fuller et al. 2010; G. Jones et al. 2013; Zeder 2006.

107. E.g., Crosby 2003; M. Jones 1988, 2009; Mintz 1985; Van der Veen 2014a; Van der Veen and Morales 2015, 2017; Viola 1991; Watson 1983.

108. Bogaard et al. 2011.

109. Ibid.

110. Bogaard et al. 2009.

111. Ibid.

112. Adler et al. 2013.

113. Morales and Gil 2014.

114. Humphrey et al. 2014.

115. Dietler 2006; Sherratt 1991; though see the work by Lewis-Williams (2004) on mind- and behaviour-altering substances.

116. E.g., Androu et al. 2013; Hamdy 2007; Jashemski 1979; Koch et al. in press; Robinson 2016; Rossignani et al. 2005; Schlumbaum et al. 2011; Wilson 2016.

117. E.g. Van der Veen, Hall and May 1993.

118. Brettell et al. 2015; see also Vermeeren and Van Haaster 2002.

119. Moreland 2001, 33–34.

References

Adler, C. J., et al. 2013. "Sequencing ancient calcified dental plaque shows changes in oral microbiota with dietary shifts of the Neolithic and Industrial revolutions." *Nature Genetics* 45 (4): 450–455. DOI: 10.1038/ng.2536.

Allaby, R.G., et al. 2014. "Using archaeogenomic and computational approaches to unravel the history of local adaptation in crops." *Philosophical Transactions of Royal Society of London* B 370: 20130377. DOI: 10.1098/rstb.2013.0377.

al-Azm, A. 2009. "The importance and antiquity of frikkeh: a simple snack or a socio-economic indicator of decline and prosperity in the ancient Near East?" In *From For-*

agers to Farmers: Papers in Honour of Gordon C. Hillman, eds. A. Fairbairn and E. Weiss. Oxford: Oxbow Books, 112–116.

Androu, S., et al. 2013. "Smelly barbarians or perfumed natives? An investigation of oil and ointment use in Late Bronze Age northern Greece." In *Diet, Economy and Society in the Ancient Greek world: Towards a Better Integration of Archaeology and Science*, eds. S. Voutsaki and S. M. Valamoti. Leuven: Peeters, 173–185.

Araus, J. L., et al. 1997. "Changes in carbon isotope discrimination in grain cereals from different regions of the western Mediterranean basin during the past seven millennia: Palaeoenvironmental evidence of a differential change in aridity during the late Holocene." *Global Change Biology* 3: 107–118. DOI: 10.1046/j.1365-2486.1997.00056.x.

Bakels, C., and S. Jacomet. 2003. "Access to luxury foods in Central Europe during the Roman period: the archaeobotanical evidence." *World Archaeology* 34 (1): 542–557. DOI: 10.1080/0043824021000026503.

Behre, K. E. 1992. "The history of beer additives in Europe—a review." *Vegetation History and Archaeobotany* 8 (1): 35–48. DOI: 10.1007/BF02042841.

Boardman, S., and G. Jones. 1990. "Experiments on the effects of charring on cereal plant components." *Journal of Archaeological Science* 17: 1–11. DOI: 10.1016/0305-4403(90)90012-T.

Bogaard, A., R. Krause, and H.-C. Strien. 2011. "Towards a social geography of cultivation and plant use in an early farming community: Vaihingen an der Enz, south-west Germany." *Antiquity* 85: 395–416. DOI: 10.1017/S0003598X00067831.

Bogaard, A., and, A. K. Outram. 2013. "Palaeodiet and beyond: stable isotopes in bioarchaeology." *World Archaeology* 45 (3): 333–337. DOI: 10.1080/00438243.2013.829272.

Bogaard, A., et al. 1999. "A FIBS approach to the use of weed ecology for the archaeobotanical recognition of crop rotation regimes." *Journal of Archaeological Science* 26: 1211–1224. DOI: 10.1006/jasc.1998.0364.

Bogaard, A., et al. 2007. "The impact of manuring on nitrogen isotope ratios in cereals: archaeological implications for reconstruction of diet and crop management practices." *Journal of Archaeological Science* 34: 335–343. DOI: 10.1073/pnas.1305918110.

Bogaard, A., et al. 2009. "Private pantries and celebrated surplus: storing and sharing food at Neolithic Çatalhöyük, Central Anatolia." *Antiquity* 83: 649–668. DOI: 10.1017/S0003598X00098896.

Bogaard, A., et al. 2013. "Crop manuring and intensive land management by Europe's first farmers." *Proceedings of the National Academy of Sciences of the United States of America* 110 (31): 12589–12594. DOI: 10.1073/pnas.1305918110.

Booth, P., et al. 2007. *Thames through Time: The Archaeology of the Gravel Terraces of the Upper and Middle Thames. The Early Historical Period: AD 1–1000*. Oxford: Oxford Archaeology.

Bottema, S., G. Entjes-Nieborg, and W. Van Zeist, eds. 1990. *Man's Role in the Shaping of the Eastern Mediterranean Landscape*. Rotterdam: Balkema.

Bouby, L., P. Boissinot, and P. Marinval. 2011. "Never mind the bottle: Archaeobotanical evidence of beer-brewing in Mediterranean France and the consumption of alcoholic beverages during the 5th Century BC." *Human Ecology* 39: 351–360. DOI: 10.1007/s10745-011-9395-x.

Bouby, L., A. Bouchette, and I. Figueiral. 2011. "Sebesten fruits (*Cordia myxa* L.) in Gallia Narbonensis (Southern France): a trade item from the Eastern Mediterranean?" *Vegetation History and Archaeobotany* 20 (5): 397–404. DOI: 10.1007/s00334-011-0285-3.

Bouby, L., and P. Marinval. 2004. "Fruits and seeds from Roman cremations in Limagne (Massif Central) and the spatial variability of plant offerings in France." *Journal of Archaeological Science* 31: 77–86. DOI: 10.1016/j.jas.2003.07.006.

Bouchaud, C. 2011. "Paysages et pratiques d'exploitation des ressources végétales en milieux semi-aride et aride dans le sud du Proche-Orient: Approche archéobotanique des périodes antique et islamique (IVe siècle av. J.-C.–XVIe siècle ap. J.-C.)." PhD dissertation, Université Paris 1 Panthéon-Sorbonne, Paris.

Bouchaud, C., M. Tengberg, and P. Dal-Prà. 2011. "Cotton cultivation and textile production in the Arabian Peninsula during Antiquity: The evidence from Madâ'in Sâlih (Saudi Arabia) and Qal'at al-Bahrain (Bahrain)." *Vegetation History and Archaeobotany* 20 (5): 405–417. DOI: 10.1007/s00334-011-0296-0.

Braadbaart, F. 2008. "Carbonisation and morphological changes in modern dehusked and husked *Triticum dicoccum* and *Triticum aestivum* grains." *Vegetation History and Archaeobotany* 17 (1): 155–166. DOI: 10.1007/s00334-011-0296-0.

Brettell, R. C., et al. 2015. " '*Choicest unguents*': molecular evidence for the use of resinous plant exudates in late Roman mortuary rites in Britain." *Journal of Archaeological Science* 53: 639–648. DOI: 10.1016/j.jas.2014.11.006.

Brown, T.A., et al. 2015. "Recent advances in ancient DNA research and their implications for archaeobotany." *Vegetation History and Archaeobotany* 24 (1): 207–214. DOI: 10.1007 /s00334-014-0489-4.

Buckland, P. C., and J. R. Magilton. 1986. *The Archaeology of Doncaster*. Volume 1: *The Roman Civil Settlement*. Oxford: British Archaeological Reports.

Bülow-Jacobsen, A. 1997. "The correspondence of Dioscorus and others (224–242)." In *Mons Claudianus: Ostraca Graeca et Latina II (O. Claud. 191–416)*, ed. J. Bingen et al. Cairo: Institut Français d'Archéologie Orientale du Caire, 43–68.

Bülow-Jacobsen, A. 2003. "The traffic on the road and the provisioning of the stations." In *La route de Myos Hormos: l'armée romaine dans le désert Oriental d'Égypte*, ed. H. Cuvigny. Cairo: Institut Français d'Archéologie Orientale du Caire, 399–425.

Bunning, S., G. Jones, and T. Brown. 2012. "Next generation sequencing of DNA in 3300-year-old charred cereal grains." *Journal of Archaeological Science* 39: 2780–2784. DOI: 10.1016/j.jas.2012.04.012.

Burger, P., et al. 2011. "Assessing the agrobiodiversity of *Prunus avium* L. (Rosaceae): a morphometric approach focussed on the stones from the archaeological site of Hôtel-Dieu (16th century, Tours, France)." *Vegetation History and Archaeobotany* 20 (5): 447–458. DOI: 10.1007/s00334-011-0310-6.

Campbell, G. 2008. "Plant utilization in the countryside around Danebury: a Roman perspective." In *The Danebury Environs Programme. A Wessex Landscape during the Roman Era*. Volume 1: *Overview*, ed. B. Cunliffe. Oxford: English Heritage and Oxford University School of Archaeology, 53–74.

Campbell, G., L. Moffett, and V. Straker. 2011. *Environmental Archaeology: A Guide to the Theory and Practice of Methods, from Sampling and Recovery to Post-excavation (2nd edition)*. Swindon: English Heritage. http://www.english-heritage.org.uk/publica tions/environmental-archaeology-2nd/.

Cappers, R.T.J. 2006. *Roman Foodprints at Berenike*: *Archaeobotanical Evidence of Subsistence and Trade in the Eastern Desert of Egypt*. Los Angeles: UCLA.

Charles, M. 1998. "Fodder from dung: the recognition and interpretation of dung-derived plant material from archaeological sites." *Environmental Archaeology* 1: 111–122. DOI: 10.1179/env.1996.1.1.111.

Charles, M., et al. 2003. "Using weed functional attributes for the identification of irrigation regimes in Jordan." *Journal of Archaeological Science* 30: 1429–1441. DOI: 10.1016 /S0305-4403(03)00038-4.

Cheddadi, R., M. Nourelbait, and O. Bouaissa. 2015. "A history of human impact on Moroccan mountain landscapes." *African Archaeological Review* 32: 233–248. DOI: 10.1007 /s10437-015-9186-7.

Ciaraldi, M. 2007. *People and Plants in Ancient Pompeii, the Use of Plant Resources at Pompeii and in the Pompeian Area from the 6th century BC to AD 79*. London: Accordia Specialist Studies on Italy 12.

Colledge, S., and J. Conolly. 2014. "Wild plant use in European Neolithic subsistence economies: a formal assessment of preservation bias in archaeobotanical assemblages and the implications for understanding changes in plant diet breadth." *Quaternary Science Reviews* 101 (1): 193–206. DOI: 10.1016/j.quascirev.2014.07.013.

Cooper, A., and H. N. Poinar. 2000. "Ancient DNA: Do it right or not at all." *Science* 289 (5482): 1139. DOI: 10.1126/science.289.5482.1139b.

Cox, A., and M. Van der Veen. 2008. "Changing foodways: watermelon (*Citrullus lanatus*) consumption in Roman and Islamic Quseir al-Qadim, Egypt." *Vegetation History and Archaeobotany* 17 (1): 181–189. DOI: 10.1007/s00334-008-0164-8.

Crosby, A.W.J. 2003. *The Columbian Exchange: Biological and Cultural Consequences of 1492*. Westport, CT: Praeger.

Cunliffe, B. 2009. "Continuity and change in a Wessex landscape." *Proceedings of the British Academy* 162, 161–210.

Cunniff, J., et al. 2010. "Was low atmospheric CO_2 a limiting factor in the origin of agriculture?" *Environmental Archaeology* 15 (2): 113–23. DOI: 10.1179/146141010X12640787648469.

Cuvigny, H. 1996. "The amount of wages paid to the quarry workers at Mons Claudianus." *Journal of Roman Studies* 86: 139–145.

Cuvigny, H. 2000. *Mons Claudianus Ostraca Graeca et Latina III. Les reçus pour avances à la familia. O Claud. 417 à 631*. Cairo: Institut Français d'Archéologie Orientale de Caire.

Davis, A. 2011. "Plant foods." In *Roman London and the Walbrook Stream Crossing; Excavations at 1 Poultry and Vicinity, City of London*, eds. J. Hill and P. Rowsome. London: Museum of London Archaeology: 409–410.

Dietler, M. 2006. "Alcohol: anthropological/archaeological perspectives." *Annual Review of Anthropology* 35: 229–249. DOI: 10.1146/annurev.anthro.35.081705.123120.

Domínguez-Delmás, M., et al. 2014. "Long-distance oak supply in mid-2nd century AD revealed: the case of a Roman harbour (Voorburg-Arentsburg) in the Netherlands." *Journal of Archaeological Science* 41: 642–654. DOI: 10.1016/j.jas.2013.09.009.

Ernst, M., and S. Jacomet, 2005. "The value of the archaeobotanical analysis of desiccated plant remains from old buildings: methodological aspects and interpretation of crop weed assemblages." *Vegetation History and Archaeobotany* 15 (1): 45–56. DOI: 10.1007/s00334-005-0077-8.

Evershed, R.P. 2008. "Experimental approaches to the interpretation of absorbed organic residues in archaeological ceramics." *World Archaeology* 40 (1): 26–47. DOI: 10.1080/00438240801889373.

Fernández, E., et al. 2013. "DNA analysis in charred grains of naked wheat from several archaeological sites in Spain." *Journal of Archaeological Science* 40: 659–670. DOI: 10.1016/j.jas.2012.07.014.

Fiorentino, G., et al. 2015. "Stable isotopes in archaeobotanical research." *Vegetation History and Archaeobotany* 24 (1): 215–227. DOI: 10.1007/s00334-014-0492-9.

Foxhall, L., M. Jones, and H. Forbes. 2007. "Human ecology and the classical landscape." *Blackwell Studies in Classical Archaeology*, eds. S. Alcock and R. Osborne. Oxford: Blackwell, 89–117.

Fraser, R. A., et al. 2011. "Manuring and stable nitrogen isotope ratios in cereals and pulses: towards a new archaeobotanical approach to the inference of land use and dietary practices." *Journal of Archaeological Science* 38: 2790–2804. DOI: 10.1016/j.jas.2011.06.024.

Fraser, R. A., et al. 2013a. "Assessing natural variation and the effects of charring, burial and pre-treatment on the stable carbon and nitrogen isotope values of archaeobotanical

cereals and pulses." *Journal of Archaeological Science* 40: 4754–4766. DOI: 10.1016/j
.jas.2013.01.032.

Fraser, R. A., et al. 2013b. "Integrating botanical, faunal and human stable carbon and nitrogen isotope values to reconstruct landuse and palaeodiet at LBK Vaihingen an der
Erz, Baden-Württemberg." *World Archaeology* 45 (3): 492–517.

Fuller, D. Q., R. G. Allaby, and C. Stevens. 2010. "Domestication as innovation: the entanglement of techniques, technology and chance in the domestication of cereal crops."
World Archaeology 42 (1): 13–28. DOI: 10.1080/00438240903429680.

Fuller, D. Q., and C. J. Stevens. 2009. "Agriculture and the development of complex societies: an archaeobotanical agenda." In *From Foragers to Farmers: Papers in Honour
of Gordon C. Hillman*, eds. A. Fairbairn and E. Weiss. Oxford: Oxbow Books, 37–57.

Fuller, D. Q., C. Stevens, and M. McClatchie. 2014. "Routine activities, tertiary refuse and
labor organisation: social inferences from everyday archaeobotany." In *Ancient Plants
and People: Contemporary Trends in Archaeobotany*, eds. M. Madella, C. Lancelotti,
and M. Savard. Tucson: University of Arizona Press, 174–217.

Giesecke, A. 2013. "Gardens, Roman Empire." In *The Encyclopedia of Ancient History*,
eds. R. S. Bagnall et al. Oxford: Blackwell, 2853–2856. DOI: 10.1002/9781444338386
.wbeah18049.

Goody, J. 1982. *Cooking, Cuisine and Class: A Study in Comparative Sociology*. Cambridge:
Cambridge University Press.

Gros-Balthazard, M., et al. 2016. "The domestication syndrome in *Phoenix dactylifera* seeds:
toward the identification of wild date palm populations." *PLOS One*. DOI: 10.1371/jour
nal.pone.0152394.

Halstead, P. 2014. *Two Oxen Ahead: Pre-Mechanized Farming in the Mediterranean*. Oxford:
Wiley-Blackwell. DOI: 10.1002/9781118819333.

Hamdy, R. 2007. "Plant remains from the intact garlands present at the Egyptian Museum
in Cairo." In *Fields of Change: Progress in African Archaeobotany*, ed. R. Cappers. Groningen: Barkhuis and Groningen University Library, 115–126.

Harris, S. A., J. P. Robinson, and B. E. Juniper. 2002. "Genetic clues to the origin of the
apple." *Trends in Genetics* 18: 426–430. DOI: 10.1016/S0168-9525(02)02689-6.

Harris, W. V. 2013. "Defining and detecting Mediterranean deforestation, 800 BCE to
700 CE." In *The Ancient Mediterranean Environment between Science and History*, ed.
W. V. Harris. Leiden: Brill, 173–194.

Heiss, A. G. 2014. "Ceremonial foodstuffs from prehistoric burnt offering places in the
Alpine region." In *Plants and People: Choices and Diversity Through Time*, eds.
A. Chevallier, E. Marinova, and L. Peña-Chocarro. Oxford: Oxbow Books, 343–353.
DOI: 10.13140/2.1.2776.5767.

Heiss, A. G., et al. 2015. "Tissue-based analysis of a charred flat bread (*galette*) from a Roman
cemetery at Saint-Memmie (Dép. Marne, Champagne-Ardenne, north-eastern France)."
Journal of Archaeological Science 55: 71–82. DOI: 10.1016/j.jas.2014.12.014.

Hillman, G. 1981. "Reconstructing crop husbandry practices from charred remains of crops."
In *Farming Practice in British Prehistory*, ed. R. Mercer. Edinburgh: University of
Edinburgh Press, 123–162.

Hillman, G. 1984. "Interpretation of archaeological plant remains: the application of ethnographic models from Turkey." In *Plants and Ancient Man: Studies in Paleoethnobotany*,
eds. W. van Zeist and W. A. Casparie. Rotterdam: Balkema, 1–41.

Humphrey, L. T., et al. 2014. "Earliest evidence for caries and exploitation of starchy plant
foods in Pleistocene hunter-gatherers from Morocco." *Proceedings of the National
Academy of Sciences of the United States of America* 111 (3): 954–959. DOI: 10.1073
/pnas.1318176111.

Ingold, T. 1993. "The temporality of landscape." *World Archaeology* 25 (2): 152–174. DOI: 10.1080/00438243.1993.9980235.

Ingold, T. 1996. "Growing plants and raising animals: an anthropological perspective on domestication." In *The Origins and Spread of Agriculture and Pastoralism in Eurasia*, ed. D. R. Harris. London: University College London Press, 12–24.

Jacomet, S. 2013. "Archaeobotany: analyses of plant remains from waterlogged archaeological sites." In *The Oxford Handbook of Wetland Archaeology*, eds. F. Menotti and A. O'Sullivan. Oxford: Oxford University Press, 497–514. DOI: 10.1093/oxfordhb /9780199573493.013.0030.

Jacomet, S., et al. 2002. "*Punica granatum* L. (pomegranates) from early Roman contexts in Vindonissa (Switzerland)." *Vegetation History and Archaeobotany* 11 (1–2): 79–92. DOI: 10.1007/s003340200009.

Jansma, E., K. Haneca, and M. Kosian. 2014. "A dendrochronological reassessment of three Roman boats from Utrecht (the Netherlands): evidence of inland navigation between the lower-Scheldt region in Gallia Belgica and the limes of Germania inferior." *Journal of Archaeological Science* 50: 484–496. DOI: 10.1016/j.jas.2014.07.019.

Jashemski, W.J. 1979. *The Gardens of Pompeii, Herculaneum and the Villas Destroyed by Vesuvius*. New York: Caratzas.

Jiang, H., et al. 2015. "Archaeobotanical evidence of plant utilization in the ancient Turpan of Xinjiang, China: a case study at the Shengjindian cemetery. *Vegetation History and Archaeobotany* 24 (1): 165–177. DOI: 10.1007/s00334-014-0495-6.

Jones, G. 1984. Interpretation of archaeological plant remains: ethnographic models from Greece. In W. Van Zeist and W.A. Casparie (ed.), *Plants and Ancient Man: Studies in Paleoethnobotany*. Rotterdam: Balkema, pp. 43–61.

Jones, G. 1991. "Numerical analysis in archaeobotany." In *Progress in Old World Palaeoethnobotany*, eds. W. van Zeist, K. Wasylikowa, and K.E. Behre. Rotterdam: Balkema, 63–80.

Jones, G. 1998. "Distinguishing food from fodder in the archaeobotanical record." *Environmental Archaeology* 1: 95–98. DOI: 10.1179/env.1996.1.1.95.

Jones, G. 2002. "Weed ecology as a method for the archaeobotanical recognition of crop husbandry practices." *Acta Archaeobotanica* 42: 185–193.

Jones, G., et al. 1999. "Identifying the intensity of crop husbandry practices on the basis of weed floras." *Annual of the British School at Athens* 94: 167–89. DOI: 10.1017 /S0068245400000563.

Jones, G., et al. 2000. "Distinguishing the effects of agricultural practices relating to fertility and disturbance: a functional ecological approach in archaeobotany." *Journal of Archaeological Science* 27: 1073–2784.

Jones, G., et al. 2010. "Crops and weeds: the role of weed functional types in the identification of crop husbandry methods." *Journal of Archaeological Science* 37: 70–77.

Jones, G., et al. 2012. "Phylogeographic analysis of barley DNA as evidence for the spread of Neolithic agriculture through Europe." *Journal of Archaeological Science* 39: 3230–3238. DOI: 10.1016/j.jas.2012.05.014.

Jones, G., et al. 2013. "Barley DNA evidence for the routes of agricultural spread into Europe following multiple domestications in Western Asia." *Antiquity* 87: 701–13. DOI: 10.1017/S0003598X00049401.

Jones, M. 1981. "The development of crop husbandry." In *The Environment of Man: The Iron Age to the Anglo-Saxon period*, eds. M. Jones and G. Dimbleby. Oxford: British Archaeological Reports, 95–127.

Jones, M. 1988. "The arable field: a botanical battleground." In *Archaeology and the Flora of the British Isles: Human Influence on the Evolution of Plant Communities*, ed. M. Jones. Oxford: Oxford University Committee for Archaeology, 86–92.

Jones, M. 1991. "Sampling in Palaeoethnobotany." In *Progress in Old World Palaeoethnobotany*, eds. W. van Zeist, K. Wasylikowa, and K. E. Behr. Rotterdam: Balkema, 53–62.

Jones, M. 2007. *Feast.: Why Humans Share Food*. Oxford: Oxford University Press.

Jones M. 2009. "Dormancy and the plough: weed seed biology as an indicator of agrarian change in the first millennium AD." In *From Foragers to Farmers: Papers in Honour of Gordon C. Hillman*, eds. A. Fairbairn and E. Weiss. Oxford: Oxbow Books, 58–63.

Kenward, H. K., and D. Williams. 1979. *Biological Evidence from the Roman Warehouses at Coney Street*. The Archaeology of York, The Past Environment of York, 14/2: 44–100.

Kislev, M. E. 1988. *Pinus pinea* in agriculture, culture and cult. In *Der Prähistorische Mensch und seine Umwelt: Festschrift für Udelgard Körber-Grohne*, ed. H. J. Küster. Stuttgart: Konrad Theiss, 73–79.

Kislev, M. E., and O. Simchoni. 2007. "Hygiene and insect damage of crops and foods at Masada." In *Masada VIII*, eds. J. Aviram et al. Jerusalem: Israel Exploration Society and The Hebrew University, 133–170.

Koch, P., et al. In Press. "Feasting in a sacred grove: a multidisciplinary study of the Roman sanctuary of Kempraten, CH." In *Bioarchaeology of Ritual and Religion*, eds. A. Livarda, R. Madgwick, and S. Riera. Oxford: Oxbow Books.

Kohler-Schneider, M., A. Caneppele, and A. G. Heiss. 2015. "Land use, economy and cult in late Iron Age ritual centres: an archaeobotanical study of the La Tène site at Sandberg-Roseldorf, Lower Austria." *Vegetation History and Archaeobotany* 24 (4): 517–540. DOI: 10.1007/s00334-014-0511-x.

Kooistra, L. I., et al. 2013. "Could the local population of the Lower Rhine delta supply the Roman army? Part 1." *Journal of the Archaeology of the Low Countries* 4 (2): 5–23. DOI: http://jalc.nl/cgi/t/text/text-idx7c37.html.

Körber-Grohne, U. 1981. "Distinguishing prehistoric grains of *Triticum* and *Secale* on the basis of their surface patterns using scanning electron microscopy." *Journal of Archaeological Science* 8: 197–204. DOI: 10.1016/0305-4403(81)90024-8.

Kreuz, A. 1990. "Searching for single activity refuse in Linearbandkeramik settlements: an archaeobotanical approach." In D. E. Robinson, ed. *Experimentation and Reconstruction in Environmental Archaeology*. Symposia of the Association for Environmental Archaeology No. 9. Oxford: Oxbow Books: 63–74.

Kreuz, A. 2004. "Landwirtschaft im Umbruch? Archäobotanische Untersuchungen zu den Jahrhunderten um Christi Geburt in Hessen und Mainfranken." *Bericht der Römisch-Germanischen Kommission* 85: 97–304.

Kron, G. 2013. "Agriculture, Roman Empire." In *The Encyclopedia of Ancient History*, eds. R. S. Bagnall, et al. Oxford: Blackwell, 217–222. DOI: 10.1002/9781444338386.wbeah18004.

Kučan, D. 1995. "Zur Ernahrung und dem Gebrauch von Pflanzen im Heraion von Samos im 7. Jahrhundert v. Chr." *Jahrbuch des Deutschen Archaologischen Instituts* 110: 1–64.

Letts, J. B. 1999. *Smoke Blackened Thatch: A Unique Source of Late Medieval Plant Remains from Southern England*. London: English Heritage and University of Reading.

Lewis-Williams, D. 2004. *The Mind in the Cave*. London: Thames & Hudson.

Li, C. X., et al. 2011. "Ancient DNA analysis of desiccated wheat grains excavated from a Bronze Age cemetery in Xinjiang." *Journal of Archaeological Science* 38: 115–119. DOI: 10.1016/j.jas.2010.08.016.

Livarda, A. 2011. "Spicing up life in north-western Europe: exotic food plant imports in the Roman and Medieval world." *Vegetation History and Archaeobotany* 20 (2): 143–164. DOI: 10.1007/s00334-010-0273-z.

Livarda, A. 2013. "Date, rituals and socio-cultural identity in the north-western Roman provinces." *Oxford Journal of Archaeology* 32 (1): 101–117. DOI: 10.1111/ojoa.12004.

Livarda, A., and M. Van der Veen. 2008. "Social access and dispersal of condiments in north-west Europe from the Roman to the Medieval period." *Vegetation History and Archaeobotany* 17 (1): 201–209. DOI: 10.1007/s00334-008-0168-4.

Lodwick, L. 2014. "Condiments before Claudius: new plant foods at the Late Iron Age oppidum at Silchester, UK." *Vegetation History and Archaeobotany* 23 (5): 543–549. DOI: 10.1007/s00334-013-0407-1.

Macaulay-Lewis, E. 2006. "The role of *ollae perforatae* in understanding horticulture, planting techniques, garden design and plant trade in the Roman world." In *The Archaeology of Crops, Fields and Gardens*, eds. J.-P. Morel, J. T. Juan, and J. C. Matalama. Bari: Centro Universitario Europeo Per I Beni Culturali, 207–219.

McGovern, P. E. et al. 1996. "Neolithic resinated wine. *Nature* 381: 480–481.

McGovern P. E., S. J. Fleming, and S. H. Katz, eds. 1995. *The Origins and Ancient History of Wine*. Amsterdam: Gordon & Breach.

Manen, J.-F., et al. 2003. "Microsatellites from archaeological *Vitis vinifera* seeds allow a tentative assignment of the geographical origin of ancient cultivars." *Journal of Archaeological Science* 30: 721–729. DOI: 10.1016/S0305-4403(02)00244-3.

Margaritis, E. 2013. "Distinguishing exploitation, domestication, cultivation and production: the olive in the third millennium Aegean." *Antiquity* 337: 746–757. DOI: 10.1017/S0003598X00049437.

Margaritis, E., and M. Jones. 2006. "Beyond cereals: crop processing and *Vitis vinifera* L. Ethnography, experiment and charred grape remains from Hellenistic Greece." *Journal of Archaeological Science* 33: 784–805. DOI: 10.1016/j.jas.2005.10.021.

Margaritis, E., and M. Jones. 2008a. "Crop processing of *Olea europaea* L.: an experimental approach for the interpretation of archaeobotanical olive remains." *Vegetation History and Archaeobotany* 17 (4): 381–392. DOI: 10.1007/s00334-007-0122-x.

Margaritis, E., and M. Jones. 2008b. "Olive oil production in Hellenistic Greece: the interpretation of charred olive remains from the site of Tria Platania, Macedonia, Greece (fourth–second century B.C.)." *Vegetation History and Archaeobotany* 17 (4): 393–401. DOI: 10.1007/s00334-008-0155-9.

Margaritis, E., and M. Jones. 2009. "Greek and Roman agriculture." In *The Oxford Handbook of Engineering and Technology in the Classical World*, ed. J. P. Oleson. Oxford: Oxford University Press, 159–174. DOI: 10.1093/oxfordhb/9780199734856.013.0008.

Marinova, E., et al. 2011. "An experimental approach for tracing olive processing residues in the archaeobotanical record, with preliminary examples from Tell Tweini, Syria." *Vegetation History and Archaeobotany* 20: 471–478. DOI: 10.1007/s00334-011-0298-y.

Marvelli, S., et al. 2013. "The origin of grapevine cultivation in Italy: the archaeobotanical evidence." *Annali di Botanica* 3: 155–163. DOI: 10.4462/annbotrm-10326.

Megaloudi, F. 2005. "Burnt sacrificial plant offerings in Hellenistic times: an archaeobotanical case study from Messene, Peloponnese, Greece." *Vegetation History and Archaeobotany* 14 (4): 329–340. DOI: 10.1007/s00334-005-0083-x.

Megaloudi, F., S. Papadopoulos, and M. Sgourou. 2007. "Plant offerings from the classical necropolis of Limenas, Thasos, northern Greece." *Antiquity* 81: 933–943. DOI: 10.1017/S0003598X00096010.8.

Mercuri, A. M., et al. 2015. "Pollen and macroremains from Holocene archaeological sites: a dataset for the understanding of the bio-cultural diversity of the Italian landscape." *Review of Palaeobotany and Palynology* 218: 250–266. DOI: 10.1016/j.revpalbo.2014.05.010.

Miller, N. F. 2008. "Sweeter than wine? The use of the grape in early Western Asia." *Antiquity* 82: 937–946. DOI: 10.1017/S0003598X00097696.

Miller, N. F., and T. Smart. 1984. "Intentional burning of dung as fuel: a mechanism for the incorporation of charred seeds into the archaeological record." *Journal of Ethnobiology* 4 (1): 15–28.

Mintz, S. W. 1985. *Sweetness and Power: The Place of Sugar in Modern History.* New York: Viking.

Mlekuž, D. 2014 "The Neolithic year." In *The Oxford Handbook of Neolithic Europe*, eds. C. Fowler, J. Harding, and D. Hofmann. Oxford: Oxford University Press. DOI: 10.1093 /oxfordhb/9780199545841.013.023.

Morales, J., and J. Gil. 2014. "Fruits as a staple food: the role of fig (*Ficus carica* L.) during the pre-Hispanic period of Gran Canaria (3rd c BC–15th c CE)." In *Plants and People. Choices and Diversity Through Time*, eds. A. Chevallier, E. Marinova, and L. Peña-Chocarro. Oxford: Oxbow Books, 182–190.

Morales, J., et al. 2014. "The archaeobotany of long-term crop storage in northwest African communal granaries: a case study from pre-Hispanic Gran Canaria (cal. AD 1000–1500)." *Vegetation History and Archaeobotany* 23 (5): 789–804. DOI: 10.1007 /s00334-014-0444-4.

Moreland, J. 2001. *Archaeology and Text.* London: Duckworth.

Motta, L. 2002. "Planting the seed of Rome." *Vegetation History & Archaeobotany* 11: 71–77.

Neef, R. 1990. "Introduction, development and environmental implications of olive culture: the evidence from Jordan." In *Man's Role in the Shaping of the Eastern Mediterranean Landscape*, eds. S. Bottema, G. Entjes-Nieborg, and W. van Zeist. Rotterdam: Balkema, 295–306.

Nesbitt, M., and D. Samuel. 1996. "From staple crop to extinction? The archaeology and history of the hulled wheats." In *Hulled Wheats: Proceedings of the First International Workshop on Hulled Wheats. Promoting the Conservation and Use of Underutilized and Neglected Crops 4*, eds. S. Padulosi, K. Hammer, and J. Heller. Rome: International Plant Genetic Resources Institute, 41–100.

Newton, C., et al. 2014. "On the origins and spread of *Olea europaea* L. (olive) domestication: evidence for shape variation of olive stones at Ugarit, Late Bronze Age, Syria—a window on the Mediterranean Basin and on the westward diffusion of olive varieties." *Vegetation History and Archaeobotany* 23 (5): 567–575. DOI: 10.1007/s00334-013-0412-4.

Nitsch, E. K., M. Charles, and A. Bogaard. 2015. "Calculating a statistically robust d13C and d15N offset for charred cereal and pulse seeds." *Science and Technology of Archaeological Research* 1 (1): 1653–1676. DOI: STAR20152054892315Y.0000000001.

Oliveira, H. R., et al. 2012. "Ancient DNA in archaeological wheat grains: preservation conditions and the study of pre-Hispanic agriculture on the island of Gran Canaria (Spain)."*Journal of Archaeological Science* 39: 828–835. DOI: 10.1016/j.jas.2011 .10.008.

Orrù, M., et al. 2013. "Morphological characterisation of *Vitis vinifera* L. seeds by image analysis and comparison with archaeological remains." *Vegetation History and Archaeobotany* 22 (3): 231–242. DOI: 10.1007/s00334-012-0362-2.

Pagnoux, C., et al. 2015. "Inferring the agrobiodiversity of *Vitis vinifera* L. (grapevine) in ancient Greece by comparative shape analysis of archaeological and modern seeds." *Vegetation History and Archaeobotany* 24 (1): 75–84. DOI: 10.1007/s00334-014-0482-y.

Palmer, S. A., et al. 2009. "Archaeogenetic evidence of ancient Nubian barley evolution from six to two-row indicates local adaptation." *PLoS ONE* 4 (7): e6301. DOI: 10.1371 /journal.pone.0006301.

Palmer, S. A., et al. 2012. "Archaeogenomic evidence of punctuated genome evolution in *Gossypium.*" *Molecular Biology and Evolution* 29 (8): 2031–2038. DOI: 10.1093/molbev /mss070.

Pals, J. P., V. Beemster, and A. Noordam, 1989. "Plant remains from the Roman castellum Praetorium Agrippinae near Valkenburg (prov. of Zuid-Holland)." In *Archäobotanik*, eds. U. Körber-Grohne, and H. Küster. Stuttgart: Cramer, 117–133.

Pals, J. P., and T. Hakbijl. 1992. "Weed and insect infestation of a grain cargo in a ship at the Roman fort of Laurium in Woerden (Prov. of Zuid-Holland)." *Review of Palaeobotany and Palynology* 73: 287–300. DOI: 10.1007/s10530-008-9339-6.

Papathanasiou, A., T. Theodoropoulou, and S.M. Valamoti. 2013. "The quest for prehistoric meals: towards an understanding of past diets in the Aegean: integrating stable isotope analysis, archaeobotany and zooarchaeology." In *Diet, Economy and Society in the Ancient Greek world: Towards a Better Integration of Archaeology and Science*, eds. S. Voutsaki and S. M. Valamoti. Leuven: Peeters, 18–31.

Pearson, E., and J. Letts. 1996. "Waterlogged plant remains and charred plant remains." In *Wavendon Gate, A Late Iron Age and Roman Settlement in Milton Keynes*, eds. R. J. Williams, P. J. Hart, and A.T.L. Williams. England: Buckinghamshire Archaeological Society, 236–255.

Pecci, A., M.A.C. Ontiveros, and N. Garnier. 2013. "Identifying wine and oil production: analysis of residues from Roman and Late Antique plastered vats." *Journal of Archaeological Science* 40: 4491–4498. DOI: 10.1016/j.jas.2013.06.019.

Pelling, R. 2008. "Garamantian agriculture: the plant remains from Jarma, Fazzan." *Libyan Studies* 39: 41–71. DOI: 10.1017/S0263718900009997.

Pelling, R., et al. 2015. "Exploring contamination (intrusion and residuality) in the archaeobotanical record: case studies from central and southern England." *Vegetation History and Archaeobotany* 24 (1): 85–99. DOI: 10.1007/s00334-014-0493-8.

Petrucci-Bavaud, M., and S. Jacomet. 1997. "Zur Interpretation von Nahrungsbeigaben in römischen Brandgrabern." *Ethnographisch-Archäologische Zeitschrift* 38: 567–593.

Pollmann, B., S. Jacomet, and A. Schlumbaum. 2005. "Morphological and genetic studies of waterlogged *Prunus* species from the Roman vicus Tasgetium (Eschenz, Switzerland)." *Journal of Archaeological Science* 32: 1471–1480. DOI: 10.1016/j.jas.2005.04.002.

Riehl, S. 2009. "Archaeobotanical evidence for the interrelationship of agricultural decision-making and climate change in the ancient Near East." *Quaternary International* 197: 93–114. DOI: 10.1016/j.quaint.2007.08.005.

Riehl, S., et al. 2014. "Drought stress variability in ancient Near Eastern agricultural systems evidenced by $d13C$ in barley grain." *Proceedings of the National Academy of Sciences of the United States* 111 (34): 12,348–12,353. DOI: 10.1073/pnas.1409516111.

Roberts, N., et al. 2004. "Holocene climate, environment and cultural change in the circum-mediterranean region." In *Past Climate Variability through Europe and Africa*, eds. R. W. Battarbee, F. Gasse, and C. E. Stickley. Dordrecht: Kluwer Academic Publishers, 627–637. DOI: 10.1007/978-1-4020-2121-3_17.

Robinson, M. 1999. "The macroscopic plant remains." *Towards a History of Pre-Roman Pompeii: Excavations beneath the House of Amarantus (I. 9.11-12), 1995-8*, eds. M. Fulford and A. Wallace-Hadrill. London: Papers of the British School at Rome 67, 95–102.

Robinson, M. 2002. "Domestic burnt offerings and sacrifices at Roman and pre-Roman Pompeii." *Vegetation History and Archaeobotany* 11 (1): 93–99. DOI: 10.1007/s003340200010.

Robinson, M. 2012. "The place of Silchester in archaeobotany." In *Silchester and the Study of Romano-British Urbanism*. ed. M. Fulford. Journal of Roman Archaeology Supplementary Series 90. Portsmouth, RI: Journal of Roman Archaeology, 213–226.

Robinson, M. 2016. "The environmental archaeology of the South Agora Pool, Aphrodisias." In *Aphrodisias Papers 5: Excavation and Research at Aphrodisias, Turkey, 2006-2012*, eds. R.R.R. Smith et al. Portsmouth, RI: Journal of Roman Archaeology, 91–99.

Robinson, M., N. Fulford, and K. Tootell. 2006. "The macroscopic plant remains." In *Life and Labour in Roman Silchester: Excavations in Insula IX since 1997*, eds. M. Fulford, A. Clarke, and H. Eckardt. London: Britannia Monograph Series 22, 206–18 and 374–379.

Robinson, M., and E. Rowan. 2015. "Roman food remains in archaeology and the contents of a Roman sewer at Herculaneum." In *A Companion to Food in the Ancient World*, eds. J. Wilkins and R. Nadeau. Chichester: Wiley-Blackwell, 105–115

Rodríguez-Ariza, M. O., and E. M. Moya. 2005. "On the origin and domestication of *Olea europaea* L. (olive) in Andalucia, Spain, based on the biogeographical distribution of its finds." *Vegetation History and Archaeobotany* 14 (4): 551–561. DOI: 10.1007/s00334-005-0012-z.

Rösch, M., S. Jacomet, and S. Karg. 1992. "The history of cereals in the region of the former Duchy of Swabia (Herzogtum Schwaben) from the Roman to the Post-medieval period: results of archaeobotanical research." *Vegetation History and Archaeobotany* 1 (4): 193–231. DOI: 10.1007/BF00189499.

Rossignani, M. P., M. Sannazaro, and G. Legrottaglie, eds. 2005. *La Signora del Sarcofago. Una Sepoltura di Rango Nella Necropoli dell'Università Cattolica*. Milan: Vita & Pensiero.

Rottoli, M., and E. Castiglioni. 2011. "Plant offerings from Roman cremations in northern Italy: a review." *Vegetation History and Archaeobotany* 20 (5): 495–506. DOI: 10.1007/s00334-011-0293-3/.

Rowan, E. 2014. "Roman diet and nutrition in the Vesuvian Region: a study of the archaeobotanical remains from the Cardo V sewer at Herculaneum." PhD dissertation, St. Cross College, University of Oxford, England.

Rowan, E. 2015. "Olive pressing waste as a fuel source in antiquity." *American Journal of Archaeology* 119: 465–482. DOI: 10.3764/aja.119.4.0465.465.

Rowley-Conwy, P. 1994. "Dung, dirt and deposits: site formation under conditions of near-perfect preservation at Qasr Ibrim, Egyptian Nubia." In *Whither Environmental Archaeology*, eds. R. Luff, and P. Rowley-Conwy. Oxford, Oxbow Monograph 38, 25–32.

Sadori, L., and M. Giardini. 2007. "Charcoal analysis, a method to study vegetation and climate of the Holocene: the case of Lago di Pergusa (Sicily, Italy). *Geobios* 40 (2): 173–180.

Sadori, L., et al. 2010. "The plant landscape of the imperial harbour of Rome." *Journal of Archaeological Science* 37: 3294–3305. DOI: 10.1016/j.jas.2010.07.032.

Sadori, L., et al. 2014. "Archaeobotany in Italian ancient Roman harbours." *Review of Palaeobotany and Palynology*. DOI: 10.1016/j.revpalbo.2014.02.004.

Samuel, D. 2000. "Brewing and baking." In *Ancient Egyptian Materials and Technology*, eds. P. T. Nicholson and I. Shaw. Cambridge: Cambridge University Press, 537–576.

Schlumbaum, A., and C. J. Edwards. 2013. "Ancient DNA research on wetland archaeological evidence." In *The Oxford Handbook of Wetland Archaeology*, eds. F. Menotti and A. O'Sullivan. Oxford: Oxford University Press, 569–583.

Schlumbaum, A., et al. 2011. "Die Ulmen und andere besondere Gehölze aus dem Römischen Tempelbezirk und der Zivilsiedlung." In *Oedenburg*. Volume 2: *L'agglomération civile et les sanctuaires. 2—Matériel et etudes*, ed. M. Reddé. *Monographien des Römisch-Germanischen Zentralmuseums* 79.2.2. Mainz: Verlag des Römisch-Germanischen Zentralmuseums, 73–102.

Sealey, P. R., and P. A. Tyres. 1989. "Olives from Roman Spain: a unique amphora find in British waters." *Antiquaries Journal* 69: 54–72. DOI: 10.1017/S0003581500043419.

Sherratt A. 1991. "Sacred and profane substances: the ritual use of narcotics in later prehistoric Europe." In *Sacred and Profane: Proceedings of a Conference on Archaeology, Ritual and Religion*, eds. P. Garwood et al. Oxford: Oxford University Committee for Archaeology, 50–64.

Smith, D., and H. Kenward. 2011. "Roman grain pests in Britain: implications for grain supply and agricultural production." *Britannia* 42: 243–262. DOI: 10.1017/S0068113X11000031.

Smith, H., and G. Jones. 1990. "Experiments on the effects of charring on cultivated grape seeds." *Journal of Archaeological Science* 17: 317–327. DOI: 10.1016/0305-4403(90)90026-2.

Smith, W. 1998. "Fuel for thought: archaeobotanical evidence for the use of alternatives to wood fuel in Late Antique North Africa." *Journal of Mediterranean Archaeology* 11: 191–205. DOI: 10.1558/jmea.v11i2.191.

Stahl, A. B. 1989. "Plant-food processing: its implications for dietary quality." In *Foraging and Farming: The Evolution of Plant Exploitation*, eds. D. K. Harris and G. C. Hillman. London: Unwin Hyman, 171–194.

Stika, H. P. 2011. "Early Iron Age and Late Mediaeval malt finds from Germany—attempts at reconstruction of early Celtic brewing and the taste of Celtic beer." *Archaeological and Anthropological Sciences* 3: 41–48. DOI: 10.1007/s12520-010-0049-5.

Straker, V. 1984. "First and second century carbonized cereal grain from Roman London." In *Plants and Ancient Man*, eds. W. van Zeist, and W. A. Casparie. Rotterdam: Balkema, 323–329.

Styring, A., et al. 2013. "The effect of charring and burial on the biochemical composition of cereal grains: investigating the integrity of archaeological plant material." *Journal of Archaeological Science* 40: 4767–4779. DOI: 10.1016/j.jas.2013.03.024.

Terral, J. F., et al. 2004. "Historical biogeography of olive domestication (*Olea europaea* L.) as revealed by geometrical morphometry applied to biological and archaeological material." *Journal of Biogeography* 31: 63–77. DOI: 10.1046/j.0305-0270.2003.01019.x.

Terral, J. F., et al. 2010. "Evolution and history of grapevine (*Vitis vinifera*) under domestication: new morphometric perspectives to understand seed domestication syndrome and reveal origins of ancient European cultivars." *Annals of Botany* 105 (3): 443–455. DOI: 10.1093/aob/mcp298.

Terral, J. F., et al. 2012. "Insights into the complex historical biogeography of date palm agrobiodiversity (*Phoenix dactylifera* L.) using geometric morphometrical analysis of modern seeds from various origins and Egyptian ancient material." *Journal of Biogeography* 39: 929–941.

Tomber, R. S. 1996. "Provisioning the desert: pottery supply to Mons Claudianus." In *Archaeological Research in Roman Egypt*, ed. D. M. Bailey. Ann Arbor, Michigan: Journal of Roman Archaeology, 39–49.

Tomlinson, P. 1987. "Plant remains." In *A Plank Tank from Nantwich*, ed. R. McNeil and A. F. Roberts. England: Cheshire, 287–296.

Valamoti, S. M. 2011. "Ground cereal food preparations from Greece: the prehistory and modern survival of traditional Mediterranean 'fast foods.'" *Archaeological and Anthropological Sciences* 3: 19–39. DOI: 10.1007/s12520-011-0058-z.

Valamoti, S. M. 2012–2013. "Healing with plants in prehistoric northern Greece—a contribution from archaeobotany." *Offa* 69/70: 479–494.

Valamoti, S. M., A. Moniaki, and A. Karathanou. 2011. "An investigation of processing and consumption of pulses among prehistoric societies: archaeobotanical, experimental and ethnographic evidence from Greece." *Vegetation History and Archaeobotany* 20 (5): 381–396. DOI: 10.1007/s00334-011-0302-6.

Valamoti, S. M. et al. 2007. "Grape pressings from northern Greece: the earliest wine in the Aegean?" *Antiquity* 81: 54–61. DOI: 0.1017/S0003598X00094837.

Valamoti, S. M., et al. 2008. "Prehistoric cereal foods in Greece and Bulgaria: investigation of starch microstructure in experimental and archaeological charred remains." *Vegetation History and Archaeobotany* 17 (1): 265–276. DOI: 10.1007/s00334-008-0190-6.

Van der Veen, M. 1984. "Sampling for seeds." In *Plants and Ancient Man: Studies in Palae-oethnobotany*, eds. W. van Zeist, and W. A. Casparie. Rotterdam: Balkema, 193–199.

Van der Veen, M. 1989. "Charred grain assemblages from Roman-period corn driers in Britain." *Archaeological Journal* 146: 302–319. DOI: 10.1080/00665983.1989.11021292.

Van der Veen, M. 1992. *Crop Husbandry Regimes: An Archaeobotanical Study of Farming in Northern England, 1000 BC–AD 500*. Sheffield: J. R. Collis.

Van der Veen, M. 1995. "Ancient agriculture in Libya: a review of the evidence." *Acta Palae-obotanica* 35 (1): 85–98.

Van der Veen, M. 1998a. "A life of luxury in the desert? The food and fodder supply to Mons Claudianus." *Journal of Roman Archaeology* 11: 101–116. DOI: 10.1017/S1047759400017219.

Van der Veen, M. 1998b. "Gardens in the desert." In *Life on the Fringe: Living in the Southern Egyptian Deserts during the Roman and Early-Byzantine Periods*, ed. O. E. Kaper. Leiden, The School of Asian, African, and Amerindian Studies, 221–242.

Van der Veen, M. 2001. "The Botanical Evidence. (Chapter 8)." In *Survey and Excavations at Mons Claudianus 1987–1993*. Volume 2: *The Excavations: Part 1*, eds. V.A. Maxfield, and D.P.S. Peacock. Cairo: Institut Français d'Archéologie Orientale du Caire, 174–247.

Van der Veen, M. 2003. "When is food a luxury?" *World Archaeology* 34: 405–427. DOI 10.1080/0043824021000026422.

Van der Veen, M. 2007. "Formation processes of desiccated and carbonized plant remains: the identification of routine practice." *Journal of Archaeological Science* 34: 968–990. DOI: 0.1016/j.jas.2006.09.007.

Van der Veen, M. 2008. "Food as embodied material culture—diversity and change in plant food consumption in Roman Britain." *Journal of Roman Archaeology* 21: 83–110. DOI: 10.1017/S1047759400004396.

Van der Veen, M. 2010. "Agricultural innovation: invention and adoption or change and adaptation?" *World Archaeology* 42 (1): 1–12. DOI: 10.1080/00438240903429649.

Van der Veen, M. 2011. *Consumption, Trade and Innovation: Exploring the Botanical Remains from the Roman and Islamic Ports at Quseir al-Qadim, Egypt*. Frankfurt: Africa Magna Verlag.

Van der Veen, M. 2014a. "Arable farming, horticulture, and food: expansion, innovation, and diversity in Roman Britain." In *The Oxford Handbook of Roman Britain*, eds. M. Millett, L. Revell, and A. Moore. Oxford: Oxford University Press. DOI: 10.1093/oxfordhb/9780199697713.013.046.

Van der Veen, M. 2014b. "The materiality of plants: plant–people entanglements." *World Archaeology* 46 (5): 799–812. DOI: 10.1080/00438243.2014.953710.

Van der Veen, M., and N. Fieller. 1982. "Sampling Seeds." *Journal of Archaeological Science* 9: 287–298.

Van der Veen, M., and R. Gale. 2011. "Woodworking and firewood—resource exploitation." In *Consumption, Trade and Innovation: Exploring the Botanical Remains from the Roman and Islamic Ports at Quseir al-Qadim, Egypt*, ed. M. Van der Veen. Frankfurt: Africa Magna Verlag, 205–226.

Van der Veen, M., A. Grant, and G. Barker. 1996. "Romano-Libyan agriculture: crops and animals." In *Farming the Desert: The UNESCO Libyan Valleys Archaeological Survey*. Vol. 1: *Synthesis*, eds. G. Barker et al. Paris: UNESCO: 227–263.

Van der Veen, M., A. J. Hall, and J. May. 1993. "Woad and the Britains painted blue." *Oxford Journal of Archaeology* 12 (3): 367–371. DOI: 10.1111/j.1468-0092.1993.tb00340.x

Van der Veen, M., A. Hill, and A. Livarda. 2013. "The archaeobotany of medieval Britain (c. AD 450–1500): identifying research priorities for the 21st century." *Medieval Archaeology* 57: 151–182. DOI: 10.1179/0076609713Z.00000000018.

Van der Veen, M., and G. Jones. 2006. "A re-analysis of agricultural production and consumption: implications for understanding the British Iron Age." *Vegetation History and Archaeobotany* 15 (3): 217–228. DOI: 10.1007/s00334-006-0040-3.

Van der Veen, M., and G. Jones. 2007. "The production and consumption of cereals: a question of scale." In *The Later Iron Age of Britain and Beyond*, eds. C. Haselgrove, and T. Moore. Oxford: Oxbow Books: 419–429.

Van der Veen, M., A. Livarda, and A. Hill. 2008. "New food plants in Roman Britain: dispersal and social access." *Environmental Archaeology* 13 (1): 11–36. DOI: 10.1179/174963108X279193.

Van der Veen, M., and J. Morales. 2015. "The Roman and Islamic spice trade: new archaeological evidence." *Journal of Ethnopharmacology* 167: 54–63. DOI: 10.1016/j.jep.2014.09.036.

Van der Veen, M., and J. Morales. 2017. "Food globalisation and the Red Sea: new evidence from the ancient ports at Quseir al-Qadim, Egypt." In *Human Interaction with the Environment in the Red Sea. Selected Papers of Red Sea Project VI*, eds. D. A. Agius et al. Leiden: Brill, 254–289.

Van der Veen, M., and T. O'Connor. 1998. "The expansion of agricultural production in Later Iron Age and Roman Britain." In *Science in Archaeology: An Agenda for the Future*, ed. J. Bayley. London: English Heritage, 127–143.

Van der Veen, M., and H. Tabinor. 2007. "Food, fodder and fuel at Mons Porphyrites: the botanical evidence." In *Survey and Excavation at Mons Porphyrites 1994–1998*. Volume 2: *The Excavations*, eds. V. A. Maxfield and D. P. S. Peacock. London: Egypt Exploration Society, 83–142.

Van Dinter, M., et al. 2015. "Could the local population of the Lower Rhine delta supply the Roman army? Part 2: Modelling the carrying capacity using archaeological, palaeoecological and geomorphological data" *Journal of the Archaeology of the Low Countries* 5: 5–50 DOI: http://jalc.nl/cgi/t/text/get-pdf6920.pdf.

Vandorpe, P., and S. Jacomet. 2011a. "Plant economy and environment." In *Oedenburg*. Vol. 2: *L'agglomération civile et les sanctuaires*. Vol. 2: *Matériel et études*, ed. M. Reddé. Mainz: Monographien des RGZM, Band 79/2, 3–72.

Vandorpe, P., and S. Jacomet. 2011b. "Remains of burnt vegetable offerings in the temple area of Roman Oedenburg (Biesheim-Kunheim, Alsace, France)—First results." In *Carpologia: Articles réunis à la mémoire de Karen Lundström-Baudais*, ed. J. Wiethold. Glux-en-Glenne: Bibracte, 87–100.

Van Zeist, W., S. Bottema, and M. Van der Veen. 2001. *Diet and Vegetation at Ancient Carthage: The Archaeobotanical Evidence*. Groningen: Groningen Institute of Archaeology.

Veal, R. J. 2012. "From context to economy: charcoal and its unique potential in archaeological interpretation: a case study from Pompeii." In *More Than Just Numbers? The Role of Science in Roman Archaeology*, ed. I. E. Schrüfer-Kolb. Portsmouth, RI, Journal of Roman Archaeology, 19–52.

Veal., R. J. 2013. "Fuelling ancient Mediterranean cities: a framework for charcoal research." In *The Ancient Mediterranean between Science and History*, ed. W.V. Harris. Leiden: Brill, 37–58. DOI: 10.1163/9789004254053_004.

Veal, R. J. 2014. "Pompeii and its hinterland connection: the fuel consumption of the House of the Vestals, 3rd c. BC to AD 79." *European Journal of Archaeology* 17 (1): 27–44. DOI: 10.1179/1461957113Y.0000000043.

Vermeeren, C. 1999. "The use of imported and local wood species at the Roman port of Berenike, Red Sea coast, Egypt." In *The Exploitation of Plant Resources in Ancient Africa*, ed. M. Van der Veen. New York: Kluwer/Plenum, 199–204.

Vermeeren, C., and H. Van Haaster. 2002. "The embalming of the ancestors of the Dutch royal family." *Vegetation History and Archaeobotany* 11 (1): 121–126. DOI: 10.1007/s003340200013.

Viola, H. J. 1991. "Seeds of change." In *Seeds of Change: A Quincentennial Commemoration*, eds. H. J. Viola and C. Margolis. Washington, DC: Smithsonian Institution Press, 11–15.

Visser, J.M. 2015. "Imperial timber? Dendrochronological evidence for large-scale road building along the Roman limes in the Netherlands." *Journal of Archaeological Science* 53: 243–254. DOI: 10.1016/j.jas.2014.10.017.

Vittori, C., et al. 2015. "Palaeoenvironmental evolution of the ancient lagoon of Ostia Antica (Tiber delta, Italy)." *Journal of Archaeological Science* 54: 374–384. DOI: 10.1016/j.jas.2014.06.017

Wallace, M., and M. Charles. 2013. "What goes in does not always come out: the impact of the ruminant digestive system of sheep on plant material, and its importance for the interpretation of dung-derived archaeobotanical assemblages." *Environmental Archaeology* 18 (1): 18–30. DOI: 10.1179/1461410313Z.00000000022.

Wallace, M., et al. 2013. "Stable carbon isotope analysis as a direct means of inferring crop water status and water management practices." *World Archaeology* 45 (3): 388–409. DOI 10.1179/1461410313Z.00000000022.

Ward, C. 2001. "The Sadana Island shipwreck: an eighteenth-century AD merchantman off the Red Sea coast of Egypt." *World Archaeology* 32 (3): 368–382. DOI: 10.1080/00438240120048680.

Ward, C. 2003. "Pomegranates in eastern Mediterranean contexts during the Late Bronze Age." *World Archaeology* 34 (3): 529–541. DOI: 10.1080/0043824021000026495.

Watson, A. M. 1983. *Agricultural Innovation in the Early Islamic World*. Cambridge: Cambridge University Press.

Wilson, A. 2016. "Water, nymphs, and a palm grove: monumental water display at Aphrodisias." In *Aphrodisias Papers 5: Excavation and Research at Aphrodisias, Turkey, 2006–2012*, eds. R.R.R. Smith et al. Portsmouth, RI: Journal of Roman Archaeology, 100–135.

Zach, B. 2002. "Vegetable offerings on the Roman sacrificial site in Mainz, Germany—a short report on the first results." *Vegetation History and Archaeobotany* 11 (1/2): 101–106. DOI: 10.1007/s003340200011.

Zech-Matterne, V., et al. 2009. "L'Agriculture du VIe au Ier siècle avant J.-C. en France: état des recherches carpologiques sur les établissements ruraux." In *Habitats et paysages ruraux en Gaule et regards sur d'autres régions du monde celtique*, eds. I. Bertrand et al. Chauvigny: Association des Publications Chauvinoises, 383–416.

Zech-Matterne, V., et al. 2014. "L'essor des blés nus en France septentrionale: systèmes de culture et commerce céréalier autour de la conquête césarienne et dans les siècles qui suivent." In *Consommer dans les campagnes de la Gaule romaine*, eds. X. Deru and R. González Villaescusa. Lille: Université Charles-de-Gaulle, 23–50.

Zeder, M. A. 2006. "Central questions in the domestication of plants and animals." *Evolutionary Anthropology* 15: 105–117. DOI: 10.1002/evan.20101.

CHAPTER THREE

Zooarchaeology

RECONSTRUCTING THE NATURAL AND CULTURAL WORLDS FROM ARCHAEOLOGICAL FAUNAL REMAINS

Michael MacKinnon

Introduction

As commodities that can embody and connect environmental, biological, and cultural aspects, animals occupy integral positions in the investigation of antiquity. Like all living beings, animals are biological entities that evolve with, and adapt to, the natural environment, in whatever manner the latter is constructed or developed across space and throughout time. But animals may also act as cultural artifacts—things (albeit biological in nature) created, modified, and manipulated by humans—an expansive concept when one considers the myriad ways in which not just animals, but a host of primary and secondary resources from them are retrieved, exploited, or otherwise incorporated into ancient life. Environmental, biological, and cultural worlds are interwoven; animals yield knowledge about each.[1] Consequently, their study forms a powerful, indeed essential, line of investigation for holistic reconstruction of the past.

Our understanding of animals in Greek and Roman antiquity fundamentally draws upon three principal means. First, a number of ancient texts, themselves spanning diverse forms and functions (e.g., commemorative and dedicatory inscriptions, literary works, histories, encyclopedic volumes, didactic manuals, poetry, legal codes, among multiple other categories) document aspects—from descriptions of various types of animals encountered and used, to details about their care, maintenance, and widespread contribution to ancient life. A second source is iconography. Images as portrayed in Greek and Roman visual culture or art allow exploration of animal morphology, characterization, and physical variation, among other concepts. Finally, archaeology, most notably zooarchaeology, supplies physical remains of animals to examine for cultural, biological, and environmental reconstructions. Excavated faunal remains reveal data about species representation, age, sex, and state of health; they may also yield evidence about cultural practices such as consumer decisions in marketing, resource acquisition and use, butchery, cooking, and eating. Archaeological analyses, consequently, can disclose much about animals themselves, as well as the cultures that kept, controlled, killed, ate, and exploited them.

While text, art, and bones provide key information about animals in antiquity, each has not shared the same degree of attention in the historic development of classical archaeology.[2] Early, established classical archaeology grew out of eighteenth-century traditions of scholarship in aesthetics of ancient art, architecture, and artifacts, alongside a strong connection with ancient Greek and Roman textual analyses and philology. Zooarchaeological remains were certainly encountered, but ceramic, architectural, artistic, and literary preeminence relegated faunal materials to lower importance, if indeed they were even saved in the first place. Intermittent scholarship permeates some reports through the late nineteenth and early twentieth century, but it is not until the 1960s and 1970s with the advent and progress of methods and theories framing "New Archaeology / Processual Archaeology" that zooarchaeological studies garner more serious attention in classical archaeology. The ensuing invigorated focus on scientific studies and reasoning across the wider scope of archaeological materials retrieved, including faunal materials, sparked greater attention to reconstructions of past environments, economies, diets, and husbandry practices, as well as interest in spatial distribution of finds, site depositional processes, and statistical assessments of samples, among other facets.

Scientific investigations and analyses of zooarchaeological materials from classical archaeological sites have grown rapidly since the 1980s, broadened further through initiatives taken not simply to understand biological and environmental components of the past (aspects that might, superficially, ally better with natural science), but to engage animals as markers of cultural complexity as well. Investigations today are increasingly conducted by zooarchaeologists who specialize in the scholarship of Greek and Roman antiquity, a tactic consequently helping to blur or dissolve traditional academic boundaries in classical studies that previously emphasized primacy to other categories of material remains, such as texts or art. Scientists now infiltrate classics, and vice versa. Multidisciplinary ventures, linking humanities, social sciences, and sciences, are increasingly commonplace in collaborative and synthetic reconstructions of the past.

The proliferation of science in classical archaeology registers boldly in the discipline of zooarchaeology, with studies continuously shaping and transforming our understanding of the people, places, events, and activities of antiquity. Much has happened over the last few decades, where today a wider reflection of the contributions, potential, and future directions as regards scientific study of faunal remains from ancient Greek and Roman sites touches upon wide-ranging aspects of interest to, important to, and in many cases, vitally integral to, our understanding of the natural and cultural worlds of antiquity. The intention of this chapter is not to overview general zooarchaeological theory, method, and practice. Numerous "how to" resources, from identification and laboratory manuals to textbooks and related publications exist for those components.[3] Nor is this chapter meant as a forum with which to summarize results obtained from zooarchaeological exploration of ancient sites. Again, tremendous scholarship exists that presents and assesses trends and patterns to this effect, including topics such as animal husbandry, the contribution of dietary meat, and the multiple roles in which animals might factor in economic and cultural life in antiquity.[4] To center investigation within this chapter, the interrogative pronouns—who, what, when, where, why—act here as guides with which to shape subsequent discussion and frame avenues of zooarchaeological exploration that might be less familiar (but no less important) within the discipline's broader contribution to classical archaeology, both today and into the future. Such terms encapsulate various categories or dimensions that comprise or underscore data collection, the necessary building blocks of scholarship from which larger reconstructions are crafted. Limitations of space prevent detailed examination of all zooarchaeological aspects that may be nestled within each term in any depth; rather the intention here is to outline the breadth of topics that can be pursued in anticipating new directions in classical archaeology overall.

When

ESTABLISHING CHRONOLOGIES

Establishing "when" events or activities occurred has been a major thrust of classical archaeology, with countless examples of chronology-building using an array of materials. Incorporation of evidence drawn from zooarchaeological materials in such schemes, however, tends to be secondary. Overall, there is often an assumed reluctance (sometimes, outright dismissal) to consider radiocarbon dating of any bones collected as correlative means to set alongside chronologies derived more astutely from ceramic and/or coin evidence. Certainly genuine cautions and limitations register in the use of radiocarbon dates into historical ages. The wiggly shape of the calibration curve during various phases across Greek and Roman antiquity, for example, imposes a very valid concern in that a ^{14}C date can potentially match to more than one calendar date. Horizontal stretches or 'plateau' episodes within the calibration curve add a second complication; here a provided ^{14}C date may correspond to a long range of calibrated dates.[5] While processing costs, survival and diagenetic issues, contamination worries, and error factors can present further hurdles in using radiocarbon dating in classical archaeology,[6] applications using Bayesian statistics have greatly improved the method's precision and accuracy.[7] Not to downplay any of these considerations, it is important to recognize that each need not apply universally nor equally across the whole of classical antiquity to warrant any blanket rejection of radiocarbon dating a priori among sites.

Notwithstanding practical issues in implementing radiocarbon dating for certain time frames in antiquity—mainly a factor of the 'wiggles' and 'plateaus' in the calibration curve that infiltrate various periods under consideration—radiocarbon dating more bones among ancient Greek and Roman sites may actually help refine chronologies for materials often rendered "less datable," typically items such as cook wares, coarse wares, tiles, and bricks—often the most plentiful categories of finds among sites. It might also help reduce the temptation to date a huge deposit of "undatable" materials (be these bones or otherwise) on the basis of small amount (perhaps even a single piece!) of well-datable ceramics or coins, which may be secondary to such deposits.

Two examples help illustrate issues of complementarity and conflict that may arise from correlation of radiocarbon-dated faunal materials with chronologies derived from more traditional means, chiefly ceramics and coins. First, recent excavations of western rooms at the theater at Corinth, Greece, unearthed a huge deposit (over 1 ton!) of bones, chiefly remains of cattle. The deposit contained surprisingly few coins or finewares; what were retrieved dated to the fourth to fifth centuries CE.[8] The zooarchaeological assemblage was heavily skewed, with practically no cattle incisor teeth, ribs, or foot bones. Rather curiously, such parts dominated another fairly massive faunal deposit recovered from the east theater, excavated in the 1980s; however, that assemblage was dated to the second century CE.[9] The temptation to view these two distinct and relatively substantial zooarchaeological assemblages as more contemporary in age as opposed to separated by over 200 years loomed large, on one level in terms of general taphonomic parallels between each bone deposit, but more important in the rather clear-cut manner that one bone deposit appeared to contain what was missing from the other. Two additional finds provoked even more suspicion: a leg bone with a unique infection in the west theater deposit displayed similar pathological markings to some foot bones found in the east theater. The interconnecting bones linking this 'leg' and 'foot' were missing from both assemblages, so a firm match between elements could not be determined. DNA signatures for each might conclusively tie the two components, but until such work can be performed, this temporal conundrum exists. Chronologies, here, might benefit from radiocarbon-dating bones from each deposit to help clarify any temporal correlation between deposits.

In another example, calibrated radiocarbon dates for a sample of five cattle bones from excavated leveling layers or fillings below the *decumanus maximus* at the site of Bir Massouda, in present-day Carthage, Tunisia, indicated with a high probability that the animals were slaughtered during the late ninth century BCE.[10] Such values, however, were older than originally expected, based on conventional chronologies of Greek Late Geometric pottery (which gave dates from the second half of the eighth century BCE) from which traditional explanations of the founding of Phoenician Carthage centered. Recognizing the repercussions of challenging or adjusting already established pottery sequences, investigators went to great lengths to argue that all of the bone samples analyzed derived from residual waste, that is, mainly early materials from the late ninth century BCE that had been redeposited in the second half or last quarter of the eighth century BCE, at a time when this area of Carthage became urbanized. While theoretically possible, this explanation prompted a host of puzzling scenarios: why did only the bones date to an older time frame; why were materials so uniformly intermixed if indeed they were only indirectly associated with one another; why were no eighth-century BCE bones found alongside the pottery; what taphonomic factors were at play; were the inconsistencies noted a matter of, or exacerbated by, human error in the earlier recording of faunal and ceramic materials; was this simply some odd stratigraphic mystery? Nevertheless, a call was made that with more radiocarbon testing, revisions to Mediterranean chronologies may be necessary, given that ^{14}C dates provided a clear indication that Phoenician contacts with cultures in the western Mediterranean appeared older than previously thought.[11]

Certainly chronology-building can benefit from greater input across disciplines and material categories. Pots, especially valuable ones, can be kept and curated over one's lifetime, but the beef ribs on one's plate are a single event, from an animal with a relatively short life span compared to many ceramics. The date of the meal, therefore, may be more precise and culturally meaningful than the pot. Extending this principle, and put another way, dating a cow bone bearing cut marks and contained within a stratified context not only dates the depositional stratum in which the find was recovered, but also dates a human presence and event, itself occurring under an ecological and climatic setting or context. Culture, biology, and environment all interact within this scenario. Such a call is not meant to detract from the incredible value and precision underscoring traditional chronologies and chronology-building in classical archaeology, but to put such into their larger perspective of the individual events (large/small, important/mundane, etc.) that occur within such periods and time frames, and the broader contexts (cultural and environmental) under which each activity or behavior is set. Zooarchaeological work can contribute to this temporal discussion.

CHARTING TEMPORAL PATTERNS

Aside from implementing more radiocarbon dating, a host of other means to extract "when" from zooarchaeological remains can assist classical archaeology. Certainly, trends associated with shifting animal husbandry schemes, dietary changes over time, or introductions and extinctions of taxa offer help. Such topics have formed traditional foci among many zooarchaeological studies for antiquity, and in essence really encompass a host of interrogative dimensions—prompting questions of who, what, where, and why, for example, in charting, detailing, and explaining patterns.

A range of biological, cultural, and environmental factors shapes and influences human behaviors. Reconstructions of ancient diets and animal husbandry practices, fundamental components to zooarchaeological research, certainly epitomize the interconnections among these aspects. Most faunal remains from ancient sites derive from consumed animals. The variety and frequency of taxa represented at a site, the environmental conditions or settings under which each lived, herd demographics, proportions of different cuts of meat, marked by the percentages of bones or skeletal parts associated with such cuts, not to mention the manner in which meats and other materials from these animals are butchered, processed, and consumed, all provide data to reconstruct more holistic and nuanced dietary assessments. Indeed, a clearer understanding of diet and subsistence practices is one area where zooarchaeology has greatly illuminated and refined, even altered, our picture of antiquity. Although assumptions that most ancient Greeks and Romans shared a basic, largely vegetarian diet are justified, zooarchaeological data affirm consumption of meat. Mammals, birds, and fish typically comprise the bulk of the faunal remains recovered and identified from ancient sites. Variation registers in the proportions of each category, however, depending upon multiple aspects, such as type of recovery scheme employed (e.g., sieved or hand-collected), time frame of investigation (e.g., domestic fowl tend to be more common among Roman sites than Greek sites), geographic and environmental setting (e.g., sites near the coast tend to have higher relative frequencies of fish), and type of site (e.g., rural/urban, rich/poor, Greek/Roman, etc.), as well as a host of other cultural issues connected with production, marketing, trade, ethnicity, religion, food taboos, and personal choice that affect access to foodstuffs and ultimately influence human diets. Even if meat comprised only a small part of the popular diet overall during antiquity, zooarchaeological data largely highlight the predominance of domestic taxa, and especially domestic livestock (i.e., cattle, sheep/goat, and pig) among the meats more commonly and universally consumed. Although wild game was eaten, it tends to be linked to more elite diets, at least within the Roman world. Fish, and especially fresh fish, are also generally connected with elite gastronomy in antiquity. Nevertheless, on average, wild game and fish normally account for less than 5% of the identified bones from most zooarchaeological assemblages from ancient Greek and Roman sites; in many instances their values are much lower than this baseline.

The available pool of zooarchaeological literature reporting upon issues such as diet and subsistence practices, animal husbandry, and food distribution for Greco-Roman antiquity is immense, so much so that broad syntheses are published. Currently, most pertain to Roman antiquity, given the geographic and cultural expanse of that empire,[12] but synthetic-style articles have also addressed more focused issues, including burnt-animal sacrifice and ritual feasting, for the Greek world.[13] Temporal scales and topics continue to expand. For example, wide-ranging diachronic investigation of zooarchaeological materials from sites in the Athenian Agora supports the hypothesis that specialized husbandry and dietary schemes focusing on domestic sheep, goats, and cattle began in Neolithic times, with some hunted game in that phase as well.[14] Subsequent periods build upon this, culminating in extensive deposits of butchery, horn and bone processing, and dietary waste within Classical levels (fifth and fourth centuries BCE). Patterns shift with Roman and Late Antique influence to slightly augmented pork consumption and even more systematic butchery patterns, before shifting back again to higher frequencies of goat pastoralism during Byzantine times, likely in response to changing cultural and environmental conditions. Overall, while broad patterns of environmental and cultural change may be postulated on the basis of the zooarchaeological evidence, deeper inspection reveals regional variation in the role of animals across prehistoric, ancient, and Byzantine Greece.

TIME FRAMES FOR EXTINCTION EVENTS, MIGRATIONS, AND MOVEMENTS OF ANIMALS

Zooarchaeology adds more to our knowledge of the concept "when" (as well as "where") through assessments of origin, migration, movement, and extinction events for various taxa. A number of cases register. Fallow deer (*Dama dama*), it seems, become more widespread across Mediterranean and continental Europe in Roman times,[15] perhaps as a commodity that helped exemplify or display a message of social and economic standing in its wild status, exotic origin, and maintenance within a private space.[16] Rabbit (*Oryctolagus cuniculus*) bones are common across Iron Age and Roman sites in Iberia, confirming remarks from ancient authors about their ubiquity in that region and general fecundity overall,[17] but available evidence indicates the species was probably not imported into Italy and the eastern Mediterranean until late antiquity.[18] Porcupines (*Hystrix cristata*) became extinct throughout Europe during the early Holocene, but were subsequently re-introduced into Sicily and southern Italy during late antique times, from North Africa, a fact now confirmed zooarchaeologically.[19]

Microfauna present a unique opportunity for zooarchaeologists to chart cultural patterns in antiquity. The distribution and spread of the house mouse (*Mus musculus*) and black rat (*Rattus rattus*) is often associated with urbanization and human movements. Zooarchaeological evidence shows Bronze Age mariners unintentionally caused the spread of mice and shrews across

Crete.[20] Phoenicians and Greeks, in turn, were further responsible for the spread of the house mouse into the western Mediterranean, while the Romans inadvertently imported and dispersed the black rat, among other pests, throughout their empire.[21] Bones of house mice are recorded from classical Greek excavations, but do not appear in Italy until the later Republic.[22] Once embedded, they take strong footing in urban locations, displacing other types of mice. In Pompeii, for example, zooarchaeological evidence registers a rise in the frequency of house mice and a concomitant decline in wood mice (*Apodemus sylvaticus*), with which the house mouse competes, coincident with the intensification of urbanization in the city.[23]

The situation involving the black rat (*Rattus rattus*) is complicated. Clear and widespread archaeological evidence for rats in Greek and Roman antiquity, be this in the form of rat gnaw marks on other bones; owl or other predator pellets containing rat pieces; or actual rat remains preserved *in situ*, is piecemeal. Black rats are of eastern origin; a record of them exists among early Near Eastern and Egyptian sites.[24] Their presence in fourth–second century BCE levels in Corsica and among second-century BCE contexts from Pompeii and Minorca suggests colonization of black rats into the western Mediterranean during Republican times,[25] probably as stowaways on trading ships. Ancient texts offer little temporal help, given that classical sources often spoke generically of mice and rats. Still, elements conducive for rat colonization and expansion (e.g., increased trade and shipping; rising population densities and urbanization; poor sanitation) existed in many parts of the ancient world, in turn promoting havens for rats. The black rat is common across large portions of the western Mediterranean during antiquity, and may be linked with plague outbreaks (e.g., bubonic plague wherein microbes are transferred via fleas carried by rats) that affected areas at this time.[26]

TEMPORAL AND SEASONAL SEQUENCING IN ZOOARCHAEOLOGICAL ASPECTS

A further means by which zooarchaeological evidence assists investigations of "when" components in classical archaeology is through the analysis of butchery practices and technologies. Temporal variations may register across antiquity, particularly in augmented use of saws, sharper blades, and bisected vertebrae in some Roman contexts.[27] A multidisciplinary assessment of butchered bones from Romano-British sites, combining data from archaeometallurgy, iconographic and literary sources, and ethnographic observation and experimentation in modern butchery, concluded that urban Romano-British butchery in fact involved great skill, despite what might at first glance seem somewhat 'slapdash' patterning in chop-mark placement on bones.[28] Results indicated that speed maximization underscored these butchery practices, thus helping promote their efficiency. Typically, such a pattern tends to imply greater incorporation of large-scale, perhaps even commercial, butchers in this process. Roman-British cleavers, moreover, performed as dual-purpose implements, both facilitating and improving carcass dismemberment. Further investigation of animal butchery patterns and equipment, from both the macroscopic and microscopic level, may help refine our knowledge of procedural variation as well as recognized chronological phases that may underscore such operations.

Zooarchaeological data also offer various means with which to establish seasonality. The ubiquity of seasonal schedules and calendrical events in antiquity, encompassing components such as festivals, rituals, sacrifices, agricultural duties, among many other behaviors, highlights its research importance. At a macroscopic level, refinements to dental aging methods and correlations of age at death sequences for animal taxa have assisted in narrowing down potential seasonal culling schedules. Such techniques work best for younger animals where age patterns can be observed in dental wear stages within the first year of life. Available evidence for Roman Italy indicates that rural sites register slightly more deaths among sheep/goat in the 3–6 month dental-age group than do urban sites, which show a higher percentage of deaths in the 7–12 month category.[29] Assuming autumn births,[30] such a pattern supports the hypothesis of enhanced late-winter or early-spring culls at rural sites, but a preference for predominantly summer- or autumn-culled ovicaprids at urban sites.

Age and season at death in animals may also be investigated by examining incremental structures in the cementum of teeth through the preparation of microscopic thin-sections. Such techniques have assisted greatly in clarifying seasonal rounds in various animal taxa,[31] but as yet have seen little application in classical archaeology, perhaps in part due to practical limitations in processing and analyzing materials. Studies of cementum banding among sheep and goats from the Roman-Byzantine site of Sagalassos, Turkey, provided greater resolution to seasonal scheduling in pastoral schemes during antiquity, and compared favorably to patterns observed among modern ovicaprids from the region.[32] Results bode well that these less established routes for getting ages and seasons from archaeological materials offer much potential and should see greater use in classical archaeology.

Where

ESTABLISHING ECOLOGICAL AND ENVIRONMENTAL SETTINGS AND CONDITIONS

"Where" is a second category where scientific studies within zooarchaeology benefit classical archaeology. Reconstructions of environmental and climatic conditions using faunal remains have long been a component of paleoenvironmental studies (see Chapters 1 and 2, for this relationship with paleoclimatic and archaebotanical data). Traditional investigations of preferred ecological habitats for various animals recovered from sites, and their corresponding implications in determining movement or migration of taxa offer some measure of assessing "where" animals originated, and "where" they ended up. Zooarchaeology thus helps establish "where" forests, fields, pastures, shallow lakes, deep lakes, and so forth may be, as well as where and how far animals traveled or were moved across different settings. Environmental and climatic tolerances for domestic livestock are generally rather broad, in turn rendering them of relatively limited use in environmental reconstruction. The fact that such taxa typically comprise the bulk of faunal remains recovered (or at least reported) from classical archaeological sites has likely tended to downplay traditional paleoecological zooarchaeological contributions for Greco-Roman antiquity. Nevertheless, future advancements in refining our assessments of ecological and environmental preferences and tolerances for domestic livestock species (both within and between such taxa) augment the value of using their bones to chart such conditions in the past.

Remains of small mammals, birds, amphibians and fish often provide better resolution to understand past environmental conditions, and factor somewhat regularly in classical archaeology. The extensive zooarchaeological corpus of these taxa from excavations at Pompeii yields remarkably vibrant and nuanced reconstructions of environmental conditions for this site—at both macro- and micro-scales—and stands as one of the more comprehensive of such studies for classical archaeology.[33]

STABLE ISOTOPE INVESTIGATION, ANIMALS, AND LANDSCAPE RECONSTRUCTION

A recently burgeoning field in investigating the concept "where" from zooarchaeological remains and consequent relations to landscape reconstruction in classical antiquity focuses on stable isotopes. This field of research has multiple contributions in biological exploration of the past, with applications across paleoclimatic (Chapter 1), archaeobotanical (Chapter 2) and human osteological (Chapter 4) fields as well. As regards their role in zooarchaeology, elements and their stable isotopes cycle through the biosphere driven by physical, chemical, and biological processes, but at different rates due to their differing atomic masses. This leads to different ratios of these substances in organisms, that in turn help provide signals for aspects such as varying diets, home ranges, breeding and foraging areas, and migration routes. Commonly used stable isotopes in archaeology today include carbon (which typically correlates with vegetation), nitrogen (which correlates with trophic levels, commonly the contribution of meat to one's diet), strontium (for geological deposits and their resulting vegetation), and oxygen (an indicator for things like temperature, altitude, and hydrology). Correlating and cross-referencing values allows placement of animals in different environmental settings on the basis of accumulated isotopes, thus helping to recognize outliers and imports. Moreover, comparing values within tooth enamel (which forms during earlier years), to those in the bone (which remodels until death) helps determine mobility in an animal, across its lifetime at one level, but seasonally as well, through correlation with dental aging analyses.

Case studies from two sites help illustrate potential. The site of Sagalassos has acted as somewhat of a pioneer for zooarchaeological stable isotope research in classical archaeology, with several influential studies. First, oxygen and strontium isotope ratios were measured in archaeological fish remains to address issues of provenance. Results from the stable oxygen studies excluded a riverine origin for these fish (carp was the taxon chosen for assessment), while those from the strontium investigation helped eliminate some local lakes as the source.[34] Second, stable carbon and nitrogen isotope results across livestock taxa at Sagalassos indicated a shift in values for sheep, suggesting that they were herded together with cattle in Roman times, but with goats during the Early Byzantine period. Such a pattern demonstrates the development of extensive stock breeding operations during Roman antiquity in the area, no doubt bolstered not simply through enhanced exploitation of available pasturage at this time, but presumably from good management of those resources as well. Finally, results from stable strontium values for ovicaprid teeth from the site document where these animals were born as well where they may have moved throughout their lives.[35]

The second study for consideration explores pasturing regimes in Neolithic Anatolia. Investigators detected, through measuring carbon and nitrogen isotopes in archaeological sheep and goat bone collagen, that flocks that provided the site of Çatalhöyük moved over a much more extensive territory (thus encountering multiple isotopically distinctive plant biomasses) than did flocks from the nearby site of Aşikli Höyük.[36] Work incorporating oxygen isotopes and dental microwear studies at Çatalhöyük added further dimensions, in concluding, among other aspects, that (1) neither long-distance, seasonal transhumance, nor fully separate, nomadic pastoralism was practiced at the site during the Neolithic; (2) flocks grazed on dedicated seasonal pastures and did not suffer from resource stress; (3) most sheep were slaughtered in early spring, after fattening on autumn grass regrowth.[37] Although these examples situate temporally outside the scope of antiquity, the results produced are of tremendous interest in framing similar questions about livestock movements in Greco-Roman antiquity. Considerable debate, for example, exists surrounding the various scales of pastoralism in antiquity.[38] Schemes can vary from localized, small, nontranshumant herding at a permanent site, to large-scale, long-distance transhumance. Implementing scientific means to track an animal's movement over the course of its lifetime, increasingly accomplished through archaeological isotopic analyses, provides an objective measure that focuses upon the actual participants in these processes (i.e., the animals themselves) as opposed to cultural recordings, reflections, or descriptions of such operations.

LANDSCAPES, TRACE ELEMENTS, POLLUTANTS, AND ANIMALS

Evaluation of trace element data from zooarchaeological remains adds a further component to environmental and cultural reconstructions in classical archaeology. Various chemical compounds (some poisonous) can infiltrate or otherwise be incorporated into skeletal tissues, including bones and teeth, through interaction with environmental conditions, dietary intake, ingestion or inhalation of pollutants, and other means. Through chemical analysis of goat bones found at Sagalassos, researchers were able to detect augmented levels of zinc, lead, magnesium, copper, and other heavy metals. Higher amounts from second-century CE contexts suggest stock were kept closer to, perhaps even within, the site, and thus exposed more to urban pollutants from workshops and manufacturing in that vicinity. Lower uptake of these pollutants during the fourth century CE can be explained by a wider catchment area that encompassed less polluted areas further afield, a concept that itself might signal more secure rural conditions for the region during this time.[39]

Clearly, research utilizing stable isotopes and trace elements can greatly enhance our knowledge of transhumant scales and routes, mobility of animals between coastal and interior areas, herding strategies, relations of stock to ecological zones, exploitation of coastal, riverine, and lacustrine resources, imports and exports of animals across regions of the ancient world, and differential feeding regimes. These are very powerful tools in classical archaeology, and areas of incredible potential for zooarchaeological input. Currently, however, more attention has focused on isotopes in human bones (and human dietary investigations);[40] what has been employed for animals has concentrated more on prehistoric and Roman contexts, predominantly at the Anatolian sites of Çatalhöyük and Sagalassos.

SPATIAL DISTRIBUTION OF ZOOARCHAEOLOGICAL REMAINS WITHIN A SITE

While understanding "where" an animal derives from, in a broad ecological, environmental, and habitat sense, is certainly a worthy topic to explore, it is important to recognize that "where" can also be pursued at a smaller level, notably in the spatial distribution of bones within a site. Such work is proving instrumental in demarcating the range of cultural and behavioral variability that might underscore "where" faunal materials are discarded, placed, or interred. One area of attention regarding zooarchaeological work in classical archaeology has focused upon spatial distribution of burnt sacrificial materials at a number of Greek sites to clarify how the composition and placement of such deposits correlates or contrasts with what otherwise might be classified as 'secular' deposits among sites.[41] This complex inter-relationship between "sacred" and "secular" aspects, as regards meat-eating in Greek antiquity, has greatly expanded zooarchaeological work in that field of classical archaeology, with attention increasingly now devoted to examination of faunal remains from "nonritual" contexts in Greek archaeology.[42]

Roman and Near Eastern archaeology are better situated in this respect, from a zooarchaeological perspective, with more studies of spatial distribution of bones (some using techniques and technology derived from GIS (geographic information system) research and applications) and what this means for site type, room function, rubbish disposal (e.g., primary, secondary, etc.) among other related topics. Detailed investigation of spatial patterning in bones from late antique San Giovanni di Ruoti, Italy, for example, was able to clarify the sequence of, and variation within, midden formation and room abandonment at the site.[43]

TAPHONOMY AND BONE DENSITY STUDIES

Taphonomy, that is, all the site formation and postdepositional processes that underlie and shape archaeological assemblages, inevitably factors in assessments of the preservation and placement of zooarchaeological materials at archaeological sites.[44] At their core, zooarchaeological samples (indeed all archaeological materials one might argue) are inextricably contingent on taphonomic agents. Greater attention to how archaeological sites are created and how things are taphonomically transformed is critical in our investigations, but taphonomy is often an issue that is overly simplified, or worse yet, ignored, among work at some classical archaeological sites. As organic remains, bones can contribute much to taphonomic studies, since in large part they represent things other organisms (be these dogs, insects, bacteria, etc.) seek or that provide a direct link between the cultural, biological, and geological worlds shaping postdepositional aspects.

While taphonomic agents might affect zooarchaeological deposits among classical archaeological sites in numerous ways, a key concern centers upon differences in bone densities. Essentially, denser bones survive better than less-dense ones, which can lead to skewing favoring remains of older animals and specific skeletal elements such as teeth and lower limbs (carpals, tarsals, and phalanges). Variability in density rates across individual bones among taxa, alongside issues such as differences in the actual number of bones deriving from each animal, differential recovery of elements across taxa, and carnivore damage to assemblages adds further biases to consider. Taphonomy of zooarchaeological materials formed an integral line of investigation at the site of San Giovanni di Ruoti.[45] Investigation here drew upon experimental work that not only took into account variability in bone density, recovery, and related issues, but measured various rates of destruction from key taphonomic factors such as carnivore damage, post-depositional breakage patterns, and soil composition and acidity, to help compensate for the range of potential biases that could affect the constitution and integrity of the faunal assemblage retrieved. Incorporation of all these components strengthened the ultimate conclusions drawn, most critically that the patterns of skeletal-part bias (in this case a very high frequency of pig cranial elements, with differing degrees of export and processing of cuts of meat) were strongly cultural, rather than taphonomic, in nature.

What

STABLE ISOTOPES AND ANIMAL DIETS

"What" denotes another interrogative pronoun shaping investigation. Stable isotopes offer help here, in determining "what" an animal has consumed, and how such might vary due to ecological and cultural factors.[46] Studies show variation in carbon and nitrogen stable isotope values from pigs from medieval York.[47] One sample registered values that were more consistent with elevated protein in the diet, perhaps a indication this animal was yard-kept, consuming a more omnivorous diet. The majority of samples, however, yielded values indicative of largely herbaceous diets, suggesting in turn these animals were raised in rural or woodland locations, as opposed to urban settings. The potential of such applications in classical archaeology is currently untapped, but incredibly promising and germane. Manipulation of animal diets for cultural tastes was known in antiquity. The ancient sources, for example, note how pork acquired different flavors depending on what the animal ingested.[48] Determining differential feeding regimes among animals in antiquity—some perhaps receiving "choicer" diets to impart distinct flavors in their meat—may in turn reflect distinct cultural meaning and purpose.

Nitrogen levels may also be affected by manuring, debatably a rather underexplored aspect of classical archaeology but something incredibly vital for crop and animal husbandry across the ages.[49] Paleobotanical research (as noted in Chapter 2) confirms the impact manuring can have on cereal crops, which respond with elevated levels of stable nitrogen values.[50] The effect among pulses is less noticeable, given these plants can fix atmospheric nitrogen, and tend towards elevated stable nitrogen values to begin with. One potential investigatory avenue stemming from such work involves broader assessments of stable nitrogen values in herbivores, in turn linking those data to aspects such as input of pulses and legumes in their diet, as well as grasses grown on well fertilized fields (i.e., manured ones), versus poorer pastures. Such would certainly augment our understanding of the integration between agricultural and pastoral schemes in antiquity. Much, at this stage, hinges upon intensifying and linking paleobotanical and zooarchaeological isotopic studies and data so that a more comprehensive trajectory can be explored.

DENTAL MICROWEAR ANALYSES AND ANIMAL DIETS

Dental microwear, that is examining scratches, pits, and striations on tooth enamel as caused by feeding regimes, adds another dimension to dietary reconstructions in animals with potential applications for classical archaeology. These marks typically remain on the enamel surface for a few weeks after formation, thus providing some record of the types of foods ingested in the latter phases of an animal's life.[51] Microwear signatures have been used to determine aspects such as pasture grazing versus foddering; tree-leaf browse versus arable by-product fodder; fresh versus dried foodstuff consumption; and quality of pasturage.[52] The underlying environmental and cultural factors that affect animal feeding regimes, moreover, amplify the relevance of such studies. For example, dental microwear analyses revealed domestic sheep, goat, pig, and cattle from "ritual" contexts at Neolithic Markriyalos, Greece, were provided with soft fodder, perhaps as a means to impart a special diet before slaughter.[53] Within the temporal frame of classical archaeology, chronological changes in ovicaprid pastoral practices at Sagalassos were charted through dental microwear analyses,[54] whereas applications of the techniques have helped distinguish feeding regimes for pigs at Sagalassos, as well as in Roman Britain.[55]

REPRESENTATION AND DISTRIBUTION
OF SKELETAL PARTS

While chemical, isotopic, and microscopic investigations clearly provide new lines of enquiry with animal bones, critical "what" questions may also be addressed through more traditional means. Specifically, greater attention to "what" parts of animals (that is "what" skeletal sections) surface at sites, and "what" this means about production, consumption, trade, movement, and overall use of animals and their resources is an important dimension for which zooarchaeological studies can contribute significantly to classical archaeology. Often, in a quest to distill zooarchaeological studies to their core, what are ultimately reported are taxonomic frequencies (% of cattle, pig, sheep, etc.), less so any variation in body parts. Was one sample basically head bones, the other chiefly tails? This latter information is far more telling of the actual processing and use of these animals, at least those critical final steps. Again, taphonomic forces are integral to understand here, since these can have a great impact on "what" bones survive.

Within the topic of distribution of skeletal parts, one area where zooarchaeological work has contributed significantly to classical archaeology, and principally here Greek archaeology, is "what" parts of an animal were selected for burnt sacrifice: *thusia* (parts, often the thigh bones), *osphys* (the tail) or *holocaust* (the entire animal). Examples of each have been identified in the zooarchaeological record.[56] Microscopic examination of the bones has even helped refine aspects such as fire temperatures, durations, fuel sources, and other practicalities in burning.[57]

Although investigation into the biases and patterns in skeleton-part representation is important to classical archaeology, greater attention should also be paid in determining from "what" side of the body materials derive. Side may not be routinely recorded in zooarchaeological analysis, but it can be meaningful. For example, left-side bias registers in *thusia* sacrifice to the hero Opheltes at the site of Nemea, Greece. Right-side biases were shown to register in a few cases for sacrifice to Apollo (e.g., Kourion), but not always.[58] Ultimately, "what" side, part, cut, or resource is chosen from an animal can carry with it deeper cultural meanings, otherwise not recorded in any other source— archaeological, literary, or iconographic. This may be especially critical in the case of samples for which side and part data were not initially recorded. Such a concept inherently endorses the application of new techniques and analytical approaches to older, curated zooarchaeological collections to address current questions about the history of human/animal interaction in antiquity. Such reassessments should not be deemed a challenge to original results, and indeed may be more common where problems and changing parameters of excavation make it difficult to collect and study new material. Given the long histories of excavations among many ancient Greek and Roman sites, potential also surfaces in analyses of older collections merged with newly recovered material from renewed excavations at a given site, where amenable.

Who

ESTABLISHING ANCESTRY AND 'IDENTITY' IN ANIMALS THROUGH DNA ANALYSES

The interrogative pronoun "who" encompasses concepts of ancestry and identity. In the same way that these aspects can be explored in assessing human biological (including skeletal) materials from the past (see Chapters 4, 6, and 7), our knowledge of such matters in ancient animals is being reshaped immensely through DNA analyses.[59] At one level DNA investigation is helpful in securing accurate identification of remains, a tactic likely to see immense future applications in archaeology, notably in determining species from fragmented or otherwise undiagnostic bones.[60] DNA work is also valuable as a means to distinguish sheep from goat, taxa frequently recovered from ancient sites and clearly integral to components such as the wool industry.[61] The skeletons of these animals are similar; while some morphological and metrical tactics can assist in identification, these are not universally applicable across all skeletal and dental components, and can be highly variable in their representation.

A second area in which DNA analyses factor in zooarchaeological work involves tracing the geographic origins of biological products to determine trade and exchange networks. An import of Nilotic fish to Roman Sagalassos was verified when mtDNA sequences for catfish bones from the site matched those with modern catfish collected from the lower Nile. Presumably these fish were preserved somehow before being traded, but they nonetheless help confirm other lines of archaeological evidence for trade among commodities and resources between these two regions.[62]

Finally, DNA research is revolutionizing our understanding of relationships between populations of animals over time and space. Indeed, tracking and mapping origin and domestication processes for species, as well as subsequent transformations, migrations, movements, and interactions that particular species may experience, dominate much of the current DNA scholarship in archaeology.[63] The potential within these methods to examine the creation and spread of new varieties or breeds of animals resonates well to classical archaeology. Greek and Roman antiquity witnessed scores of what today may be best considered as varieties, strains, or landraces,[64] if one assumes the range of references in the ancient sources describing different morphological types of animals within one species is an indication of breeding manipulation, whether conscious or unconscious, among ancient farmers and herders. Inevitably the result was variation in traits among livestock, with selection for certain characteristics deemed of interest for specific local and regional natural and cultural factors of interest, or under consideration. The ancient sources mention several types of cattle in Roman Italy, which are often grouped by geographic zones and accompanying general traits of interest for each spot (e.g., small Ligurian cattle; thick-set, powerful cattle of Etruria and Latium, etc.).[65] Similar distinctions among pigs and sheep surface within ancient texts to suggest creation of "breeds" or varieties among these taxa as well.[66]

MORPHOMETRIC INVESTIGATIONS
AND ANIMAL 'BREEDS'

While DNA investigations, as they develop, will inevitably shed new light on our understanding of animal breeds in Greek and Roman antiquity, recent work that utilizes osteometrics is already expanding possibilities.[67] Varieties or "breeds" can be explored through detailed studies using bone measurements, a huge collection of which already exists among zooarchaeological studies for classical archaeology. Incorporating measurements of various dimensions—bone lengths, widths, depth, etc.—as well as ratios among such dimensions provides a better calculation of morphology. Such techniques are not unique to animal bone investigation, and indeed can be applied within the fields of archaeobotany (Chapter 2) and human osteology (Chapters 4) as well—in the case of the latter, for example, most notably in the study of growth and stature (Chapter 5). As regards animals, research indicates that widespread increases in cattle sizes throughout Roman Italy do not occur until Republican and Imperial times, coincident with marked agricultural and demographic changes in that region.[68] Distinct clusters of cattle "breeds" develop during these times, which see representation in both zooarchaeological and textual databases. Differential selection of traits is evident with aspects such as stockiness, leanness, height, longer legs, among other aspects variously manipulated to suit specific cultural and environmental needs and conditions. Several factors interplay to cause size and shape changes, including an augmented market and military demand for grain and other foodstuffs, local demands for more powerful plow and traction oxen, as well as the import and export of cattle brood-stock into and out of Roman Italy.

In another example, size and shape variation, as revealed from zooarchaeological metric data for sheep indicate an increase in height as a consequence of manipulation of livestock across much of the larger Mediterranean world during Roman times; however, tremendous variation is noted.[69] Smaller breeds are often never eliminated, while the introduction and spread of taller, slender types, heavier, thicker-set types, smaller, rustic types, among other varieties of sheep attests to the shrewd, productive breeding tactics during antiquity.

Morphometric assessments of domestic fowl bones present further work of interest. Investigators were able to recognize three "breeds" of domestic fowl at Sagalassos, but through limiting samples to females (as determined from the presence of medullary bone, a build-up of calcium in egg-laying birds) could avoid problems of sexual size variation that might otherwise complicate results.[70]

SKELETAL HISTOLOGY AND ANIMAL 'IDENTITY'

"Who" is further reflected in individualized patterns of bone development. Bone adapts to everyone's unique circumstances, which include environment, health, and activity. At the microstructural, histological level one can record individual markers in bone fiber alignment, crystalline orientation, rate and extent of replacement and remodeling of bone tissues, thickness of deposits, among other features that ally with variation in muscular activity and associated stresses from compressive forces (as would be encountered in any weight-bearing activity). Researchers have used histological indicators to investigate which caprines may have walked more, and which may have encountered tougher terrain in such transhumant journeys among Neolithic contexts in Iran.[71] Results showed a tendency for the degree of ovalness of the bone shaft (the humerus bone was used) and the thickness of the trabeculae within the bone to be negatively correlated, an expected outcome if the trabeculae correlated with weight transfer-ence from more compressive forces. Wild sheep showed more of this tendency than domestic sheep, a finding that supported their link to greater activity, in this case greater compressive stress on their bones, presumably from trekking longer distances over more difficult terrain. Although the theoretical basis from which to expect variations among histological elements for animals is sound, the potential of such work in archaeological practice has never caught on to any degree, most likely due to the nascent stages of development for this type of research overall, coupled with practical costs related to sample preparation and microscopy. Nevertheless, such techniques provide a virtually untapped domain of investigation for classical archaeology, especially in understanding relational effects of animals to their ecosystems.

PALEOPATHOLOGY AND ANIMAL 'IDENTITY'

Health is another aspect that individualizes an animal, both in specifics of the condition, and in any investment made to treat that condition. Paleopatho-logical analysis forms a crucial component of zooarchaeological research (as it does human osteology as well, see Chapters 4 and 5), yielding information, from the level of the individual up to the population, which can be set against a broader investigation of all the cultural and natural factors that might link to the incidence and prevalence of different pathological conditions, as well as the various ways in which disease and illness may be treated.

At the individual level, care of an animal often reveals some emotional attachment or investment, in turn an indicator of human identity. Common pathological conditions noted in an investigation of skeletal health in dogs from Roman sites across the Mediterranean include dental complications, es-pecially premortem tooth loss, healed limb fractures, osteoarthritis, and infec-tion, in patterns and frequencies similar to dog samples from other temporal and spatial contexts.[72] Generally, Roman dogs seem to be in good condition, as regards skeletal health, with minimal osteological evidence for human abuse or maltreatment, but also no conclusive data for splinting any broken bones. Smaller 'toy' breeds of dogs in Roman times appear more susceptible to mul-tiple pathological conditions, but also display signs of greater human care, es-pecially in terms of pampering and feeding.

Correlation of bone pathologies observed on modern and ancient cattle al-lowed for a better understanding of disease etiology coincident with draught exploitation.[73] Not only did this work connect the range of deformations noted on the bones with associated activities, enhancing our knowledge about animal traction, but further established a methodological basis to score, describe, and interpret various skeletal pathologies in cattle. Investigations of pathological conditions among cattle at Tiel-Passewaaij, a Roman site in the Netherlands, revealed the use of both cows and oxen for draught purposes, thereby debunking some traditional assumptions that only oxen were used for traction.[74] The implications and potential of such studies for classical archaeology, a period in which draft animals see great use and development, is especially marked.

Enamel hypoplasia, a commonly occurring defect that reflects physiological stresses encountered by the animal during dental development, represents another pathological condition of interest in classical antiquity. The condition can shed light on aspects including dietary deficiencies and general health in animals, which in turn can be influenced by environmental and cultural factors. Protocols to record enamel hypoplasia in different species, including major domesticates, are established,[75] with several noteworthy applications in classical archaeology. Enamel hypoplasia data among pigs at Sagalassos, for example, revealed two important conclusions. First, consistent patterning in the condition across phases of occupation suggested that environmental conditions remained relatively stable throughout time, perhaps an indication that the forests, where such pigs might be herded, were not heavily affected by exploitation of timber.[76] Second, evidence correlated with a single farrowing episode, in spring, a husbandry tactic that aligns better with extensive herding as opposed to household rearing.[77]

Why

Lastly, attention turns to the interrogative pronoun "why." This is a huge topic, which deserves far more than the brief overview afforded here. "Why" questions generally involve synthesis, which inherently draws upon theory. Many people tend to stigmatize theory—"oh that's too complex for me"—until they realize they use it constantly, and have internalized many aspects. Theories inherently involve connections of concepts, pattern recognition, shared trajectories of associations, and related frameworks of knowledge that guide our thinking. Many components of zooarchaeological work arguably might situate more towards the processual (i.e., 'science') end of the archaeological theoretical spectrum due to the emphasis on pattern and data analysis to craft explanations and interpretations, not to mention the complementarity with scientific testing of materials. Nevertheless, in studies of complex societies, like those of the ancient Greeks and Romans, "processual-plus" or "postprocessual" (i.e., more 'humanistic') theoretical viewpoints may be a better assessment since zooarchaeologists use scientific methods, taphonomy, and patterns of data to address questions of individual agents within the development of the archaeological record, such as ethnic groups, diasporas, social classes, and so forth. Thus, zooarchaeologists often address postprocessual questions but with processual methods. Such may supposedly blur theoretical lines, but also proves the importance of zooarchaeology in examining all parts inherent in archaeological reconstruction, and in advocating shared agendas and goals. Moreover, scientific methods often carry with them a level of procedural thinking and rigor that strives for objective reasoning. Although staunch objectivity in deciphering the often subjective-laden realm of human behavior in the past is not always achievable, incorporation of scientific thinking in this quest can be helpful. As Morris remarks:

> Archaeological interpretations will never achieve the same level of confidence as those in the natural sciences—there are far too many interacting and unknowable factors to assess—but by making testing procedures explicit, archaeologists and natural scientists can share common ground and language of scientific inquiry.[78]

To conclude, a sentiment, whether expressed outwardly or not, that "classical archaeology is a place to avoid science" is clearly not reflective of classical archaeology today. Such insularity, moreover, is not helpful for scholarship within the discipline as it ventures forward. While some of the procedures to extract and process scientific data from archaeological remains (within this chapter the focus being zooarchaeological finds) may seem complex, such is really not the case when their basic components are dissected. Indeed, as Chapter 1 in this volume makes clear (and is important here to emphasize), what can be learned from zooarchaeological materials, and scientific tests of these, adds immensely to our shared reconstructions of the past. The operative word here is "shared"—all manner of data contribute to our understanding of antiquity. Cultivating that ethos certainly serves classical archaeology well.

Notes

1. The premise of interconnections between natural and cultural worlds underlines investigation of archeobotanical remains as well (see Chapter 2).

2. See MacKinnon 2007 for a more detailed overview of the historical development of zooarchaeology in classical archaeology.

3. E.g., (for textbook-style works) Hesse and Wapnish 1985; Davis 1987; Reitz and Wing 1999; O'Connor 2003.

4. For the Roman world, see, for example, the regional studies of King 1978, 1984, 1999, 2005; Luff 1982; Lauwerier 1988; Peters 1998; Lepetz 1996; Columeau 2002; Fernández Rodríguez 2003; Colominas 2013; Bökönyi 1974; MacKinnon 2004, 2017; among others. Various handbook and companion volumes related to animals in antiquity provide further points of interest (e.g., Campbell 2014; Albarella et al. 2017).

5. One such horizontal stretch in the calibration curve, known as the "Halstatt plateau" (Becker and Kromer 1993) spans the time range 800–400 BCE, an arguably key era of interest for classical archaeology.

6. van der Plicht et al. 2009.

7. Bronk Ramsey 2009.

8. Williams 2013.

9. Reese 1987.

10. Docter et al. 2004.

11. van der Plicht et al. 2009, 227.

12. For zooarchaeological syntheses of Roman Britain and the Roman northwestern provinces in general, see King 1978, 1984, 2005; Luff 1982. For the Netherlands, see

Lauwerier 1988. For the Roman Germanic provinces, see Peters 1998. For Roman Gaul, see Lepetz 1996; Columeau 2002. For Roman Iberia, see Fernández Rodríguez 2003; Colominas 2013. For Central and Eastern Europe, see Bökönyi 1974. For Roman Italy, see MacKinnon 2004. For Roman North Africa, see MacKinnon, 2017. Audoin-Rouzeau (1993) and King (1999) provide generalized, empire-wide zooarchaeological studies.

13. For a zooarchaeological overview of fish and fishing in Greek antiquity, see Mylona 2008. For ritual feasting, see e.g., Dietler and Hayden 2001; Dabney et al. 2004. For burnt-animal sacrifice, see Ekroth 2014; 2017.

14. MacKinnon 2014.

15. Davis and MacKinnon 2009.

16. Sykes et al. 2006.

17. Strabo, 3.2.5; Pliny the Elder, *Natural History* 8.217–18

18. Masseti 2003, 58; MacKinnon 2004, 213.

19. Masseti et al. 2010.

20. Papayiannis 2012.

21. Cucchi and Vigne 2006.

22. King 2002, 435.

23. Holt and Palazzo 2013.

24. Collins 2002.

25. McCormick 2003.

26. King 2002, 443. For the exploration of plagues in antiquity from the vantage of DNA evidence, see Chapter 6.

27. MacKinnon 2004, 163–172, 177–184.

28. Seetah 2006.

29. MacKinnon 2004, 108.

30. As prescribed in the ancient texts: Columella, *On Agriculture* 7.3.12; Varro, *On Agriculture* 2.1.19, 2.2.14; Pliny the Elder, *Natural History* 8.72.187.

31. Lieberman 1994.

32. Buels 2004. The site of Sagalassos stands at the forefront of many scientific applications of zooarchaeology in classical archaeology. For a reflective overview of work, see van Neer and de Cupere 2013.

33. Jashemski and Meyer 2002.

34. Dufour et al. 2007.

35. The latter two examples derive from: W. Van Neer et al. "Herding practices inferred from multiple isotopic and heavy metal analyses of faunal remains from a classical site in Turkey," in BoneCommons, Item #1622, http://alexandriaarchive.org/bonecommons/items /show/1622 (accessed May 25, 2014).

36. Pearson et al. 2007.

37. Henton 2010, 406–409.

38. Halstead 1996; Lemak 2006.

39. Degryse et al. 2004.

40. As detailed in Chapter 4 (this volume).

41. Ekroth 2014; MacKinnon 2013.

42. MacKinnon 2014.

43. MacKinnon 2002.

44. Lyman 1994 offers an excellent overview of vertebrate taphonomy.

45. MacKinnon 2002.

46. The same concepts apply in the use of these techniques to investigate diet in humans; see Chapter 4.

47. Hammond and O'Connor 2013.

48. Varro, *On Agriculture* 2.4.3; Pliny the Elder, *Natural History* 8.209.

49. Forbes 2012.

50. Fraser et al. 2011.

51. Solounias and Hayek 1993.

52. Mainland 2006; Henton 2010.

53. Mainland and Halstead 2002.

54. Beuls 2004.

55. Vanpoucke et al. 2009 (for Sagalassos); Wilkie et al. 2007 (for Britain).

56. Ekroth 2014; 2017.

57. Hincak et al. 2007; Ubelaker and Rife 2007; MacKinnon 2013.

58. MacKinnon 2013.

59. Matisoo-Smith and Horsburgh (2012, 145–152) provide a good overview of such applications.

60. Newman et al. 2002.

61. Buckley et al. 2010.

62. Arndt et al 2003.

63. Lira et al. (2010) trace lineages of Iberian Neolithic and Bronze Age horses in relation to modern horses; Larsen et al. (2007) investigate pig domestication and spread in Europe throughout the Neolithic. Colominas et al. (2014) explore cattle ancestry in Iberia using mtDNA. For further examples, and an overview of DNA studies in classical archaeology in general, see Chapter 6 (this volume).

64. Strict "breed" cannot be truly verified without proper genetic study, as genotype and phenotype manipulation frame its modern definition.

65. See MacKinnon 2004 and 2010 for more comprehensive lists of ancient sources and their connection to cattle "breeds" in Roman Italy.

66. MacKinnon 2001; 2015.

67. Enhanced 3-D imaging will inevitably follow as an exploratory tool in zooarchaeology; see Owen et al. 2014 for an application using pig crania. Imaging, additionally, has multiple uses in archaeobotany and human osteology (see Chapter 2 and 4).

68. MacKinnon 2010.

69. MacKinnon 2015.

70. De Cupere et al. 2005.

71. Zeder 1978.

72. MacKinnon 2010.

73. Bartosiewicz et al. 1997.

74. Groot 2005.

75. E.g., Dobney and Ervynck 1998 (for pigs); Upex et al. 2014 (for caprines); Kierdorf et al. 2006 (for cattle).

76. M. Waelkens—"Stressed pigs?"—http://mill.arts.kuleuven.be/IPA-V-09/faunal.html

77. Vanpoucke et al. 2007.

78. Morris 1995: 82.

References

Albarella, U., M. Rizzetto, H. Russ, K. Vickers, and S. Viner-Daniels, eds. 2017. *The Oxford Handbook of Zooarchaeology*. Oxford: Oxford University Press.

Arndt, A., et al. 2003. "Roman trade relationships at Sagalassos (Turkey) elucidated from mtDNA of ancient fish remains." *Journal of Archaeological Science* 30: 1095–1105. DOI: 10.1016/S0305-4403(02)00204-2.

Audoin-Rouzeau, F. 1993. *Homme et animaux en Europe de l'époque antique aux temps modernes. Corpus de données archéozoologiques et historiques.* Paris: Centre National de la Recherche Scientifique.

Bartosiewicz, J., W. Van Neer, and A. Lentacker. 1997. *Draught Cattle: Their Osteological Identification and History.* Tervuren: Koninklijk Museum voor Midden-Afrika.

Becker, B., and B. Kromer. 1993. "The continental tree-ring record—absolute chronology, 14C calibration and climatic change at 11ka." *Palaeogeography, Palaeoclimatology, Palaeocology* 103: 67–71. DOI: 10.1016/0031–0182(93)90052-K.

Bökönyi, S. 1974. *A History of Domestic Mammals in Central and Eastern Europe.* Budapest: Akadémiai Kiadó.

Bronk Ramsey, C. 2009. "Bayesian analysis of radiocarbon dates." *Radiocarbon* 51 (1): 337–360. DOI: 10.2458/azu_js_rc.51.3494.

Buckley, M., et al. 2010. "Distinguishing between archaeological sheep and goat bones using a single collagen peptide." *Journal of Archaeological Science* 37: 13–20. DOI: 10.1016/j.jas.2009.08.020.

Beuls, I. 2004. "Design of Odontological Tools to Elucidate Small Ruminant Herd Management at Sagalassos (SW-Turkey) in the Roman-Byzantine Period (0–650 AD)." PhD dissertation, University of Leuven, Belgium.

Campbell, G. L., ed. 2014. *The Oxford Handbook of Animals in Classical Thought and Life.* Oxford: Oxford University Press.

Collins, B. J., ed. 2002. *A History of the Animal World in the Ancient Near East.* Leiden: Brill.

Colominas, L. B. 2013. *Arqueozoología y Romanización. Producción, distribución y consume de animals en el nordeste de la Península Ibérica entre los siglos V ane-V dne.* Oxford: Archaeopress.

Colominas L., A. Schlumbaum, and M. Saña. 2014. The impact of the Roman Empire on animal husbandry practices: Study of the changes in cattle morphology in the north-east of the Iberian Peninsula through osteometric and ancient DNA analyses. *Archaeological and Anthropological Sciences* 6: 1–16.

Columeau, P. 2002. *Alimentation carnée en Gaule du sud (VIIe s. av. J.-C.-IIVe s.).* Aix-en-Provence: Université de Provence.

Cucchi, T., and J.-D. Vigne. 2006. "Origins and diffusion of the House Mouse in the Mediterranean." *Human Evolution* 21: 95–106. DOI: 10.1007/s11598–006–9011-z.

Dabney, M. K., P. Halstead, and P. Thomas. 2004. "Mycenaean feasting at Tsoungiza at ancient Nemea." *Hesperia* 73 (2): 197–215. DOI: 10.2972/hesp.2004.73.2.197.

Davis, S. J. M. 1987. *The Archaeology of Animals.* London: Batsford.

Davis, S. J. M., and M. MacKinnon. 2009. "Did Romans Bring Fallow Deer to Portugal?" *Environmental Archaeology* 14: 15–26. DOI: 10.1179/174963109X400646.

De Cupere, B. 2001. *Animals at Ancient Sagalassos: Evidence of the Faunal Remains.* Turnhout, Belgium: Brepols.

De Cupere, B., et al. 2005. "Ancient breeds of domestic fowl (*Gallus gallus* f. domestica) distinguished on the basis of traditional observations combined with mixture analysis." *Journal of Archaeological Science* 32: 1587–1597. DOI: 10.1016/j.jas.2005.04.015.

Degryse, P., et al. 2004. "Statistical treatment of trace element data from modern and ancient animal bone: Evaluation of Roman and Byzantine environmental pollution." *Analytical Letters* 37: 2819–2834. DOI: 10.1081/AL-200032082.

Dietler, M., and B. Hayden, eds. 2001. *Feasts: Archaeological and Ethnographic Perspectives on Food, Politics, and Power.* Washington, D.C.: Smithsonian Institution Press.

Dobney, K., and A. Ervynck. 1998. "A protocol for recording linear enamel hypoplasia on archaeological pig teeth." *International Journal of Osteoarchaeology* 8: 263–273. DOI: 10.1002/oa.2227.

Docter, R. F., et al. 2004. "Radiocarbon dates of animal bones in the earliest levels of Carthage." *Mediterranea* 1: 557–577.

Dufour, E., et al. 2007. "Oxygen and strontium isotopes as provenance indicators of fish at archaeological sites: the case study of Sagalassos, SW Turkey." *Journal of Archaeological Science* 34: 1226–1239. DOI: 10.1016/j.jas.2006.10.014.

Ekroth, G. 2014. "Animal sacrifice in antiquity." In *The Oxford Handbook of Animals in Classical Thought and Life*, ed. J. L. Campbell. Oxford: Oxford University Press, 324–354.

Ekroth, G. 2017. "Bare bones: osteology and Greek sacrificial ritual." In *Animal Sacrifice in the Ancient Greek World*, eds. S. Hitch and I. Rutherford. Cambridge: Cambridge University Press.

Fernández Rodríguez, C. 2003. *Ganadería, caza y animals de compañia en la Galicia Romana: Estudio arqueozoológico*. Coruña: Museo Arqueolóxico e Histórico.

Forbes, H. 2012. "Lost souls: Ethnographic observations on manuring practices in a Mediterranean community." In *Manure Matters: Historical, Archaeological and Ethnographic Perspectives*, ed. R. Jones. Burlington, Vermont: Ashgate, 159–172.

Fraser, R. A. et al. 2011. "Manuring and stable nitrogen isotope ratios in cereals and pulses: Towards a new archaeobotanical approach to the inference of land use and dietary practices." *Journal of Archaeological Science* 38: 2790–2804. DOI: 10.1016/j.jas.2011.06.024.

Groot, M. 2005. "Palaeopathological evidence for draught cattle on a Roman site in the Netherlands." In *Diet and Health in Past Animal Populations*, eds. J. Davies et al. Oxford: Oxbow, 52–57.

Halstead, P. 1996. "Pastoralism or household herding? Problems of scale and specialization in early Greek animal husbandry." *World Archaeology* 28: 20–42.

Hammond, C., and T. O'Connor. 2013. "Pig diet in medieval York: Carbon and nitrogen stable isotopes." *Archaeological and Anthropological Sciences* 5: 123–127. DOI: 10.1007/s12520-013-0123-x.

Henton, E. 2010. "Herd Management and the Social Role of Herding at Neolithic Çatalhöyük: An Investigation Using Oxygen Isotope and Dental Microwear Evidence in Sheep." PhD dissertation, University College London, England.

Hesse, B., and P. Wapnish. 1985. *Animal Bone Archaeology: From Objectives to Analysis*. Washington, DC: Taraxacum.

Hincak, Z., Mihelic, D., and A. Bugar. 2007. "Cremated human and animal remains of the Roman period—microscopic method of analysis." *Collegium Antropologicum* 31: 1127–1134.

Holt, E., and S. Palazzo. 2013. "The role of rodents in the disease ecology of the Roman city." *Archaeological Review from Cambridge* 28 (2): 132–154.

Jashemski, W. F., and F. G. Meyer, eds. 2002. *The Natural History of Pompeii*. Cambridge: Cambridge University Press.

Kierdorf, H., J. Zeiler, and U. Kierdorf. 2006. "Problems and Pitfalls in the diagnosis of linear enamel hypoplasia in cheek teeth of cattle." *Journal of Archaeological Science* 33: 1690–1695. DOI: 10.1016/j.jas.2006.03.001.

King, A. C. 1978. "A comparative study of bone assemblages from Roman sites in Britain." *Bulletin of the Institute of Archaeology (University of London)* 15: 205–232.

King, A. C. 1984. "Animal bones and the dietary identity of military and civilian groups in Roman Britain, Germany and Gaul." In *Military and Civilian in Roman Britain*, eds. Blagg, T. F.C., and A. C. King. Oxford: Archaeopress, 187–217.

King, A. C. 1999. "Diet in the Roman world: A regional inter-site comparison of the mammal bones." *Journal of Roman Archaeology* 12: 168–202. DOI: 10.1017/S1047759400017979.

King, A. C. 2002. "Mammals: Evidence from wall paintings, sculpture, mosaics, faunal remains, and ancient literary sources." In *The Natural History of Pompeii*, eds. W. F. Jashemski and F. G. Meyer. Cambridge: Cambridge University Press, 401–450.

King, A. C. 2005. "Animal remains from temples in Roman Britain." *Britannia* 36: 329–369. DOI: 10.3815/000000005784016964.

Larsen, G., et al. 2007. "Ancient DNA, pig domestication, and the spread of the Neolithic into Europe." *Proceedings of the National Academy of Sciences of the United States of America* 104 (39): 15276–15281. DOI: 10.1073/pnas.0703411104.

Lauwerier, R. C. G. M. 1988. *Animals in Roman Times in the Dutch Eastern River Area.* Amersfoot: Rijksdienst voor het Oudheidkundig Bodemonderzoek.

Lemak, J. 2006. "Pastoralism in the Roman Empire: A Comparative Approach." PhD dissertation, State University of New York at Buffalo.

Lepetz, S. 1996. *L'animal dans la société Gallo-Romaine de la France du Nord.* Amiens: Revue Archéologique de Picardie.

Lieberman, D. E. 1994. "The biological basis for seasonal increments in dental cementum and their application to archaeological research." *Journal of Archaeological Science* 21: 525–539. DOI: 10.1006/jasc.1994.1052.

Lira, J., et al. 2010. "Ancient DNA reveals traces of Iberian Neolithic and Bronze Age lineages in modern Iberian horses." *Molecular Ecology* 19 (1): 64–78. DOI: 0.1111/j.1365 –294X.2009.04430.x.

Luff, R. 1982. *A Zooarchaeological Study of the Roman North-Western Provinces.* Oxford: Archaeopress.

Lyman, D. 1994. *Vertebrate Taphonomy.* Cambridge: Cambridge University Press.

MacKinnon, M. 2001. "High on the hog: Linking zooarchaeological, literary and artistic data for pig breeds in Roman Italy." *American Journal of Archaeology* 105: 649–673. DOI: 10.2307/507411.

MacKinnon, M. 2002. *The Excavations of San Giovanni di Ruoti 3. The Faunal and Plant Remains.* Toronto: University of Toronto Press.

MacKinnon, M. 2004. *Production and Consumption of Animals in Roman Italy: Integrating the Zooarchaeological and Textual Evidence.* Portsmouth, RI: Journal of Roman Archaeology.

MacKinnon, M. 2007. "State of the discipline: osteological research in classical archaeology." *American Journal of Archaeology* 111: 473–504. DOI: 10.3764/aja.111.3.473.

MacKinnon, M. 2010. "Cattle 'breed' variation and improvement in Roman Italy: Connecting the zooarchaeological and ancient textual evidence." *World Archaeology* 42 (1): 55–73. DOI: 10.1080/00438240903429730.

MacKinnon, M. 2013. "'Side' matters: animal offerings at ancient Nemea, Greece." In *Bones, Behaviours, and Belief: The Osteological Evidence as Source of Greek Ritual Practice*, eds. G. Ekroth, and J. Wallensten. Athens: Swedish Institute Athens, 125–143.

MacKinnon, M. 2014. "Animals, economic and culture in the Athenian Agora: Comparative zooarchaeological investigations." *Hesperia* 83 (3): 189–255. DOI: 10.2972 /hesperia.83.2.0189.

MacKinnon, M. 2015 "Changes in animal husbandry as a consequence of changing social and economic patterns: Zooarchaeological evidence from the Roman Mediterranean context." In *Ownership and Exploitation of Land and Natural Resources in the Roman World*, eds. P. Erdkamp, K. Verboven, and A. Zuiderhoek. Oxford: Oxford University Press, 249–276.

MacKinnon, M. 2017. "Animals, acculturation and colonization in ancient and Islamic North Africa." In *The Oxford Handbook of Zooarchaeology*, eds. U. Albarella et al. Oxford: Oxford University Press, 466–478.

Mainland, I. 2006. "Pasture lost? A dental microwear study of ovicaprine diet and management in Norse Greenland." *Journal of Archaeological Science* 33: 238–252. DOI: 10.1016/j.jas.2005.07.013.

Mainland, I., and P. Halstead. 2002. "The diet and management of domestic sheep and goats at Neolithic Makriyalos." In *Diet and Health in Past Animal Populations*, eds. J. Davies et al. Oxford: Oxbow, 104–112.

Masseti, M. 2003. "Holocene endemic and non-endemic mammals of the Aegean islands." In *Zooarchaeology in Greece: Recent Advances*, eds. E. Kotjabopoulou et al. London: British School at Athens Studies 9, 53–64.

Masseti, M., U. Albarella, and J. De Grossi Mazzorin. 2010. "The Crested Porcupine, *Hystrix cristata* L., 1758, in Italy." *Anthropozoologica* 45 (2): 27–42. DOI: 0.5252/az2010n2a2.

Matisoo-Smith, E., and K. A. Horsburgh. 2012. *DNA for Archaeologists*. Walnut Creek CA: Left Coast Press.

McCormick, M. 2003. "Rats, communications, and plague: Towards an ecological history." *Journal of Interdisciplinary History* 34: 1–25. DOI: 10.1162/002219503322645439.

McGovern, P. E. 1995. "Science in archaeology: A review." *American Journal of Archaeology* 99: 79–142. DOI: 10.2307/506880.

Mylona, D. 2008. *Fish-Eating in Greece from the Fifth Century B.C. to the Seventh Century A.D.* Oxford: Archaeopress.

Newman, M. E., et al. 2002. "Identification of archaeological animal bone by PCR/DNA analysis." *Journal of Archaeological Science* 29: 77–84. DOI: 10.1006/jasc.2001.0688.

O'Connor, T. 2003. *The Archaeology of Animals Bones*. Stroud: Sutton.

Owen, J., et al. 2014. "The zooarchaeological application of quantifying cranial shape differences in wild boar and domestic pigs (*Sus scrofa*) using 3D geometric morphometrics." *Journal of Archaeological Science* 43: 159–167. DOI: 10.1016/j.jas.2013.12.010.

Papayiannis, K. 2012. "The micromammals of Minoan Crete: Human intervention in the ecosystem of the island." *Paleobiology and Palaeoenvironment* 92: 239–248. DOI: 10.1007/s12549-012-0081-9.

Pearson, J. A., et al. 2007. "New light on early caprine herding strategies from isotope analysis: A case study from Neolithic Anatolia." *Journal of Archaeological Science* 34: 2170–2179. DOI: 10.1016/j.jas.2007.09.001.

Peters, J. 1998. *Römische Tierhaltung und Tierzucht: Eine Synthese aud archäozoologischer Untersuchung und schriftlich-bildicher Überlieferung*. Rahden: Marie Leidorf.

Reese, D. 1987. "A bone assemblage at Corinth of the second century after Christ." *Hesperia* 56 (3): 255–274.

Reitz, E., and E. Wing. 1999. *Zooarchaeology*. Cambridge: Cambridge University Press.

Seetah, K. 2006. "Multidisciplinary approach to Romano-British cattle butchery." In *Integrating Zooarchaeology*, ed. M. Maltby. Oxford: Oxbow, 109–116.

Solounias, N., and L. A. Hayek. 1993. "New methods of tooth microwear analysis and application to dietary determination of extinct antelopes." *Journal of Zoology, London* 229: 421–445. DOI: 10.1111/j.1469-7998.1993.tb02646.x.

Sykes, N. J., et al. 2006. "Tracking animals using strontium isotopes and teeth: The role of fallow deer (*Dama dama*) in Roman Britain." *Antiquity* 80: 948–959. DOI: 10.1017/S0003598X00094539.

Towers, J., et al. 2011. "A calf for all seasons? The potential of stable isotope analysis to investigate prehistoric husbandry practices." *Journal of Archaeological Science* 38: 1858–1868. DOI: 0.1016/j.jas.2011.03.030.

Trantalidou, K. 2000. "Animal bones and animal representations at Late Bronze Age Akrotiri." In *The Wall Paintings of Thera*. Vol. 2, ed. S. Sherratt. Athens: The Thera Foundation, 709–735.

Ubelaker, D. H., and J. L. Rife. 2007. "The practice of cremation in the Roman-era cemetery at Kenchreai, Greece: the perspective from archaeology and forensic science." *Bioarchaeology of the Near East* 1: 35–57.

Upex, B., and K. Dobney. 2012. "More than just mad cows: Exploring human-animal relationships through animal paleopathology." In *A Companion to Paleopathology*, ed. A. L. Grauer. Chichester: Blackwell, 191–213. DOI: 10.1002/9781444345940.ch11.

Upex, B., et al. 2014. "Protocol for recording enamel hypoplasia in modern and archaeological caprine populations." *International Journal of Osteoarchaeology* 24: 79–84. DOI: 10.1002/oa.2227.

Van der Plicht, J., H. J. Bruins, and A. J. Nijboer. 2009. "The Iron Age around the Mediterranean: A high chronology perspective from the Groningen radiocarbon database." *Radiocarbon* 51 (1): 213–242. DOI: 10.1017/S0033822200033786.

Van Neer, W., and B. De Cupere. 2013. "Two decennia of faunal analysis at Sagalassos." In *Exempli Gratia: Sagalassos, Marc Waelkens and Interdisciplinary Archaeology*, ed. J. Poblome. Leuven: Leuven University Press, 51–58. DOI: 10.2143/AS.27.0.632400.

Vanpoucke, S., B. De Cupere, and M. Waekens. 2007. "Economic and ecological reconstruction at the Classical site of Sagalassos, Turkey, using pig teeth." In *Pigs and Humans, 10,000 Years of Interaction*, eds. U. Albarella et al. Oxford: Oxford University Press, 269–282.

Vanpoucke, S., et al. 2009. "Dental microwear study of pigs from the classical sites of Sagalassos (SW Turkey) as an aid for the reconstruction of husbandry practices in ancient times." *Environmental Archaeology* 14: 137–154. DOI: 10.1179/146141009X124817099 28328.

Wilkie, T., et al. 2007. "A dental microwear study of pig diet and management in Iron Age, Roman-British, Anglo-Scandinavian, and Medieval contexts in England." In *Pigs and Humans, 10,000 Years of Interaction*, eds. U. Albarella et al. Oxford: Oxford University Press, 241–254.

Williams, C. K. 2013. "Corinth, 2011: Investigation of the west hall of the theatre." *Hesperia* 82 (3): 487–549. DOI: 10.2972/hesperia.82.3.0487.

Yannouli, E., and K. Trantalidou. 1999. "The Fallow Deer (*Dama dama* Linnaeus, 1758): Archaeological presence and representation in Greece." In *The Holocene History of the European Vertebrate Fauna: Modern Aspects of Research*, ed. N. Benecke. Rahden/Westf: Verlag Marie Leidorf GmbH, 247–282.

Zeder, M. 1978. "Differentiation between the bones of caprines from different ecosystems in Iran by the analysis of osteological microstructure and chemical composition." In *Approaches to Faunal Analysis in the Middle East*, eds. R. H. Meadow, and M. Zeder. Cambridge, MA: Peabody Museum, 69–84.

CHAPTER FOUR

Bones, Teeth, and History

Alessandra Sperduti, Luca Bondioli,
Oliver E. Craig, Tracy Prowse, & Peter Garnsey

Introduction

Human bones and teeth are the primary databank for biological anthropologists, but have aroused little interest among historians of antiquity. The beginnings of an explanation of this disparity are to be sought in the fact that human skeletal remains have no obvious relevance as a source of information for politics, political institutions, political thought, government, law, religion, warfare: in brief, for the traditional concerns of ancient historians. A second consideration is that biological anthropology is rooted in prehistory; its practitioners are characteristically involved in the exploration of the origins of humanity. Fortunately (for our present purposes), some anthropologists have allowed themselves to stray into historical periods, including the classical world of Greece and Rome. For example, the laboratory of bioarchaeology at the *Museo Nazionale Preistorico Etnografico 'Luigi Pigorini'* in Rome houses the skeletons of Isola Sacra, the burial ground of classical Rome's harbor-town of Portus, and over the last three decades or so has produced a lengthy sequence of articles and dissertations deriving from the study of this large sample. In the meantime, historians of antiquity are showing increased interest in social, economic, and cultural history, and are displaying a new willingness to expose themselves to other disciplines, including the natural and social sciences. Thus, the time seems ripe for fruitful communication between historians and anthropologists. Specifically, health and demography (mortality, fertility, mobility) hold promise as fields for constructive dialogue and collaborative research. Initial contacts have already been made, though not always with happy results.

The challenge awaiting historians is to provide contextualization, to put the results of scientific analysis into a historical setting, and to bring other evidence to bear—while being fully conscious of the limitations of that other evidence. As regards the central question of health and well-being—how healthy were the Romans?—the anthropological evidence appears to have a clear advantage over the conventional source material. An enquiry as to health inevitably begins with diet. The question of what people normally ate and in what quantities is in fact impossible to answer from the traditional sources with respect to past societies, at least prior to the nineteenth century. This is because quantitative data is unobtainable from the conventional source material on food and diet. And yet, without such data, any historical account of food consumption can at best be only impressionistic. One can, of course, derive from a variety of sources, for the most part literary and archaeological, a list of foods that were in principle available to and utilized by inhabitants of a given society in the past, usually members of the upper classes, towards whom the evidence is skewed, and this has been achieved for Roman society.[1] Such a catalogue is not without interest or utility, but historians of food should have higher ambitions, and should be prepared to ask what proportion of the diet came from what source, among any particular group of people, including groups representative of the mass of ordinary people, and not just the elite, in a given society. The best hope of progress in this area lies with the scientific analysis of human skeletal remains, which alone provide data which are quantitatively and qualitatively significant, and are also cross-class. In particular, through the analysis of the stable isotopes of carbon and nitrogen, one can move towards a reconstruction of cumulative diet, that is to say, diet over the last ten years of so of the life of an individual. Those (few) historians who are aware of stable isotope analysis are apt to regard it with skepticism and to underestimate its potential.

Meanwhile a number of historians have shown a tendency to progress from judgments about the diet of a people to conclusions (typically optimistic) about their health. In so doing they leave out of account the factor of disease. Diet and nutrition are not the same thing as health or nutritional status. Nutritional status equals nutrition minus the claims made by disease (and workload). Exposure to disease, endemic or episodic, undermines even the best of diets. While we are fully entitled to be enthusiastic about the impressive range and variety of foods consumed by the residents of an apartment block at Herculaneum, as revealed by the contents of their sewers, we should not forget that there are other variables to be brought into play in any judgment concerning health. On this particular issue, cooperation between historians and anthropologists can produce significant results. Historians have two main cards up their sleeve: historical demography and epigraphy. One theme of the historical demographers (whose focus of interest, inevitably, is on the early modern and modern historical periods) is the negative impact of urban concentrations on health, especially in prescientific societies. One thinks first of course of ancient Rome itself and its environs, but the Bay of Naples was also a significant center of population. At Herculaneum perhaps 4000 people were densely packed into a space of ca. 20 hectares, at ca.250 per hectare. High infant mortality implying low life expectancy at birth—something in the 20s, and maybe the lower 20s—can be assumed for urban agglomerations in Central Italy in the early Roman Empire, other things being equal. As it happens, on the basis of a remarkable inscription (of which significant fragments survive), one can surmise that the population of the town of Herculaneum was sustained at a stable level not by the natural reproduction of the existing inhabitants, but by the forced migration of slaves, a significant proportion of whom passed through manumission into the ranks of the free population. We now look to the anthropologists to fill out the picture of morbidity and mortality (and indeed geographical mobility) at Herculaneum through the various state-of-the-art techniques that they have developed and are developing—including the use of dental histology—to provide a close-grained map of the experience of stress at the most perilous stage of the life cycle, namely, early infancy.

In one particular area, related to health, historians have moved too fast, too soon, and with too little. Some economic historians who have become interested in health and nutrition are looking to bones to provide support for their thesis that the early Roman Empire witnessed substantial economic growth. Specifically, it has been asserted that the Romans enjoyed better health and nutrition than most Europeans up to the nineteenth and early twentieth century, and that this is reflected in their stature. In general, establishing a link-up between health status and economic development is not likely to be a straightforward procedure.[2] As to stature in particular, it can be agreed—and has long been familiar to historical demographers—that heights (and weights, where available) are an important index of health and nutritional status (see Chapter 5). The problem is that the figures for stature that have been derived from the anthropological literature are estimates, arrived at by regression analysis from long-bone measurements. And anthropologists have used a variety of regression formulae, in the absence of any consensus as to which is the most appropriate. Worse still, many anthropologists in their publications have failed to provide the raw data (long-bone lengths), and even neglected to identify the regression formula they have applied.

This is one indication of the fact that it is not easy for historians to make constructive use of the work of anthropologists. If historians have been blind to the opportunities provided by anthropology, anthropologists have been uninterested in reaching out to historians. They have been preoccupied with developments within their own fast-evolving discipline, which have been debated at every turn. As will soon appear, trails have been identified as false, methodologies have been found wanting, and the best way to advance is disputed.

As indicated above, in recent years the idea that bones and teeth are an essential source of information for the historical reconstruction of ancient populations has begun to be taken up by archaeologists and historians. At the same time anthropologists are now becoming aware that the reconstruction of biocultural adaptations and lifestyles of past populations relies on integrated analyses, involving biological, ecological, historical, and archaeological evidence brought together in the collaborative work of researchers from various disciplines.[3] More specifically, social institutions, subsistence strategies, and mortuary rites—and the ways they materialize in the bioarchaeological record—represent the main focus of current anthropological research. Nevertheless, data extraction and interpretation are not simple and straightforward processes: the community of the dead very rarely reflects that of the living,[4] but more often is the outcome of the interplay of cultural, environmental, and biological phenomena that are not always quantifiable.

In fact, the quality and reliability of biocultural reconstructions from skeletal populations have been much debated over the last three decades among anthropologists, since the publication of two major (and now famous) contributions that provided a serious challenge to the assumption that a good basic skeletal biology is capable of reconstructing paleodemographic profiles and describing in detail the living conditions of ancient communities.[5] Since the appearance of these papers, a new generation of studies, mostly focused on critical reassessment of both theoretical and methodological issues, has been launched, and today we can avail ourselves of an ever-increasing number of pertinent contributions in skeletal anthropology, acknowledging fundamental problems and pitfalls, and at the same time pointing to new and more promising approaches of analysis and interpretation.

Any anthropological intervention in cemetery contexts that is to contribute to an integrative, collaborative study with archaeologists and historians should be directed at asking the following, basic questions, of the buried individuals: Who were they? What was their physiological condition? What were their occupations? What did they eat? Where were they from?

Who Were They? The Basics: Sex, Age and Paleodemography

No anthropological analysis carried out on bones and teeth, whether simple or complex and based on advanced investigatory techniques, can neglect two fundamental items of data: the sex and the age at death of the individual under examination. At the same time, no reconstruction, if it is to have demographic significance, can fail to consider as a whole the demographic profile of the set of skeletal remains under scrutiny. For this reason, a large part of skeletal research, from its very beginnings, has been dedicated to the problem of how to arrive at best estimates of sex and age at death, and how to construct a coherent demographic picture out of the data on individuals.

SEX DETERMINATION

As regards the determination of sex from bone and tooth remains, every individual part of the skeleton has been subjected to analysis directed at the calculation of sexual dimorphism and its potential application for the determination of sex, on the basis of examination carried out on skeletal series whose sex and age is established. Numerous morphological and metric criteria have been proposed and published in the literature, together with geometric morphometric elaborations, mainly with reference to the skull and hip bone.[6] Moreover, the ever more concrete possibility of isolating, amplifying and sequencing DNA extracted from ancient skeletons has made available an additional diagnostic tool, although the time and expense involved means that this is a procedure which is difficult to adopt on a large scale.

There are fundamental issues connected with sex-determination. In the front line are the morphological criteria, whether whole bones are involved (usually the skull and the hip bone), or parts thereof. At the same time, as the assessments are made on the basis of visual scoring, they require that the observer have received specialized training, and they are susceptible to subjective judgment, thus undermining intra-and-interobserver replicability.[7] The metric criteria have certain advantages, in that they are applicable even to individuals presented in a highly fragmented state (as is the case with cremated skeletons); again, they are objective and are rigorously defined.[8] Nevertheless, they often issue in a relatively high percentage of misclassifications, and are strongly dependent on the genetic background, not to mention the living conditions, of the population under examination. Numerous studies have demonstrated that metrical and morphological standards derived from a population of known sex are not equally reliable when applied to skeletal series of a different origin.[9]

Still, skeletal sex estimates can attain very high levels of consistency: from 90% for the cranium alone, to 95% for the pelvis alone, to 98% for both in combination or for the pelvis alone.[10] A recent study has shown similar values for some postcranial metrical variables, whether used as a single measurement (epiphyseal breadth of long bones) or within multivariate functions.[11] The calculation of volumes and areas from 3-D models of specific bones has obtained a very high reliability in sex-prediction.[12] However, this procedure cannot be used routinely in archaeological contexts, since it requires bones that are perfectly intact.

The skeletal sexual dimorphism of subadult individuals has often been described, quantitatively and qualitatively. Nevertheless, few attempts have been made to set standards for sexing subadults.[13] Essentially, attention has been directed toward the development of diagnostic criteria based on the dimensional variations of primary and secondary dentition and on morphological aspects of the pelvis and mandible. Notwithstanding the efforts dedicated to this subject, there is now a general consensus that the determination of sex in the prepubic period on the basis of skeletal morphology is unattainable. The only viable option is aDNA analysis. This has proved to be particularly useful in providing better definition of cases and behavioral patterns of infant disposal in ancient Roman communities.[14]

AGE-AT-DEATH DETERMINATION

Age at death assessment from bones and teeth is one of the most investigated and debated topics in skeletal biology. The search for the best and most trustworthy odontoskeletal characteristics for the determination of age at death, and their constant validation, goes back to the middle of the nineteenth century. To warrant serious consideration, an age-marker should satisfy the following criteria: strong correlation with the chronological age; progressive and unambiguously identifiable aging pattern; continuous change through an extended period in the life span; wide applicability; no, or little, influence from environmental factors (pathologies, nutrition, work-load etc.).[15]

The several aging criteria that have been formulated for infant and juvenile individuals are for the most part reliant on developmental stages in the growth of teeth (the stages of formation and eruption) and of bones (the fusion of bone-ends, general and specific dimensions).[16] For adult skeletons, the indicators that are taken into account are mainly linked with postgrowth processes (mainly wear and physiological degeneration) of specific segments of the skeleton. Macroscopic morphological techniques are routinely adopted, followed by the application of radiological techniques and histological observations.[17] Because the growth and maturation process is more regular and constant across populations than the degenerative phenomena, infants and subadults are more easily and precisely diagnosed for age at death than adults.[18]

Despite the plethora of works that have appeared in this research field, there is a widespread awareness that the criteria routinely applied may not ensure high levels of accuracy (in achieving a result that comes close to the chronological age) and precision.[19] One major source of error in estimating age-at-death which is intrinsic and hard to control for, consists in the dissociation between the chronological and the biological age,[20] the point being that the biological age is influenced by the genetic background as well as by environmental factors, such as type and level of physical activity, general health, nutrition, specific pathologies that may alter the aging rate of bones and teeth. Experience shows that the skeletal maturation and aging process are not constant, regular or homogeneous among different anatomical districts of a single individual,[21] and may vary significantly across individuals and populations. Moreover, body size can influence age-at-death determination and result in misclassification.[22]

A second pitfall relates to the reference series: how far are criteria that are derived from a modern population, where sex and age are known, applicable to skeletal samples that come from different geographical regions and chronological settings? This issue has been investigated time and time again, with divergent results.[23]

A third issue was first raised by Bocquet-Appel and Masset in a famous paper. They argue that skeletal age-at-death estimates tend to mimic the age distribution of the reference sample by which the criteria were assessed, and that in consequence the mortality tables that are produced are merely "random fluctuations" and reflect "erroneous methodology."[24] Their paper has the merit of having launched a new generation of studies, mostly focused on the critical reassessment of both theoretical and methodological issues.[25]

The problem of the influence of the reference sample is partly overcome by a shift from single-traits to multiple-traits assessment procedures.[26] Multifactorial standardized procedures for the determination of skeletal age at death include the "combined" method, the summary age method, and the transition analysis.[27] The above-cited methods have had some (partial) success, but it should be acknowledged that they are not routinely used or usable, due to the incomplete preservation of skeletons. In general, research in this area suffers from inconsistency among researchers and across laboratories in the choice of aging techniques and age category definition.[28] Add to this the differences between researchers in terms of the level of their individual experience, and we can see why outcomes can be divergent.[29] The assertion of Maples is still valid,[30] that the process of age determination is "an art, not a precise science." How to react to this issue? At the least, we should continue to explore and test age indicators,[31] with the end in view of making our analyses more scientific and less subjective, through constant checking and standardization of methods, and searching for quantifiable aging procedures.[32]

The determination of infant age at death is a specialist area in itself. Age-estimation is mostly based on the development and eruption of the primary and secondary dentition, following a number of standards derived from radiographic surveys of modern populations.[33] If these estimates are imperfect, this is because of the intrinsic individual variability in the pattern of dental formation. In recent years advances in dental enamel and dentine histology are offering a more precise method of assessing the age at death of infants, calculated as the time that occurred from birth to form an incomplete deciduous tooth.[34] Dental enamel is a highly mineralized tissue that is formed of hydroxyapatite prisms produced through a complex process (amelogenesis), which rhythmically deposits a protein matrix (during the so-called secretion phase), that becomes mature enamel in a second phase (the maturation phase). This process leaves unambiguous traces inside the enamel microstructure in the form of daily striations of the enamel prism (cross-striations) and bands through the enamel thickness (Retzius lines), that correspond to a fixed time interval (in the range of 6–11 days) for each individual. Moreover, when the amelogenesis is disturbed by a stressing agent, the Retzius lines become accentuated (more marked). The first accentuated Retzius line forms at birth. The transition from an intra- to extrauterine environment leaves its mark in deciduous teeth (and first permanent molars) as an accentuated enamel incremental ring called the neonatal line (NL).[35] At birth all the deciduous teeth are present, and very often the first molar is already in the process of formation. Therefore the age at death of an infant can be determined by counting the time markers on the developing crowns from birth till death. Similarly, it is possible to estimate the dentine extension rates in the forming dental roots and extend the range of age at death estimation till the last moment of the first permanent molar formation (~9.5 years).[36] Smith and co-authors, working with macaques, estimated that standard histological techniques yield an average 3.5% overestimate and a 7.2% average absolute difference from the known age.[37] Therefore, dental histology methods offer a much more precise estimate of the infant age at death in comparison with the most widely used morphological methods, and should be used particularly when estimating mortality profiles and weaning-related studies.

DEMOGRAPHIC ESTIMATES

Skeletal data have the potential to provide valuable information on the structure, size, and biological dynamics of past human populations. At the same time their limitations are fully recognized and debated.[38] In addition to the issues relating to age and sex determination discussed above, the data are problematic in other ways, both practical and theoretical. For example, partial recovery of the skeletal material, influence of selective funerary practices, and seasonality in the use of burial are all factors that may profoundly influence the structure of the sample, introducing biases that can affect the reconstruction of the characteristics of the population.

Wood and colleagues in their landmark paper of 1992 questioned the validity of paleodemographic and paleopathologial reconstruction by adding other sources of error: the nonstationary nature of populations, selective mortality and heterogeneity of death-risk among individuals from the same age group. For the calculation of demographic parameters, in fact, a sample must be considered as a single cohort of individuals with the same birth interval. Thus the reference population should be stationary and closed, that is, it should be characterized by the absence of migration flows and events (biological and/

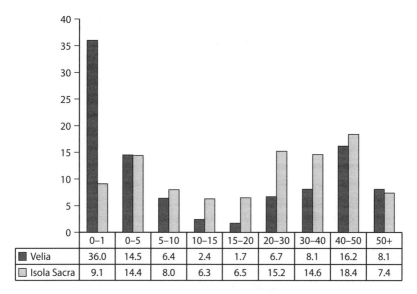

FIGURE 4.1. Age-at-death distributions of Velia Porta Marina
(I–II cent. CE; N=297) and Isola Sacra (I–III cent. CE; N=526).

or cultural) that may have altered mortality/birth rates and growth. These conditions are difficult to observe in real life, and even if some scholars think that there was greater stability in past than in present communities,[39] this is certainly not the case with many towns of the Roman period, for which large migration flows of people of both slave and free status are historically, archaeologically, and biologically attested.[40]

Thus far there have been very few paleodemographic studies of any substance of ancient Roman communities, and even fewer published in scientific journals.[41] In some cases the skeletal series proved to be too biased to yield reasonable results—for example, *Lucus Feroniae* with its skewed sex ratio of 0.79 and virtual lack of infants aged 0–1, and *Isola Sacra* with its oversized class of young adult males.[42]

IS THERE HOPE?

A striking exception to this scenario, is Velia (second century CE, Campania, Italy). Almost 300 burials were excavated at Velia during the 2003–2006 field campaigns, in a necropolis immediately adjacent to the ancient port, just outside the south entrance to the city (Porta Marina Sud). The context of the burials and finds from the excavations indicate that the cemetery was used between the first and second centuries CE. As clearly indicated by the age at death distribution compared with Isola Sacra (Fig. 4.1), the proportion of

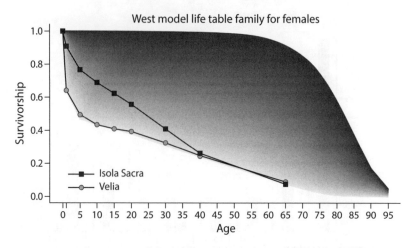

FIGURE 4.2. The survival trend through the age-cycle follows the theoretical model at Velia but not at Isola Sacra. The gray shading shows different levels of mean life expectancy at birth from e_0=20 (lightest) to e_0=80 (darkest).

infants aged 0–1, in Velia is more than 30%, and the whole subadult subsample almost reaches 60%. Velia demographic parameters fit the Western Female model no.1.[43] As shown in Figure 4.2, the survivorship trend through ages in Velia follows the theoretical model, while that in Isola Sacra does not.

As stated above, the age and sex structure of cemetery samples are not normally a reliable guide to the demographic parameters of the living community. Nevertheless, these same data are very useful for the detection of the conditions that have skewed the profile. Anomalous sex ratio and under-/over-representation of particular age classes have been variously interpreted as the outcome of selective burials, slavery, infanticide, warfare, epidemics, migration flows, and other dynamics of sample formation.[44] As for the constant and ubiquitous lack of infant remains from archeological samples, contrary to the differential preservation hypothesis, Lewis suggests that "the absence of infant remains from cemetery sites at different periods is probably revealing more about their status within the society, than their ability to 'dissolve in the ground'."[45]

For the Roman world we have a clear case of how large-scale natural disasters can affect the demographic profile: the eruption of Vesuvius in 79 CE. The skeletal data from Pompeii show a lower proportion of deaths of adult males than of women and children, and this was to be expected.[46] In the Herculaneum sample, however, the sex and age distribution is significantly different. Males outnumber females, and very young individuals are few. Further, if we set the demographic data and the provenance of the skeletons side by side, we obtain a very striking result. The skeletons were found mainly on the beach and within the *fornici* (vaults) set back from the beach. As Figure 4.3 shows, males predominate on the beach but are outnumbered in 7 out of 9 *fornici*. Conversely, infants and most of the juveniles were found in the *fornici* together with a large number of females. All this is suggestive of a pattern of social behavior, and the operation of a particular escape strategy, which remain to be uncovered and elucidated. In this case, furthermore, there is the additional informative but complicating factor, that the population of the town is known through epigraphy to have contained a substantial number of ex-slaves (and, a fortiori, slaves). This inevitably raises further issues, as to how representative the beach sample is of the population of the town as a whole, and as to how the peculiar demographic structure of the town influenced the distribution of individuals on the beach and in the *fornici*.[47]

In conclusion: we have spelled out the problems that beset paleodemography.[48] At the same time, we firmly believe that age-and-death-profile analysis is the initial, fundamental, and mandatory phase of all anthropological research, for the following reasons. First, it is possible, as demonstrated by the case of Velia, to come across a cemetery sample that meets most of the requirements of a reliable mortality profile. Second, the very same deviations from the expected norm, which appear to undermine the credibility of paleodemographic estimates (in particular, the under- or over-representation of specific categories), can serve as a highly informative source of information as to specific events and patterns of social behavior that have produced the biases in particular samples. Third, if we hold back from describing a cemetery sample in terms of (at least) the sex ratio and gross age distribution, then we cannot proceed to consider other data relevant to the population in question, especially such data as are closely related to the sex and age of individuals, such as skeletal markers of work activities, specific and aspecific health indicators, pathological changes, diet.

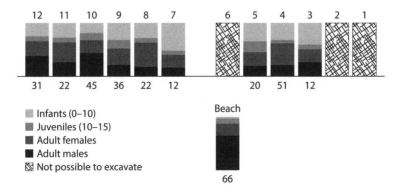

FIGURE 4.3. Spatial distribution by sex and age-at-death at Herculaneum.

How Were They Doing?

For the Roman period, scholars have been collecting paleopathological data, and integrating them with evidence from archaeological and written sources, in order to contribute to the understanding of, first, the origin, diffusion, and evolution of diseases; second, the epidemiological aspects related to the presence and spreading of the diseases, mainly in their relation with sex, age, social standing, diet, growth, mortality, and occupation, and so on; and third, medical knowledge and skills in the treatment of the diseases. With regard to the last of these items, the most striking evidence of surgical expertise is provided by one of the oldest cases of amputation with individual recovery and survival.[49]

Researchers have also published epidemiological studies involving the systematic surveys of skeletal health markers on large Roman age samples,[50] while others have sought to trace variations in health conditions through different historical ages,[51] or between Rome and the periphery of the Empire.[52] Some studies focus on selected issues, such as the presence of aspecific metabolic stress markers.[53] Others investigate the correlation of diet with health or the osteopenic/osteoporotic processes through life.[54]

A number of scholarly contributions offer description and diagnosis of rare pathological conditions. *Hyperostosis frontalis interna*, pituitary gigantism, and dwarfism are hormonal diseases located in various Roman age skeletal series.[55] Some data have been also gathered for specific infectious diseases. Rubini and colleagues isolated *Mycobacterium leprae* DNA from two skeletons showing several bony changes indicative of leprosy.[56] Evidence of tuberculosis was found in Herculaneum and in a first-century CE necropolis in Rome.[57] Among metabolic diseases, one case of scurvy and one of rickets have been described.[58] In the Collatina necropolis (Rome, first century BCE to third century CE), a case of ankylosing spondylitis and a case of gout were found.[59] These papers give valuable information in the fields of medical and disease history,[60] but contribute little to an understanding of the overall health status of the ancient Romans. It is now acknowledged that historians and anthropologists have a common interest in the study, over a broad canvas, of the health and physical well-being of the inhabitants of past communities, as an index of their adaptability to environmental resources and constraints. An additional benefit of such research is that it can provide useful insight into past medical knowledge, social organization, and resource distribution.

It hardly needs emphasizing that skeletal data are meaningful only if discussed within a framework that brings together clinical, historical, and paleoenvironmental evidence, and gives attention to their mutual influence. For instance, it is of interest to know how far ancient parasitoses may have influenced the distribution of settlements, productive activity, socioeconomic levels, and demographic trends in ancient populations.[61]

Paleopathological surveys, in past generations, have done useful ground-work in detecting, describing, and quantifying diseases, and setting them in their temporal context, through macroscopic, radiological, and microscopic analyses of bones and teeth.[62] Similarly, the integration of skeletal with histopathological and biomolecular evidence is crucial to further significant advance on the scientific front.[63] Improved techniques for aDNA detection and sequencing are now providing interesting results with regard to human hereditary diseases and ancient pathogens infections, *Treponema pallidum* (syphilis), *Brucella* sp. (brucellosis) *Mycobacterium tuberculosis* (tuberculosis) among the latter.[64] Interest in pathogens is also focused on their evolutionary trajectories in the interaction with the human host.[65] Besides DNA, other biomolecules (i.e., hemoglobins, human leukocyte antigens, hemozoin) are also providing further means of diagnosis. The search for ancient molecules extends beyond the teeth and the bone, even resorting to dental calculus and coprolites as sources of data.[66]

Furthermore, the field has been moving out of descriptive research on individual cases, into multidisciplinary approaches applied to whole populations, marshalling evidence from histology, anatomy, microbiology, physiology, biochemistry, medicine, archaeology, history, and ecology. At the same time, much effort has been dedicated to validating new research methods and addressing limitations in data assessment and interpretations.[67]

An example of how technical innovations, along with an integrated approach, are providing useful results, is the identification of malarial parasitosis in ancient bones, a subject of particular interest for the Roman period. Traditional means of diagnosis—that is, gross examination of bone manifestation of *cribra orbitalia* and porotic hyperostosis—have failed to provide concrete and secure evidence,[68] in the main because they can in fact be associated with several different conditions, related to other kind of parasites, malnutrition, or inherited hemolytic anemias.[69] Meanwhile, sequencing ancient DNA has provided some, but limited, results.[70] Finally, a quicker and less expensive diagnostic method has been tested on skeletal material in the effort to identify the biomolecules related to malarial infection. Hemozoin, a plasmodium catabolite, is indeed currently used in clinical research as a malaria biomarker.[71] Very recently this biomolecule was detected—with fluorescence microscopy—on the spongy bone tissue of a Late Roman Age sample of children,[72] confirming the diagnosis of malaria previously obtained through the isolation of *Plasmodium falciparum* DNA.[73]

However, a few methodological pitfalls and theoretical limitations in paleopathological analyses should be acknowledged and addressed.

Issue 1. Diagnosis and etiology. The first constraint in the information potentiality of paleopathological studies is that only a limited number of diseases actually manifest on the skeleton and on the teeth. Moreover, as stressed by several authors, in a skeletal sample, unaffected individuals may actually represent the most affected group within the population, given the possibility that they represent those who have died in a short time period, before the disease could leave permanent markers on the bones. For specific conditions, mostly infections and epidemics, molecular data can make a decisive contribution. The presence of *Yersinia pestis*,[74] and of other pathogens has been detected in skeletal series from particular depositional contexts (mass graves) and are characterized by biased demographic profiles, but without presenting any gross pathological changes.[75] Still, even with these techniques, a false negative can occur, that is, the disease is present, but the tests fail to detect it.

Secondly, the bone tissue responds in a very homogeneous way to a broad spectrum of afflictions, and this complicates, or even prevents, differential diagnosis. Lesions on the visceral surface of the ribs, for instance, are generally linked to tuberculosis,[76] but the association is not ubiquitous,[77] and other pathological conditions come into the reckoning,[78] like rickets. Furthermore, many diseases have a complex etiology. Since their mechanisms, as well as the role of every single causative factor, are not fully understood, nor weighted, we may fail to comprehend what we are really measuring in our surveys. In this respect, the case of osteoarthritis is emblematic. Osteoarthritis (or degenerative joint disease, DJD) is a multifactorial condition linked to hereditary factors, endocrine agents, age and sex, and nutrition, as well as functional stress, related to traumas, body weight, and articular loading and movements.[79] Nevertheless, in bioanthropological studies, the analysis of DJD has been exclusively used to assess type and extent of working activities, ignoring a more complex interpretation, or even a simple association with the ageing process.

Thirdly, skeletal analysis may reveal co-occurrences of conditions whose significance is not always clear. For instance, a correlation between enamel hypoplasia and reduced longevity has been registered.[80] However, several possible alternative explanations are available: repeated metabolic stress during growth may increase adult susceptibility to pathologies;[81] hypoplasia and mortality may both be the effect of poor dietary and sanitary conditions, or of congenital individual frailty.[82]

Association between disease, mortality, and diet gives a multifaceted picture which lends itself to complex interpretation.[83] For instance, in Velia, the presence of *cribra orbitalia* does not correlate with diets characterized by low protein content (Fig. 4.4). In the same sample, cases of diffuse idiopathic skeletal hyperostosis (DISH), a pathological condition characterized by a complex etiology, is more frequently associated with rich in proteins diets (Fig. 4.5).

Issue 2. Methodology. There is an absence of shared diagnostic standards. In order fruitfully to share and compare results, researchers should follow the same diagnostic criteria and adopt the same survey protocols. This procedure should be mandatory in the case of the most common and widespread afflictions (such as caries, degenerative joint diseases, traumas) or nonspecific

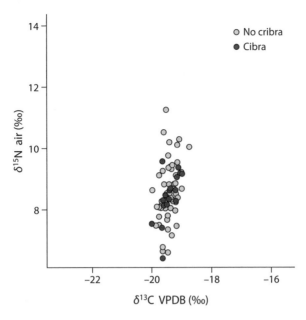

FIGURE 4.4. Nitrogen and carbon isotopic delta values and presence of cribra orbitalia in the adult sample from Velia (I–II cent. CE; N=74) do not show a significant correlation.

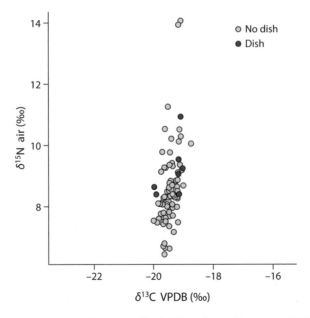

FIGURE 4.5. Nitrogen and carbon isotopic delta values and presence of diffuse idiopathic skeletal hyperostosis (DISH) in the adult sample from Velia (I–II cent. CE; N=85) show a positive correlation.

stress indicators (enamel hypoplasia, *cribra orbitalia*, porotic hyperostosis, periosteal bone reaction). Lack of standardization in scoring the trait, but also in the presentation of results, one for all, the extension of the age classes, is a frustrating situation, experienced by anyone who has attempted to conduct meaningful comparisons with data from the literature.[84]

Even within the same laboratories, adopting the same protocol, the interobserver error may reach statistically significant levels.[85] In this respect, Ortner and Buikstra and Ubelaker call for, and have been working toward, a more extensive standardization in data collection and diagnosis, associated with a more objective descriptive terminology.[86]

There is another aspect that should not be overlooked: since we are dealing with skeletal material of archaeological provenance, individuals are never complete. How is the partial recovery of bones to be reflected in the representation of pathologies? Boldsen and Milner discuss the high risk of obtain a "false negative" in fragmentary skeletons compared to those that are intact.[87]

In the specific case of the studies of osteoarthritis, the preservation of archaeological remains often precludes the recovery of complete joints. This consideration led us to investigate the effective informative value of the scoring procedure customarily adopted in the relevant literature. We used for this purpose the archaeological sample of Velia (Italy, first–second century CE) as our test case. In a sample of 103 adult individuals of both sexes, every single bone of the 6 major joint complexes (shoulder, elbow, wrist, hip, knee, and ankle) was scored separately for osteoarthritic changes. Our first objective was to evaluate the effective informative value of single bone elements within each articular complex. The results indicate that each joint shows to some extent intra-articular differences in DJD percentage, with a presence of what we have called a "bone leader" (a bone showing more DJD affection than the others within the same joint). The most frequently and most severely affected bones are: the scapula for the shoulder complex; the os coxale for the hip; the femur for the knee; and the tibia for the ankle.

A simulation was developed in order to quantify the influence of incomplete joints on the estimation of the mean scores of DJD simulating a set of hypothetical archaeological sub-samples (extracted from our sample with all complete joints) at different levels of preservation. This was achieved by simply randomly eliminating a given percentage of bones from the main sample. The preservation levels are expressed as the percentage of complete joints over the total number of joints scored, ranging between 90% and 10%. For each level, a computer routine eliminates in the data base the DJD score of at least one bone per joint, and the DJD prevalence was calculated again. For the prevalence calculation, a joint was considered affected if the score of any indicator (lipping, porosity, and eburnation) was greater than 1. The simulation was run 1000 times for each level of completeness resulting in a distribution of possible prevalence values.

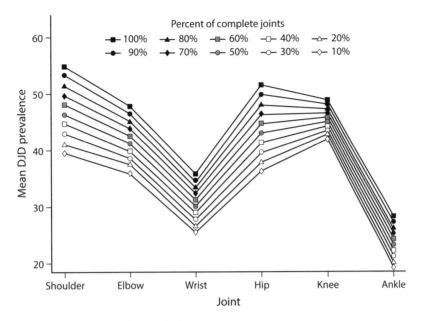

FIGURE 4.6. Decreasing levels of bone representation affect the osteoarthritic frequency, with different grade of bias across the joints. For instance, with 100% of complete joints, the hip prevalence is higher than in the knee, while, at 70% of completeness the result is reversed.

The results show that for all the 6 joints analyzed, the decrease of joint completeness (from 90% to 10%) is associated with a constant and progressive lowering of the mean DJD prevalence. At the same time, they show a progressive increase of variability of the prevalence values (standard deviation and range values), as obtained by the 1000 simulation runs. This tendency is clearly illustrated in Figure 4.6 where the results of the joints are presented, with the kernel density estimates of the simulated distributions plotted for lipping.

The results confirm that most of the joints have a bone that is more susceptible to the mechanical stress, so that joint incompleteness strongly affects the results by underestimating the pathological conditions and by neglecting the differences across the joint which consistently occur.

Issue 3. Theoretical questions. Besides the methodological issues concerning diagnosis and compatibility in data collections, scholars are currently addressing the problem of translating the results into consistent and meaningful interpretations of past life conditions, through the development and implementation of new theories and approaches. Once again, we have to cite the "Osteological Paradox" of Wood and colleagues and the three key issues hindering our understanding of the past health: demographic nonstationarity, selective mortality, and hidden heterogeneity in the morbidity and mortality risk.[88]

Demographic nonstationarity. Among the demographic variables, migration has the effect of altering the pathological profile of a population, since the health status and life history of newcomers are not those of the host community but are the product of a different biocultural environment. There is no easy solution to this problem. It has to be accepted that the disease regime is likely to be highly complex in the case of communities where there is a considerable degree of population mobility, notably Rome itself, but also cities on or near the coast in, for example, Central/South Italy.

Hidden heterogeneity in the risk of disease and mortality. There is a plethora of factors influencing the probability of falling ill and dying, ranging from genetic background to acquired susceptibility. These factors, and their interaction, are nonquantifiable, except in rare cases. One promising approach is to single out smaller homogeneous groups within a broader population, possibly identified by age, sex, working conditions, or social status, and pathological afflictions.[89] Individual life history, in so far as it can be reconstructed, should be helpful in tracing and identifying the sources of frailty. Another recommended line of action consists of analyzing the possible relationships between different health indicators, diet, and mortality.[90]

Selective mortality. In approaching once-living populations, we should always remember that we are dealing with the people who did not survive, and who are likely to embody higher frequencies of pathological afflictions and other skeletal signals related to harsher life-styles: growth disruption, poor nutritional intake, heavy working loads, amongst others. In some instances, skeletal series derived from catastrophic events can give us a glimpse of the living community, even if they certainly cannot represent an exact cross-section of the population. As stated above, the demographic parameters of the Herculaneum and Pompeii Vesuvius victims, rather than being a proxy of the city census,[91] reflect behavioral responses to the crisis.[92] Nevertheless, these samples provide us with an interesting opportunity to compare "living" Romans with dead ones. It is indeed rewarding to compare dietary habits as derived from cemetery samples (Isola Sacra and Velia) with those from the Herculaneum catastrophic assemblages.[93] Preliminary finds suggest that the variability observable in mortuary samples might be inflated when set against the more homogeneous nutritional pattern of the "living" community of Herculaneum.

Conversely, also using a comparative perspective, stature (calculated using the Pearson's regression formulae),[94] shows a remarkable overlap among the samples; the implication is that this derived trait is less affected by the "Osteological Paradox" and more dependent on the homogeneous environmental conditions prevailing in central Italy under the Roman Empire.

An integrated perspective As suggested above, a way of confronting the challenge of the "Osteological Paradox," specifically heterogeneity in the risk of disease and mortality, and selective mortality, is to isolate a subgroup within a given population and subject it to multivariate analysis. Infancy suggests it-self as a promising candidate for such an approach and within it, more partic-ularly, the weaning period in which the interplay of diet, disease, and mortality can be studied in depth by a combination of diverse methodologies.[95] The timing and procedures of weaning influence infant mortality, the demographic structure of the population, and the health status of infants and adults. Wean-ing is a critical phase of childhood, and the levels of morbidity and mortality associated with it are an indirect measure of the nursing practices and sanita-tion levels of the community. In the case of weaning, we can study the inter-play of diet, disease, and mortality by a combination of diverse methodologies. Specifically, one can combine the isotopic, mortality, and morbidity data for infants, with microscopic analysis of the dental enamel in infants, juveniles, and young adults (unfortunately older individuals can be included only with difficulty, because of dental wear).

Dental enamel histology (see above 2.2) produces more precise age deter-minations, and consequently results in a better definition of the weaning time, when studied through the change in trophic level, using the carbon and nitro-gen stable isotopes.[96] Moreover, laser ablation–inductively coupled plasma-mass spectrometry (LA-ICPMS) techniques can integrate the isotopic data with high resolution chronological variation of strontium concentration.[97] Fi-nally, the quantification and chronology of possibly weaning-related metabolic stress using the prevalence of enamel microdefects during growth,[98] provides a means of measuring the weaning time frame.

Dental histology, infancy, and multivariate analysis all feature in the lat-est investigations of the effects of exposure to and consumption of lead in the Roman world. It is well known that lead was widely used in Roman society.[99] In scholarly discussions the pendulum has swung, over time, from exagger-ating, to underrating, the extent and the consequences of lead poisoning. A recent study of sediments from the harbor at Portus and the Tiber river con-cluded that Roman drinking water was polluted with lead, but not to a danger-ous degree.[100] The authors make no reference to an earlier study which showed persuasively that Pompeians must have taken in high levels of lead through the water supply: regular repair and maintenance of the water distribution system reduced the protective crust of sinter in the pipes within the city.[101] The au-thors' use of skeletal evidence in relation to Herculaneum—an excellent idea in principle, because what we really need to know is how far lead built up in the bodies of consumers—is less successful, in that they take over an earlier assessment of a skeletal sample,[102] which ignored the postmortem alteration of bones. The way forward may lie with the examination of tooth enamel, which is less vulnerable to environmental contamination than are buried bones. In any case, it is clear that lead cannot yet be written off as a major health risk for Roman populations, while enamel lead concentration and isotopic ratio can contribute to the topic of mobility during the Roman Age.[103]

What Were They Doing?

Economic activities within ancient communities and workload distribution across different social groups are important indicators of past biocultural adaptation to the environment. Complementing historical accounts and archaeological evidence, skeletons can be informative as to occupational tasks. The underlying assumption is that working activities are reflected in both skeletal morphology and pathology. Repeated gestures may be responsible for skeletal overloading and biomechanical bone response.[104] A number of jobs or occupations may lead to specific pathologies, or may increase the risk of skeletal traumatism.[105] Craft production involving the dentition as "third hand" leave permanent signs on teeth.[106]

Many odontoskeletal indicators of occupational activities are presented and discussed in the literature. In some cases there is corroboration from contemporary observations from epidemiology, sports medicine, or ethnography. In more detail, one can list among the markers of odontoskeletal activities the following: degenerative joint disease; enthesopathies; cross-sectional bone geometry, specific anatomical variants (such as Allen's fossa on the femur or kneeling facets), extramasticatory unintentional modifications of teeth (that is dental grooves, notches, chipping and specific patterns of attrition).

This line of research, after an enthusiastic beginning, has to some extent lost its momentum. Meanwhile, the informative value of enthesopathic and osteoarthritic changes has come under critical scrutiny as they are skeletal markers, whose etiology is multifactorial and complex. Many researchers now reject the claim that a specific pattern of bone changes corresponds to a precise working activity.[107] Conversely, cross-sectional bone geometry has proved to be a more reliable indicator, as is indicated by some clinical studies and bioarchaeological applications.[108] Variation in the cross-sectional properties of long bones is related to mechanical stress induced by habitual body postures and movements. This approach has produced good results, especially in detecting asymmetric mechanical loads. But even with this approach the risk of misinterpretation is high if some important aspect of the individual condition and life history is neglected. For instance, Figure 4.7 shows an adult male femur, anatomically normal, from Velia, whose cortical thickness distribution is anomalous and resembles that of a great ape.[109] The discrepancy between the external and internal (and virtually never accessible) morphology signals is striking but explicable in terms of the presence of a completely healed tibia fracture that drove the anomalous remodeling of the femur shaft bone.

In spite of the great interest aroused by the analysis of occupational stress markers on past populations, very little is known, in this respect, about the ancient Roman world.[110] The most fully investigated topic is water-related activities, as detected at the two coastal towns of Portus and Velia, in the presence

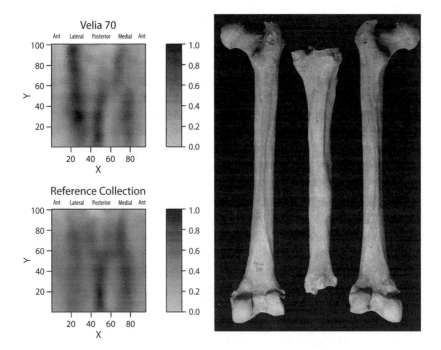

FIGURE 4.7. Cortical thickness of the femoral diaphysys of the
individual Velia 70 (above left; right femur), compared with data from
a reference collection (bottom left). The anomalous cortical pattern is
explained by the presence of a healed fracture on the right tibia.

of external auditory exostosis, an ear pathology caused by frequent and prolonged contact of the auditory canal with cold water.[111] These were port cities,
and it was only to be expected that a certain proportion of their populations
would have been engaged in aquatic activities, namely, the procurement and
processing of marine foods, and, especially at Portus which was the entrepôt of
Rome, the maritime trade, which employed sailors, divers, porters, and ship-
and wharf-workers. The incidence of the pathology is high: at Portus, 21%
among adult males (aged over 15) and at Velia 35.3% suffered from EAE, and
these figures certainly underestimate the percentage of the populations regularly involved in aquatic activities. Significantly, the condition does not occur
among females, which points clearly to an occupational divide between the
sexes. Companion studies of diet through carbon and nitrogen stable isotope
analysis highlight the contribution made by marine foods to the diet of both
populations. In addition, within a subgroup of Velians, a conspicuously higher
consumption of food of a higher trophic level, mostly marine in origin, and a
higher incidence of both EAE and skull traumas, provide strong supporting
evidence of their sphere of employment.

What Did They Eat?

Of the many sources pertaining to diet in the Roman world, direct information from skeletal tissues in the form of stable isotopes is often thought to eclipse all. Theoretically, measurements of isotopes of the atoms of carbon, nitrogen, oxygen, and hydrogen preserved in the organic and mineral portion of bone and teeth directly reflect what an individual consumed over an extended period prior to death. The power of an isotope approach is therefore its ability to provide direct, deep, and broad dietary reconstructions, independent of other sources of evidence that often appear anecdotal in comparison. As such, stable isotope analysis has been widely applied to prehistoric human remains, transforming our understanding of dietary change through time.[112] With such high credentials, it is apt to ask what exactly stable isotope analysis brings to the table in later periods—particularly one already richly furnished with historical accounts of food production and consumption, and extensive assemblages of food remains.

While there has been no shortage of stable isotope studies of Roman populations in Italy, Britain, and elsewhere,[113] the impact of these studies is arguably lessened compared to prehistoric examples. Expectations too have been higher. After all, dietary reconstruction implies data that may be fruitfully compared with other premodern, developing, and modern societies and economies in order to properly contextualize the Roman diet. For example, we may want to know the proportion of proteins, lipids (fats), and carbohydrates in an individual's diet or the contribution by weight of different foodstuffs. A second expectation is to understand how diets vary both within populations: for example, according to age, sex, and social standing, and between populations with different geographic, demographic, and social dimensions. Here, we will briefly review progress to each of the goals and attempt to define the limits of this technology in an attempt to curb expectations. Numerous reviews and texts describe both the method and the rationale for inferring diet from stable isotope data; the reader is directed to these.[114]

HOW ACCURATE ARE DIETARY RECONSTRUCTIONS USING STABLE ISOTOPES?

Most commonly, stable isotope analysis involves measuring the carbon ($\delta^{13}C$) and nitrogen ($\delta^{15}N$) stable isotopes in collagen, the major protein in bone and tooth dentine. Carbon isotopes have also been widely measured in the mineral part of bone (bioapatite). With these two (or three) measurements inferring precise dietary composition is as challenging as it may seem. Firstly, the sources of the different elements that are measured isotopically need to be considered. For carbon and nitrogen in collagen, these are largely derived from dietary proteins; however, an additional contribution of carbon from nonprotein sources (carbohydrate and perhaps lipids) has also been demonstrated through animal feeding experiments.[115] Conversely, the carbon present in bone apatite is assumed to be derived equally from all dietary carbon although, in this case, alteration of the biomineral fraction of bone during burial will affect the $\delta^{13}C$ values. Therefore, one needs to be cautious of bioapatite isotope data, without extensive tests to check the integrity of the biomineral, which are seldom carried out.[116]

Secondly, the carbon and nitrogen isotope values for the sources of protein, lipids, and carbohydrate in reference foodstuffs need to be determined. Collagen from animal bone found in association with the humans is the most obvious source for this information, provided this is available, which is not always the case in funerary contexts. Also, as the bones from animals were not actually consumed, corrections based on assumptions to the corresponding consumed tissues need to be made. Notably missing, however, are isotope values for carbohydrate and proteins in plants, values for lipids in meat and fish, and values for other foods of potential dietary significance, olive oil, *garum* etc. The third source of uncertainty is how the carbon and nitrogen isotope values change (fractionate) as they pass through the food chain, that is, from the foodstuff to its consumer. Standard values of enrichment of the heavier isotope are often assumed but rarely scrutinized and may even vary between the types of food consumed.[117]

The list of uncertainties and assumptions involved in accurately reconstructing diets is often seen as insurmountable and thus it is not often attempted. Researchers often prefer to discuss comparative isotope difference between individuals or describe diets in the broadest terms, for example as "predominantly" terrestrial or marine, with little attention to differentiating the protein or carbohydrate parts of diets. When investigating the diet at large coastal necropolises, such as Velia or Isola Sacra and at Herculaneum, this approach has been particularly problematic since the data show a large variation in $\delta^{15}N$ encompassing expected ranges for consumption of both terrestrial foods at one extreme, and marine fish at the other, but the values $\delta^{13}C$ are entirely consistent with terrestrial foods. These seemingly mixed signals have been interpreted as reflecting dietary complexity, where substantial amounts of cereal carbohydrate are combined with various amounts of protein from marine fish and terrestrial animal products.[118] This interpretation is partially informed by our knowledge of the extent of grain supply and consumption in the Roman world, so it is not without some circularity. Indeed, similar isotope values from humans at a riverine hunter-gatherer site have been interpreted as reflecting a diet rich in freshwater fish,[119] and in Rome itself, this interpretation has been proposed for humans from the catacombs of St. Callixtus.[120]

Radiocarbon dating of human bone has been used to investigate the fish contribution to diet, since carbon from a marine source contains a known amount of "old" carbon derived from the marine reservoir. The discrepancy in radiocarbon dates between bones that are assumed to be of the same age can therefore reveal the extent of fish consumption. This approach has been applied to humans from both Herculaneum and the catacomb of Sts. Peter and Marcellinus in Rome leading to an estimation of up to 30% marine carbon in diet, which is equivalent to about 30% marine foods by dry weight.[121] At Herculaneum, it is estimated that the fish contribution to dietary protein, reflected in $\delta^{15}N$ values, is almost double given its relatively high protein content. Consequently, here, it is estimated that the remainder of the bulk diet (50–70% by weight) consists almost entirely of cereal grain, consistent with estimates based on historical records.[122] The degree of freshwater fish consumption is more difficult to assess using this approach. Consumers of freshwater fish will also incorporate "old" carbon in their bone collagen, but the reservoir age may vary greatly depending on the environment. Without knowledge of this "age," estimates are fruitless.

For most cemeteries, the luxury of dating individuals of known age is not afforded. In these cases, another approach is to use a Bayesian statistical approach that is able to take into account the uncertainties in the assumptions mentioned above.[123] Such models provide a range of estimations of the proportional contribution by weight of different foodstuffs sources to an individual's diet. The efficacy of this approach was demonstrated at Herculaneum, where the model outputs supported the interpretation from the radiocarbon dating regarding the marine contribution.[124] Nevertheless, as with all other approaches, without knowledge of the reference food ranges, and without isotopic discrimination of these, we will not be able reconstruct diet with any useful precision.

Looking to the horizon, analysis of different isotope systems, such as sulfur ($\delta^{34}S$), hydrogen (δD), or even oxygen ($\delta^{18}O$), in collagen hold some promise as additional variables for dietary discrimination. Perhaps more exciting, though, is the measurement of $\delta^{15}N$ and $\delta^{13}C$ in the individual amino acids that make up collagen. As individual amino acids are obtained from different dietary components (either through biosynthesis or routing) in fairly predictable ways, we should be able to reconstruct diet at much greater precision than with bulk analysis. This approach has already shown to be useful for quantifying the terrestrial and marine contributions to diet and, although more methodologically challenging, is surely set to transform the field of paleodietary study.[125]

WHAT DIFFERENCES EXIST IN DIETS
IN THE ROMAN WORLD?

Even if we are unable to quantify the Roman diet with absolute certainty, absolute differences in isotope values between individuals still offer exceptional insights. Large Roman necropolises provide excellent contexts for such studies of intrapopulation dietary variability. Differences in carbon and nitrogen isotope values between fish and terrestrial foods mean that isotope differences are more evident at coastal sites compared to inland settings. At both Velia and Isola Sacra there are significant differences in $\delta^{15}N$ values between males and females,[126] reflecting increased access to higher trophic-level marine foods by males. These differences may reflect differential access to fish, meat and cereals at the household level or are perhaps related to occupation outside the home. In contrast, there is very little evidence that individuals afforded different burials had different diets, at least, at the Italian Imperial Roman sites.

Larger isotopic differences are noticeable between archaeological sites, but here caution needs to be exercised so that like may be compared with like. This is because the marine and terrestrial fauna available at different locations, and also perhaps the grain supplied, might also vary isotopically therefore influencing the values measured in humans. Such changes in the faunal baseline means that ideally the "isoscape" needs to be carefully defined by measuring animal bones and plant materials so that isotope values in humans can be meaningfully compared through space and time. Clearly, this is no easy task at complex urban centers such as Rome, where food may be supplied from a range of sources, not all of which are available for analysis. At Isola Sacra and Velia both the humans and the terrestrial fauna show significant differences between sites. However, when the faunal differences are taken into account, it becomes clear the individuals buried at Isola Sacra still generally show higher $\delta^{15}N$ values reflecting higher consumption of marine fish than at Velia, which may be related to differences in the social standing and occupation of the populations that these cemeteries represent.

As mentioned in Section 3, the problem of demographic bias and the osteological paradox is still crucially important when interpreting stable isotope data in attritional death assemblages skewed to the frail, the old, and the young. As collagen is formed over a period of time prior to death, the data obtained from stable isotope analysis offers an integrated dietary signal over this period. Therefore, additional complexity relates to the time taken for collagen to "turn-over" in bone. The most detailed study of this phenomenon was carried out on femurs from individuals who acquired ^{14}C into their bone during the nuclear bomb testing of the 1960s and 1970s.[127] These bomb tracer experiments show differences in collagen turn-over between males and females, related to physiology. It was also seen that collagen is much more rapidly replaced in earlier than in later life (see Table 4.1). For example, a 60-year-old man still retains 25% of his collagen in his femur from before he was were 19 years old. Interestingly therefore, the sex-related differences noted above may relate to access to foods earlier in adolescence rather than in later life. Furthermore, that age-related isotopic differences should be observable at all would be remarkable given such an attenuated signal.

Nevertheless, at Isola Sacra, it has been observed that the collagen from older individuals is significantly enriched in ^{15}N compared with younger,[128]

Table 4.1. Approximate collagen formation times in femoral bone based on the radiocarbon tracer experiments (Hedges et al. 2007)

	Age at death	% femoral collagen formed since age X		
Male		25%	50%	75%
	20	17	15	12
	40	30	21	15
	60	48	35	19
Female				
	20	14	11	8
	40	33	24	12
	60	52	42	26

implying, at face value, greater access to fish with age. Alternatively, considering the rates of collagen turnover, it is perhaps more likely that individuals who had more access to fish over their life simply lived longer. Resolving these two interpretations, although far from trivial, has consequences for our interpretation of how food is distributed within a population and/or the effects of diet on health. Understanding both demography and tissue turnover holds the key. Two future approaches may help in this respect. First, by microsampling incrementally growing tissues that do not turn-over, such as tooth dentine, we may be able to more meaningfully compare diet across cohorts of children and adolescents.[129] Hair keratin would be another useful tissue were it readily available. A second approach is to focus attention on catastrophic death assemblages, such as Herculaneum, Oplontis, and Pompeii, which record snapshots of living Roman populations avoiding many of the demographic issues that plague the field. One thing is clear, as the application of stable isotopes broadens, equal research effort needs to be directed to understanding some of these fundamental issues.

Where Were They From?

Migration, the movement of individuals, families, and populations, was a feature of the Roman world. In addition to being one of the main demographic variables, migration has an obvious relevance to significant areas of enquiry such as slavery, the economy, and identity—not to mention the dissemination of cultural and religious practices and ideas.[130] Historical, epigraphic, and archaeological sources are useful for the identification of both voluntary and involuntary migrants, although each kind of evidence has its limitations.[131] Isotopic analysis, in addition to its role in the study of diets in the past, is an important method of investigating mobility of individuals and groups, providing key evidence for the presence of migrants at a site and information about their possible places of origin.

Isotopic analysis of oxygen ($^{18}O/^{16}O$) and strontium ($^{87}Sr/^{86}Sr$) in human bones and teeth is employed to investigate mobility and geographic origins in skeletal samples, because the ratios of these isotopes are primarily determined by characteristics of local water (for oxygen) and underlying geology (for strontium). Humans acquire these isotopic signatures during life by consuming local food and water. Thus, people born and raised in a given area will have a chemical "signature" in their tissues that is consistent with baseline $\delta^{18}O$ and $^{87}Sr/^{86}Sr$ values for that region. The "delta" (δ) symbol is used to represent the ratio of a heavier (e.g., ^{18}O) to lighter (e.g., ^{16}O) isotope, measured in relation to an international standard. Strontium isotope values are usually represented as a ratio only, not with the delta symbol, because one of the isotopes (^{87}Sr) is radiogenic, that is, it is the decay product of rubidium (^{87}Rb). Carbon ($\delta^{13}C$) and nitrogen ($\delta^{15}N$) isotopes can be used together with oxygen and strontium isotopes to investigate mobility, based on the premise that recent immigrants to an area will display dietary signals that are different from the local isotopic signature.[132] One limitation is that different diets may have similar isotopic values, so $\delta^{13}C$ and $\delta^{15}N$ values should not be used on their own to infer mobility.

Bones and teeth provide different information about mobility over the life course. Teeth retain isotopic information about diet from infancy through late childhood, while bones preserve the average isotopic signal from the last few decades of an adult's life. Tooth enamel forms early in life, and once enamel mineralization is complete the isotopic signal remains unchanged. Permanent teeth start to form around birth, and the crown of the last tooth to erupt (the third molar) is complete by 12 to 16 years of age. Tooth enamel is also one of the hardest substances in the human body and preserves well in the archaeological record. Bone, in comparison, goes through rapid growth and development during childhood and adolescence, and once growth is complete it continues to remodel at a relatively constant rate, so it is a long-term signal. The rate of bone remodeling is not well understood, although it is estimated to be anywhere from 3–8% per year, and this rate decreases with advanced age.[133] Other factors that can affect the *in vivo* isotopic signals of bone are chemical changes that occur in the burial environment (i.e., diagenesis). Strontium is particularly problematic since strontium ions readily substitute with calcium and other trace elements in archaeological soils. Thus, isotopic studies of strontium mainly use teeth, as this exchange occurs less readily, given that enamel is much harder and more chemically stable. There is also the possibility of postmortem alteration in the case of oxygen isotopes, but there are ways of detecting when the sample has been compromised by diagenetic changes.[134]

OXYGEN AND STRONTIUM ISOTOPES

Oxygen in bones and teeth is largely determined by $\delta^{18}O$ values in drinking water, which approximates $\delta^{18}O$ in local precipitation, although a small portion of the signal can come from diet. Oxygen isotope composition of rainwater fluctuates in relation to local air temperature, humidity, distance from coastlines, latitude, and elevation.[135] Oxygen consumed as drinking water is incorporated into the body's tissues, although there is a slight offset between the values in drinking water and human tissues due to metabolic processes. Formulae exist to convert $\delta^{18}O$ values in bones and teeth to drinking water $\delta^{18}O$ values.[136] These adjusted $\delta^{18}O$ values can then be compared to published regional and global maps of annual average $\delta^{18}O$ in modern precipitation, which show variability due to differences in climate and geography. Modern precipitation maps can be compared to $\delta^{18}O$ data from Roman period skeletal samples, because temperature changes of global precipitation have been minimal over the past few millennia.

The strontium isotope signals in local food and water consumed by humans vary in relation to the $^{87}Sr/^{86}Sr$ values of the underlying bedrock.[137] Generally very old geological formations (>100 mya) have higher $^{87}Sr/^{86}Sr$ ratios, while younger formations (<1–10 mya) have lower ratios.[138] In contrast to oxygen, strontium undergoes little change as it passes through the food chain, and readily substitutes for calcium in the mineral portion of bones and teeth. If the underlying geology of the region is known, then regional maps of $^{87}Sr/^{86}Sr$ variation can be used to establish a local strontium range for comparison with human values. However, not all regions have detailed geological maps with known $^{87}Sr/^{86}Sr$ signals, and although strontium values are correlated with characteristics of the underlying bedrock, soil values can vary slightly due to factors such as soil transportation, which in turn affect the strontium signal of the foods consumed.[139]

IDENTIFYING MIGRANTS IN ROMAN ITALY

Oxygen isotope analysis of teeth from the inhabitants of Portus Romae reveals that approximately one-third of the individuals in the sample were not from Rome and its environs, including both males and females. Further, a comparison of early- and late-forming teeth from the same individuals suggested that some of these people migrated as children.[140] A subsequent critique by Bruun, which raised concerns about the use of isotopes for studying human mobility in the Roman world, and underlined the limitations of the evidence, produced responses from Killgrove and Prowse.[141] Bruun correctly noted that oxygen isotopes do not explain the varied motivations behind human mobility, as between voluntary migration of individuals and families, or involuntary mobility linked to slavery.[142] It is clearly essential to integrate isotopic data with contextual information from historical, literary, and archaeological evidence if we are to try to understand why men, women, and/or children migrated.

Strontium and oxygen isotopic analysis of skeletal samples from two suburban Roman sites, Casal Bertone and Castellaccio Europarco, found proportionately fewer people born outside of Rome and its suburbs than at Portus, suggesting a lower rate of migration at the former sites; in addition, women and children were underrepresented in the suburban samples.[143] We clearly need many more, and much larger, samples from Roman sites before we can attempt to estimate migration rates to Rome and its suburbs on the basis of skeletal evidence. As with other isotopic studies, the precise geographic origins of migrants to these places could not be identified because different regions have similar baseline values. It is likely enough that there was no standard pattern of migration nor source of migrants, even within the comparatively limited region of Rome and its environs; each cemetery sample will provide individualized histories of mobility, which can be explored in relation to age, gender, status, and other aspects of social identity.

Isotopic analyses of origins in Roman Italy to-date have for the most part studied samples from Rome itself or places in the region, where high levels of mobility are to be expected. Chenery and coworkers noted the need for studies on populations where lower levels of mobility can be anticipated.[144] The rural Roman estate of Vagnari in southern Italy is one such place. Archaeological evidence from this site indicates the presence of a substantial settlement, probably an Imperial estate, along with a necropolis mainly in use from the first to third centuries CE.[145] A preliminary study integrating oxygen isotope and ancient DNA (aDNA) analysis found evidence of a small number of nonlocals.[146] More recently, an integrated mortuary and $\delta^{18}O$ analysis found that only 8% of the sample were nonlocal, but of those 5 individuals 2 were children under the age of 10 years and 1 was a 6-month-old infant, suggestive of mobility throughout the life course.[147]

LIMITATIONS AND PROSPECTS OF ISOTOPE ANALYSIS

Isotopes can be used to identify nonlocals in a skeletal sample, but they are limited in their ability to account for short-term residency or repeated movement, as people may live in multiple locations prior to death. Due to the slow turnover of bone, it will take years for the isotopic signature of the new location to be registered in bone. In oxygen isotope studies, there are a number of factors that may contribute to the consumption of water in areas that do not reflect the regional $^{18}O/^{16}O$ of rainwater. For example, Rome's extensive aqueduct system brought in drinking water from topographically higher regions inland, an area where oxygen isotope composition of rain is significantly lighter.[148] Food preparation activities, such as brewing, boiling, and stewing, may cause a slight increase in $\delta^{18}O$ values of liquids consumed.[149] Breastfeeding has also been shown to shift the isotopic signal towards heavier values in teeth that develop during this period of time. Breast milk is enriched in the heavier isotope (^{18}O) in relation to the water consumed by the mother, so teeth forming during infancy will have slightly higher $\delta^{18}O$ values.[150] The isotope data can be adjusted to account for this effect, or teeth that formed after weaning can be used. A final limitation of oxygen isotope analysis is that it does not pinpoint a specific geographic location associated with a particular $\delta^{18}O$ value. One possible way to deal with this issue is to use modern human values as "proxies" for local oxygen values,[151] although obtaining sufficient modern samples for this type of analysis can be challenging. In any case, the integration of oxygen and strontium isotope data will improve our chances of identifying possible geographic origins.

With respect to strontium, Slovak and Paytan stressed that there must be sufficient geologic variability between regions under study in order to detect mobility.[152] Further, if marine foods are a significant component of a population's diet, the strontium values in teeth may reflect seawater $^{87}Sr/^{86}Sr$ values and confound the interpretations of mobility between coastal and inland regions.[153] Killgrove and Montgomery have published estimates for strontium isotope ratios in and around the city of Rome based on geological maps of Italy, and Tafuri and coworkers for Neolithic sites in the Gargano region of southern Italy.[154] There are, however, no large-scale regional studies of bioavailable strontium in the Mediterranean region, with the exception of work done by Nafplioti for Greece.[155] It is possible to have multiple regions with similar strontium isotope signals—hence the advantage of using oxygen and strontium together to better define possible geographic origins of migrants. Finally, regular food items may come from different regions (with different $^{87}Sr/^{86}Sr$ signals), so the strontium isotope signals in human tissues represent an average of these sources. To control for these issues with strontium, isotope studies use 'bioavailable' $^{87}Sr/^{86}Sr$ values from local, nonmigratory animals or modern plant remains from the same region represent local strontium values.[156] It is assumed that these values reflect the average $^{87}Sr/^{86}Sr$ values of food available to humans in the region.[157]

The number of isotopic studies of mobility in the Roman world is rapidly increasing, particularly on samples from Roman Britain.[158] More work is needed on comparable human isotope data from all geographic areas in the Mediterranean region and other areas of the Roman Empire, as well as more information on local drinking water values (for oxygen) and baseline strontium signatures from local plants and animals in order to refine the ranges for specific geographic areas. The integration of dietary (C, N) and mobility (O, Sr) isotopic data may help to further clarify the geographic origins of individuals.[159] Lead (Pb) is also being used to identify migrants in Roman Britain in relation to the increased postconquest use of lead in both industrial and household contexts.[160] Only a small number of studies have integrated isotopic and aDNA evidence to explore geographic origins and biological relationships,[161] but this is another possible future avenue of research.

These are destructive analyses, even though the amounts needed for isotopic analysis are relatively small, so research questions should be carefully designed before testing takes place. If our ultimate aim is to understand how and why people (adult males, women, children, or families) migrated, then isotopic data needs to be interpreted within the social and historical framework obtained from literary, epigraphic, and archaeological evidence. Each kind of evidence has its own strengths and weaknesses, but when these apparently disparate elements and diverse methodologies are brought together, we gain a much more nuanced picture of human mobility in the past.

Conclusion

Bones and teeth are a uniquely valuable source of information on population structure, diet, disease, health, social behavior, migration. The data on offer are quantitatively impressive and cross-class, and there is more to come.

The data do not speak for themselves: they have to be interpreted. This statement, of course, applies also to other kinds of evidence from antiquity. There are however specific problems with the interpretation of skeletal data. The most vocal and trenchant criticisms have come from within the anthropological community itself. One famous article declared paleodemography defunct; another hammered some more nails into the coffin of paleodemography, and, into the bargain, proclaimed that a prominent, current approach to paleopathology is a false trail.

Biological anthropology is an imperfect science. It is, however, a science, and also one that is on the advance, constantly learning from its mistakes, and assessing and improving the analyses and methodologies that it employs. Is paleodemography dead? The report of its death is greatly exaggerated.[162] True, one must lower one's sights and have modest objectives. One cannot expect to arrive at a plausible reconstruction of the demographic shape of a community from a particular skeletal sample. This is not actually impossible, as the case of Velia in South Italy demonstrates; rather it is not normally possible. But if that is the case, why not? What are the factors that produce skewed profiles? A negative result of an experiment is not worthless, as any scientist knows; it may lead to productive enquiries and, eventually, plausible explanations. Meanwhile there are other dimensions of paleodemography that remain thus far relatively unexplored, such as the comparative demography of the dead of the cemeteries and the "living" victims of catastrophes, not to mention the bone evidence for a key demographic variable, migration.

The critics themselves hint at the way forward: isolate a particular age group, study it in depth, bring into play the latest techniques from a whole range of scientific disciplines, and make comparisons, ideally between communities from similar environments. Thus, paleopathology is moving away from a previous, narrow focus on individual cases of eccentric diseases, and on some specific but problematic markers of stress in the form of bone lesions, and advancing towards a broad reconstruction of the morbidity and mortality regime of whole communities.

Our discussion has highlighted an age cohort in the case of which promising research is emerging, and converging: infants and juveniles. Age at death is more easily diagnosed here than is the case with adults. Dental histology has made a breakthrough in enabling us to count the time-markers on the crowns as they develop from birth to death. In addition, state-of-the-art techniques make possible a close analysis of the impact of disease-induced stress, and the infusion of potentially toxic minerals, centering on, but extending beyond, the critical period of weaning.

Stable isotope analysis is a relatively new methodology, and it is all too easy to focus on its present limitations and deficiencies while failing to appreciate its singular contribution—in providing an integrated dietary signal over an extended period of time, or identifying the nonlocal element in a population—and underestimating its potential to reconstruct diets and uncover patterns of migration. We can expect steady progress to be made, for example, in quantifying the terrestrial and marine contributions to the diet, and in locating the geographic origins of migrants.

A multidisciplinary approach holds the key to further advances in our knowledge. On the scientific side, histology, microbiology, biochemical medicine, among other disciplines, are combining forces, for example, in the identification of infectious diseases, such as malarial parasitosis, in skeletal samples. Archaeology is the handmaid of anthropology. What of history? History at present stands on the sidelines, although historians and anthropologists have overlapping interests, most obviously in the areas of diet, health, the structure and movement of populations, and the urban environment. For historians (or archaeologists) to ignore, or write off, the contribution of biological anthropology and adjacent sciences to our knowledge of ancient society is not a sensible option.

Notes

1. André 1981.
2. Scheidel 2012.
3. Larsen 2006.
4. Bondioli and Sperduti 2011.
5. Bocquet-Appel and Masset 1982; Wood et al. 1992.
6. See, e.g., Ferembach et al. 1979; Buikstra and Mielke 1985; Buikstra and Ubelaker 1994, Krogman and İşcan 1986; İşcan and Kennedy 1989; White and Falkens 2000; Garvin 2012.
7. Walrath et al. 2004.

8. Weiss 1972.

9. Walker 2008; Guyomarc'h and Bruzek 2011; Garvin 2012.

10. Krogman and İşçan 1986; Rogers and Saunders 1994.

11. Spradley and Jantz 2011.

12. Lee et al. 2015.

13. See review in Sutter 2003.

14. Mays and Faerman 2001; Hassan et al. 2014. See Chapter 6.

15. Spirduso 1995; Schmitt et al. 2002; Cox 2000; Corsini et al. 2005.

16. For a review, see Schaefer et al. 2009.

17. Radiological: Macchiarelli et al. 1990. Histological: see synthesis in Kemkes-Grottenhaler 2002.

18. Klepinger 2006.

19. Soomer et al. 2003; Baccino et al. 1999; Garvin et al. 2012.

20. See Nawrocki 2010 for a full discussion.

21. Spirduso 1995.

22. Merritt 2015.

23. Katz and Suchey 1989; Schmitt 2004; Falys et al. 2006; Hens et al. 2008; Merritt 2014; Mays 2014.

24. Bocquet-Appel and Masset 1982, 329.

25. See, for instance, Gowland and Chamberlain 2002; Hoppa and Vaupel 2002; Soomer et al. 2003; Gowland 2007; Jackes 2011.

26. Kemkes-Grottenhaler 2002.

27. Acsàdi and Nemeskéri 1970 (combined); Lovejoy et al. 1985 (summary age); Boldsen et al. 2002 (transition).

28. Falys and Lewis 2011; Garvin and Passalacqua 2012.

29. Kimmerle et al. 2008.

30. Maples 1989, 323.

31. Falys and Prangle 2015; Tang et al. 2014.

32. Stoyanova et al., 2015.

33. See most recently AlQahtani 2009.

34. Antoine et al. 2009; Birch and Dean 2009; Mahoney 2011.

35. Sabel et al. 2008.

36. Macchiarelli et al. 2006; Dean 2009; Dean and Cole 2013.

37. Smith et al. 2006.

38. Bocquet-Appel and Masset 1982, 1985; Van Gerven and Armelagos 1983; Konigsberg and Frankenberg 1994, 2002; Bocquet-Appel and Masset 1996; Paine and Harpending 1998; Hoppa and Vaupel 2002; Jackes 2011.

39. Green et al. 1974; Johansson and Horowitz 1986.

40. McNeill 1979; Scheidel 2001a; Sperduti 1995; Paine and Storey 2006.

41. Farwell and Molleson 1993. Journals: Molleson 1989; Lewis 2010; Redfern and Dewitte 2011a.

42. Sperduti 1995.

43. Coale and Demeny 1983; on the viability of model life tables, see Scheidel 2001b; Hin 2013.

44. Mays and Faerman 2001; Hassan et al. 2014; Sperduti et al. in press.

45. Lewis 2011, 4.

46. Luongo et al. 2003; Frey et al. 2010.

47. Fattore et al. 2012.

48. Cf. Jackes 2011.

49. Weaver et al. 2000.

50. See for instance Bisel 1991; Capasso 2001; Paine et al. 2009; Gowland and Garnsey 2010; Petrone et al. 2011.

51. Cucina et al. 1998; Belcastro et al. 2007; Manzi et al. 1999.

52. Gowland and Redfern 2010.

53. Manzi et al. 1989; Ricci et al. 1997; Salvadei et al. 2001; Facchini et al. 2004; Fitz-Gerald et al. 2006.

54. Bisel 1988; Bonfiglioli et al. 2003; Prowse et al. 2008 (diet); Cho and Stout 2011; Beauchesne and Agarwal 2014 (processes).

55. Sperduti and Manzi 1990; Ottini et al. 2001; Lazer 2011; Minozzi et al. 2012a.

56. Rubini et al. 2014.

57. Capasso and Di Tota 1999 (Herculaneum); Canci et al. 2005 (Rome).

58. Lunardini et al. 2005; Minozzi et al. 2012b.

59. Lunardini et al. 2008 (spondylitis); Minozzi et al. 2013 (gout).

60. For a detailed description of orthopedic pathologies in Roman Imperial age, see Piccioli et al. 2015.

61. Sallares 2002.

62. Ortner and Putschar 1981.

63. Ragsdale and Lehmer 2012; Zuckerman et al. 2016.

64. For an updated review, see Spigelman 2012 and below, Chapter 6.

65. Bos et al. 2011; Schuenemann et al. 2011; Bouwman et al. 2012.

66. Jaeger and Iñiguez 2014; Warinner et al. 2014; Weyrich et al. 2015.

67. Ortner 1991; Wood et al. 1992; Zuckerman et al. 2012, 2015; DeWitte and Stojanowski 2015.

68. Walker et al. 2009; Setzer 2014.

69. Stuart-Macadam 1989, 1992.

70. Hawass et al. 2010; Sallares and Gomzi 2001; Marciniak et al. 2016. See below, Chapter 6.

71. Mirdha et al. 1999.

72. Lugnano in Teverina, Inwood, J., personal communication.

73. Abbott 2001; Sallares et al. 2004.

74. Drancourt et al. 1998; Zink et al. 2006; Bos et al. 2012; Wagner et al. 2014.

75. DeWitte and Wood 2008.

76. Santos and Roberts 2001.

77. Mays et al. 2001, 2002.

78. Molto 1990.

79. Hough 2001; Weiss and Jurmain 2007; Musumeci et al. 2015.

80. Palubeckaitė et al. 2002.

81. Duray 1996; Armelagos et al. 2009.

82. Rothman and Greenlander 1998.

83. Bondioli et al. 2016.

84. Bridges 1993.

85. Macchiarelli et al. 1994; Davis et al. 2013; Jacobi and Danforth 2002; Waldron and Rogers 1991.

86. Ortner 1991, 2012; Buikstra and Ubelaker 1994.

87. Boldsen and Milner 2012.

88. Wood, et al. 1992.

89. Storey 1997; DeWitte and Bekvalac 2010; Boldsen et al. 2015.

90. DeWitte 2014.

91. Capasso 2001; Lazer 2011.

92. Fattore et al. 2012.

93. Crowe et al. 2010; Craig et al. 2013 (Isola Sacra and Velia); Martyn et al. 2015 (Herculaneum).

94. See Giannecchini and Moggi-Cecchi 2008.

95. Sandberg et al. 2014.

96. For a review see Fulminante 2015.

97. Humphrey et al. 2007; Bondioli et al. 2011.

98. FitzGerald et al. 2006.

99. E.g., Retief and Cilliers 2005.

100. Delile et al. 2014.

101. Keenan-Jones et al. 2012.

102. Bisel and Bisel 2002.

103. Montgomery et al. 2010.

104. Ruff et al. 1984; Ruff 1987; Chenorkian et al.1990.

105. Crowe et al. 2010 (pathologies); Brasili et al. 2004; Kanz and Grosschmidt 2006; Van der Merwe et al. 2010 (traumatism).

106. Frayer and Russell 1987; Erdal 2008; Waters-Rist et al. 2010; Lorkiewicz 2010; Sperduti et al. 2011.

107. Jurmain et al. 2012; Villotte and Knüsel 2013.

108. Shaw and Stock 2009a, b (clinical studies); Maggiano et al. 2008; Sládek et al. 2007 (archaeology).

109. See Bondioli et al. 2010 for the methodology.

110. Farwell and Molleson 1993; Petrone 1993; Sperduti 1997; Catalano et al. 2010.

111. Manzi et al. 1991; Crowe et al. 2010; Sperduti et al. 2012. These articles also provide clinical and epidemiological references for the etiology of the pathology.

112. Richards et al. 2003; Richards et al. 2005; Richards et al. 2000.

113. Prowse et al. 2004; Craig et al. 2009; Craig et al. 2013; Müldner et al. 2011; Richards and Hedges 1998; Keenleyside et al. 2009.

114. E.g., Katzenberg 2000; Sealy 2001; van Klinken et al. 2000.

115. Hedges 2006.

116. Salesse et al. 2014.

117. O'Connell et al. 2012; Hedges and Reynard 2007.

118. Craig et al. 2009; Prowse et al. 2004; Craig et al. 2013.

119. Cook et al. 2001.

120. Rutgers et al. 2009.

121. Craig et al. 2013; Salesse et al. 2014.

122. Foxhall and Forbes 1982.

123. Fernandes et al. 2014.

124. Fernandes 2015.

125. Colonese et al. 2014; Naito et al. 2010.

126. Craig et al. 2009; Prowse et al. 2004.

127. Hedges et al. 2007.

128. Prowse et al. 2005.

129. Beaumont et al. 2013.

130. E.g., Noy 2000; Scheidel 2004, 2005, 2007; de Ligt and Northwood 2008; Eckardt 2010; Holleran 2011; de Ligt 2012; Hin 2013; de Ligt and Tacoma 2016.

131. Eckardt et al. 2014.

132. E.g., Dupras and Schwarcz 2001; Chenery et al. 2010, 2011; Müldner et al. 2011.

133. Parfitt 2004; Hedges et al. 2007; Stepańczak et al. 2014.

134. Hedges 2002.

135. Yurtsever and Gat 1981; Gat 1996.

136. Daux et al. 2008; Pollard et al. 2011a; Chenery et al. 2012.

137. Slovak and Paytan 2012.

138. Bentley 2006.

139. Chenery et al. 2010, 2011.

140. Prowse et al. 2007.

141. Bruun 2010; Killgrove 2010a; Prowse 2016.

142. Bruun 2010.

143. Killgrove and Montgomery 2016; cf. Killgrove 2010b; 2013.

144. Chenery et al. 2010.

145. Small 2011.

146. Prowse et al. 2010.

147. Prowse 2016.

148. Lightfoot et al. 2014.

149. Brettell et al. 2012; Lightfoot et al. 2014.

150. Roberts et al. 1998; Wright and Schwarcz 1998.

151. E.g., Prowse et al. 2007.

152. Slovak and Paytan 2012.

153. Prowse et al. 2007; Slovak and Paytan 2012.

154. Killgrove and Montgomery 2016; Tafuri and coworkers 2015.

155. Nafplioti 2011.

156. Bataille and Bowen 2012; Chenery et al. 2011.

157. For a detailed discussion of the use of strontium isotopes in archaeology, see Bentley 2006 and Slovak and Paytan 2012.

158. E.g., Evans et al. 2006; Chenery et al. 2010, 2011; Montgomery et al. 2010; Müldner et al. 2011; Eckardt et al. 2014.

159. E.g., Pollard et al. 2011b.

160. E.g., Montgomery et al. 2010; Shaw et al. 2016.

161. E.g., Prowse et al. 2010; Sofeso et al. 2012.

162. Mark Twain, adapted; see Van Gerven and Armelagos 1983.

References

Abbott, A. 2001. "Earliest malaria DNA found in Roman baby graveyard." *Nature* 412 (6850): 847. DOI: 10.1038/35091226.

Acsàdi, G., and J. Nemeskéri. 1970. *History of Human Life Span and Mortality*. Budapest: Akadémiai Kiadò.

AlQahtani, S. J. 2009. *Atlas of Human Tooth Development and Eruption*. London: Queen Mary and Westfield College.

André, J. 1981. *L'alimentation et la cuisine à Rome*. Paris: Belles Lettres.

Antoine, D., S. Hillson, and M. C. Dean. 2009. "The developmental clock of dental enamel: a test for the periodicity of prism cross-striations in modern humans and an evaluation of the most likely sources of error in histological studies of this kind." *Journal of Anatomy* 214: 45–55. DOI: 10.1111/j.1469-7580.2008.01010.x.

Armelagos, G., et al. 2009. "Enamel hypoplasia and early mortality: bioarcheological support for the Barker hypothesis." *Evolutionary Anthropology* 18: 261–271. DOI: 10.1002/evan.20239.

Baccino, E., et al. 1999. "Evaluation of seven methods of estimating age at death from mature human skeletal remains." *Journal of Forensic Sciences* 44: 931–936.

Bataille, C., and G. Bowen. 2012. "Mapping $^{87}Sr/^{86}Sr$ variations in bedrock and water for large scale provenance studies." *Chemical Geology* 304–305: 39–52.

Beauchesne, P., and S. C. Agarwal. 2014. "Age-related cortical bone maintenance and Loss in Imperial Roman population." *International Journal Osteoarchaeology* 24: 15–30. DOI: 10.1002/oa.1303.

Beaumont, J., et al. 2013. "Childhood diet: a closer examination of the evidence from dental tissues using stable isotope analysis of incremental human dentine." *Archaeometry* 55 (2): 277–295. DOI: 10.1111/j.1475-4754.2012.00682.x.

Belcastro, G., et al. 2007. "Continuity or discontinuity of the life-style in central Italy during the Roman Imperial Age–Early Middle Ages transition: diet, health, and behavior." *American Journal of Physical Anthropology* 132: 381–394. DOI: 10.1002/ajpa.20530.

Bentley, R. A. 2006. "Strontium isotopes from the earth to the archaeological skeleton: a review." *Journal of Archaeological Method and Theory* 13: 135–187. DOI: 10.1007/s10816-006-9009-x.

Birch, W., and M. C. Dean. 2009. "Rates of enamel formation in human deciduous teeth." In *Comparative Dental Morphology: Frontiers of Oral Biology* 13, eds. T. Koppe, G. Meyer, and K. W. Alt. Basel: Karger, 116–120. DOI: 10.1159/000242402.

Bisel, S. 1988. "Nutrition in 1st Century Herculaneum." *Anthropologie Brno* 26: 61–66.

Bisel, S. 1991. "The human skeletons of Herculaneum." *International Journal of Anthropology* 6: 1–20. DOI: 10.1007/BF02447284.

Bisel, S., and J. Bisel. 2002. "Health and Nutrition at Herculaneum: an examination of human skeletal remains." In *The Natural History of Pompeii*, eds. W. F. Jashemski and F. G. Meyer. Cambridge: Cambridge University Press, 451–475.

Bocquet-Appel, J.-P., and C. Masset. 1982. "Farewell to paleodemography." *Journal of Human Evolution* 11 (4): 321–333.

Bocquet-Appel, J.-P., and C. Masset. 1985 "Paleodemography: resurrection or ghost?" *Journal of Human Evolution* 14 (2): 107–111.

Bocquet-Appel, J.-P., and C. Masset. 1996. "Palaeodemography: expectancy and false hope." *American Journal of Physical Anthropology* 89: 235–256. DOI: 10.1002/(SICI)1096-8644(199604)99:4<571::AID-AJPA4>3.0.CO;2-X.

Boldsen, J. L., and G. R. Milner. 2012. "An epidemiological approach to paleopathology." In *A Companion to Paleopathology*, ed. A. L. Grauer Malden, MA: Wiley-Blackwell, 114–132. DOI: DOI: 10.1002/9781444345940.ch7.

Boldsen, J. L., G. R. Milner, and S. Weise. 2015. "Cranial vault trauma and selective mortality in medieval to early modern Denmark." *Proceedings of the National Academy of Sciences of the United States of America* 112 (6): 1721–1726. DOI: 10.1073/pnas.1412511112.

Boldsen, J. L., et al. 2002. "Transitional analysis: a new method for estimating age from skeletons." In *Paleodemography: Age Distributions from Skeletal Samples*, eds. R. D. Hoppa, and J. W. Vaupel. New York: Cambridge University Press, 73–106.

Bondioli, L., W. Müller, and P. F. Rossi. 2001. "Evaluation of secretion vs. maturation in human dental enamel from LA-ICPMS compositional profiles." American Association of Physical Anthropologists Supplement 144 (S52): 93.

Bondioli, L., and A. Sperduti. 2011. "Comunità dei morti ed individui scheletrici: dallo studio di popolazioni alla ricostruzione della storia biologica individuale" In *Dalla nascita alla morte: antropologia e archeologia a confronto. Atti dell'Incontro Internazionale di studi in onore di Claude Lévi-Strauss*, ed. V. Nizzo. Roma: Museo Nazionale Preistorico Etnografico Luigi Pigorini, 431–460.

Bondioli L., et al. 2010. "Technical note: morphometric maps of long bone shafts and dental roots for imaging topographic thickness variation." *American Journal of Physical Anthropology* 142: 328–334. DOI: 10.1002/ajpa.21271.

Bondioli, L., et al. 2016. "Diet and health in Central-Southern Italy during the Roman Imperial time." *ACTA IMEKO* 5 (2): 19–25.

Bonfiglioli, B., P. Brasili, and M. G. Belcastro. 2003. "Dental-alveolar lesions and nutritional habits: paleopathology of Roman skeletons of a Roman Imperial Age population (1st–4th c.AD): Quadrelli (Molise, Italy)." *Homo* 54: 35–56.

Bos, K. I., et al. 2011. "A draft genome of *Yersinia pestis* from victims of the Black Death." *Nature* 478 (7370): 505–510. DOI: 10.1038/nature10549.

Bos, K. I., et al. 2012. "*Yersinia pestis*: new evidence for an old infection." *PLoS ONE* 7 (11): e49803. DOI: 10.1371/journal.pone.0049803.

Bouwman, A. S., et al. 2012. "Genotype of a historic strain of *Mycobacterium tuberculosis*." *Proceedings of the National Academy of Sciences of the United States of America* 109 (45): 18511–18516. DOI: 10.1073/pnas.1209444109.

Brasili, P., E. Bianchi, and A. R. Ventrella. 2004. "Traumatic events and life-style in ancient Italian populations." *Collegium Antropologicum* 28: 179–191.

Brettell, R., J. Montgomery, and J. Evans. 2012. "Brewing and stewing: the effect of culturally mediated behaviour on the oxygen isotope composition of ingested fluids and the implications for human provenance studies." *Journal of Analytic Atomic Spectrometry* 27: 778–785. DOI: 10.1039/C2JA10335D.

Bridges, P. S. 1993. "The effect of variation in methodology on the outcome of osteoarthritic studies." *International Journal of Osteoarchaeology* 3: 289–295. DOI: 10.1002/oa.1390030407.

Bruun, C. 2010. "Water, oxygen isotopes, and immigration to Ostia-Portus." *Journal of Roman Archaeology* 23: 109–132. DOI: 0.1017/S1047759400002324.

Buikstra, J. E., and J. H. Mielke. 1985. "Demography, diet and health." In *The Analysis of Prehistoric Diets*, eds. R. I. Gilbert and J. H. Mielke. Orlando: Academic Press, 359–422.

Buikstra, J. E., and D. H. Ubelaker. 1994. *Standards for Data Collection from Human Skeletal Remains*. Fayetteville: Arkansas Archeological Survey Research.

Canci, A., et al. 2005. "A case of healing spinal infection from Classical Rome." *International Journal of Osteoarchaeology* 15: 77–83. DOI: 10.1002/oa.734.

Capasso, L. 2001. *I fuggiaschi di Ercolano*. Rome: L'Erma Di Bretschneider.

Capasso, L., and G. Di Tota. 1999. "Tuberculosis in Herculaneum (79 AD)." In *Tuberculosis: Past and Present*, eds. G. Palfi et al. Golden Book Publishers and Tuberculosis Foundation: Budapest, 463–467.

Catalano, P., et al. 2010. "Health status and life style in Castel Malnome (Rome, I–II cent. A. D.)." *Medicina nei Secoli* 22: 11–28.

Chenery, C., H. Eckardt, and G. Müldner. 2011. "Cosmopolitan Catterick? Isotopic evidence for population mobility on Rome's northern frontier." *Journal of Archaeological Science* 38 1525–1536. DOI: 10.1016/j.jas.2011.02.018.

Chenery, C., et al. 2010. "Strontium and stable isotope evidence for diet and mobility in Roman Glouchester, UK." *Journal of Archaeological Science* 37: 150–163. DOI: 10.1016/j.jas.2009.09.025.

Chenery, C. A., et al. 2012. "The oxygen isotope relationship between the phosphate and structural carbonate fractions of human bioapatite." *Rapid Communication in Mass Spectrometry* 26: 309–319. DOI: 10.1002/rcm.5331.

Chenorkian, R., et al. 1990. "Pour une archéologie du geste." *Travaux du Lapmo*: 147–151.

Cho, H., and S. D. Stout. 2011. "Age-associated bone loss and intraskeletal variability in the Imperial Romans." *Journal of Anthropological Science* 89: 109–125. DOI: 10.4436/jass.89007.

Coale, A. J., and P. Demeny. 1983. *Regional Model Life Tables and Stable Populations*. Princeton: Princeton University Press.

Colonese, A. C., et al. 2014. "Long-Term Resilience of Late Holocene Coastal Subsistence System in Southeastern South America." *PLoS ONE* 9 (4): e93854. DOI: 10.1371/journal.pone.0093854.

Cook, G. T., et al. 2011. "A freshwater diet-derived C-14 reservoir effect at the stone age sites in the iron gates gorge." *Radiocarbon* 43 (2): 453–460. DOI: 10.2458/azu_js_rc.43.3985.

Corsini, M., A. Schmitt, and J. Bruzek. 2005. "Ageing process variability on the human skeleton: artificial network as an appropriate tool for age at death assessment." *Forensic Science International* 148: 163–167. DOI: 10.1016/j.forsciint.2004.05.008.

Cox, M. 2000. "Ageing adults from the skeleton." In *Human Osteology in Archaeology and Forensic Science*, eds. M. Cox and S. Mays. London: Greenwich Medical Media, 61–81.

Craig, O. E., et al. 2009. "Stable isotopic evidence for diet at the imperial Roman coastal site of Velia (1st and 2nd Centuries AD) in Southern Italy." *American Journal of Physical Anthropology* 139: 572–583. DOI: 10.1002/ajpa.21021.

Craig, O. E., et al. 2013. "Evaluating marine diets through radiocarbon dating and stable isotope analysis of victims of the AD 79 eruption of Vesuvius." *American Journal of Physical Anthropology* 152: 345–352. DOI: 10.1002/ajpa.22352.

Crowe, F., et al. 2010. "Water-Related occupations and diet in two Roman coastal communities (Italy, first to third century AD): Correlation between stable carbon and nitrogen isotope values and auricular exostosis prevalence." *American Journal of Physical Anthropology* 142: 355–366. DOI: 10.1002/ajpa.21229.

Cucina, A., D. Mancinelli, and A. Coppa. 1998. "Demography, nutrition and stress in the Italian Peninsula from the Copper Age to the Roman Imperial Age." *Rivista di Antropologia* 76: 135–138.

Daux, V., et al. 2008. "Oxygen isotope fractionation between human phosphate and water revisited." *Journal of Human Evolution* 55 (6): 1138–1147. DOI: 10.1016/j.jhevol.2008.06.006.

Davis, C. B., et al. 2013. "Patterns of interobserver error in the scoring of entheseal changes." *International Journal of Osteoarchaeology* 23: 147–151. DOI: 10.1002/oa.2277.

Dean, M. C. 2009. "Extension rates and growth in tooth height of modern human and fossil hominin canines and molars." In *Frontiers of Oral Biology: Interdisciplinary Dental Morphology*, 13, eds. T. Koppe, G. Meyer, and K.W. Alt. Karger: Basel, 68–73. DOI: 10.1159/000242394.

Dean, M.C., and T. J. Cole. 2013. "Human life history evolution explains dissociation between the timing of tooth eruption and peak rates of root growth." *PLoS ONE* 8 (1): e54534. DOI: 10.1371/journal.pone.0054534.

Delile, H., et al. 2014. "Lead in ancient Rome's city waters." *Proceedings of National Academy of Sciences of the United States of America* 111 (18): 6594–6599. DOI: 10.1073/pnas.1400097111.

DeWitte, S. N. 2014. "Differential survival among individuals with active and healed periosteal new bone formation." *International Journal of Paleopathology* 7: 38–44. DOI: 10.1016/j.ijpp.2014.06.001.

DeWitte, S. N., and J. Bekvalac. 2010, "Oral health and frailty in the medieval English cemetery of St Mary Graces." *American Journal of Physical Anthropology* 142: 341–354. DOI: 10.1002/ajpa.21228.

DeWitte, S. N., and C. M. Stojanowski. 2015. "The Osteological Paradox 20 years later: past perspectives, future directions." *Journal of Archaeological Research.* DOI 10.1007/s10814-015-9084-1.

DeWitte, S. N., and J. W. Wood. 2008. "Selectivity of Black Death mortality with respect to pre-existing health." *Proceedings of the National Academy of Sciences of the United States of America* 105 (5): 1436–1441. DOI: 10.1073/pnas.0705460105.

Drancourt, M., et al. 1998. "Detection of 400-year-old *Yersinia pestis* DNA in human dental pulp: an approach to the diagnosis of ancient septicemia." *Proceedings of the National*

Academy of Sciences of the United States of America 95 (21): 12637–12640. DOI: 10.1073 /pnas.95.21.12637.

Dupras, T. L., and H.P. Schwarcz. 2001. "Strangers in a strange land: Stable isotope evidence for human migration in the Dakhleh Oasis, Egypt." *Journal of Archaeological Science* 28: 1199–1208. DOI: 10.1006/jasc.2001.0640.

Duray, S. 1996. "Dental indicators of stress and reduced age at death in prehistoric native Americans." *American Journal of Physical Anthropology* 99, 275–286. DOI: 10.1002 /(SICI)1096-8644(199602)99:2<275::AID-AJPA5>3.0.CO;2-Y.

Eckardt, H., ed. 2010. *Roman Diasporas: Archaeological Approaches to Mobility and Diversity in the Roman Empire.* Portsmouth, RI: Journal of Roman Archaeology. Supplementary volume 78.

Eckardt, H., G. Mülnder, and M. Lewis. 2014. "People on the move in Roman Britain." *World Archaeology* 46 (4): 534–550. DOI: 10.1080/00438243.2014.931821.

Erdal, Y.S. 2008. "Occlusal grooves in anterior dentition among Kovuklukaya inhabitants (Sinop, Northern Anatolia, 10th Century AD)." *International Journal of Osteoarchaeology* 18: 152–166. DOI: 10.1002/oa.925.

Evans, J., N. Stoodley, and C. Chenery. 2006. "A strontium and oxygen isotope assessment of a possible fourth century immigrant population in a Hampshire cemetery, England." *Journal of Archaeological Science* 33: 265–272. DOI: 10.1016/j.jas.2005.07.011.

Facchini, F., E. Rastelli, and P. Brasili. 2004. "*Cribra orbitalia* and *cribra cranii* in Roman skeletal remains from the Ravenna area and Rimini (I–IV Century AD)." *International Journal of Osteoarchaeology* 14: 126–136. DOI: 10.1002/oa.717.

Falys, C. G., H. Schutkowski, and D. Weston. 2006. "Auricular surface ageing: worse than expected? A test of the revised method on a documented historic skeletal assemblage." *American Journal of Physical Anthropology* 130: 508–513. DOI: 10.1002 /ajpa.20382.

Falys, C. G., and M. E. Lewis. 2011. "Proposing a way forward: a review of standardisation in the use of age categories and ageing techniques in osteological analysis (2004–2009)." *International Journal of Osteoarchaeology* 21: 704–716. DOI: 10.1002/oa.1179.

Falys, C. G., and D. Prangle. 2015. "Estimating age of mature adults from the degeneration of the sternal end of the clavicle." *American Journal of Physical Anthropology* 156: 203–214. DOI: 10.1002/ajpa.22639.

Farwell, D. E., and T. I. Molleson. 1993. *Excavations at Poundbury 1966–80. Volume II: The Cemeteries.* Dorchester: Dorset Natural History and Archaeology Society Monograph 11.

Fattore, L., et al. 2012. "The Roman skeletal remains from Herculaneum: new evidence from the excavation of the fornici 7, 8, 9, 10, and 11." *American Association of Physical Anthropologists Suppl. 81st Annual Meeting.* Portland, Oregon: 11–14, April 2012. 142 DOI 10.1002/ajpa.22032.

Ferembach, D., I. E., Schwidetzky, and M. Stloukal. 1979. "Raccomandazioni per la determinazione dell'età e del sesso sullo scheletro." *Rivista di Antropologia* 60: 5–51.

Fernandes, R. 2015. "A simple(r) model to predict the source of dietary carbon in individual consumers." *Archaeometry*: 500–512. DOI: 10.1111/arcm.12193.

Fernandes, R., et al. 2014. "Food reconstruction using isotopic transferred signals (FRUITS): a bayesian model for diet reconstruction." *PLoS ONE* 9 (2): e87436. DOI: 10.1371/journal .pone.0087436.

FitzGerald, C., et al. 2006. "Health of infants in an Imperial Roman skeletal sample: perspective from dental microstructure." *American Journal of Physical Anthropology* 130: 179–189. DOI: 10.1002/ajpa.20275.

Foxhall, L., and H. A. Forbes. 1982. "Sitometria: The role of grain as a staple food in classical antiquity." *Chiron* 12: 41–90.

Frayer, D. W., and M. D. Russell. 1987. "Artificial grooves in the Krapina Neanderthal teeth." *American Journal of Physical Anthropology* 74: 393–405. DOI: 10.1002/ajpa.1330740311.

Frey, B. S., D. A. Savage, and B. Torgler. 2010. "Interaction of natural survival instincts and internalized social norms exploring the Titanic and Lusitania disasters." *Proceedings of the National Academy of Sciences of the United States of America* 16 (11): 4862–4865. DOI: 10.1073/pnas.0911303107.

Fulminante, F. 2015. "Infant feeding practices in Europe and the Mediterranean from prehistory to the Middle Ages: a comparison between the historical sources and bioarchaeology." *Childhood in the Past* 8: 24–47. DOI: 10.1179/1758571615Z.00000000026.

Garvin, H. M. 2012. "Adult sex determination: methods and application." In *A Companion to Forensic Anthropology*, ed. D. C. Dirkmaat. Malden, MA: Wiley-Blackwell, 239–247. DOI: 10.1002/9781118255377.ch12.

Garvin, H. M., and N. V. Passalacqua. 2012. "Current practices by forensic anthropologists in adult skeletal age estimation." *Journal of Forensic Sciences* 57: 427–433. DOI: 10.1111/j.1556-4029.2011.01979.x.

Garvin, H. M., et al. 2012. "Developments in forensic anthropology: age at-death estimation." In *A Companion to Forensic Anthropology*, ed. D.C. Dirkmaat. Malden, MA: Wiley-Blackwell, 202–223. DOI: 10.1002/9781118255377.ch10.

Gat, J. R. 1996. "Oxygen and hydrogen isotopes in the hydrologic cycle." *Annual Review of Earth and Planetary Sciences* 24: 225–262. DOI: 10.1146/annurev.earth.24.1.225.

Giannecchini, M., and J. Moggi-Cecchi. 2008. "Stature in archeological samples from central Italy: methodological issues and diachronic changes." *American Journal of Physical Anthropology* 135: 284–292. DOI: 10.1002/ajpa.20742.

Gowland, R. L. 2007. "Age, ageism and osteological bias: the evidence from late Roman Britain." *Journal of Roman Archaeology Supplementary* 65: 153–169.

Gowland, R. L., and A. T. Chamberlain. 2002. "A Bayesian approach to ageing perinatal skeletal remains: implications for the evidence of infanticide in Roman Britain." *Journal of Archaeological Science* 29: 677–685. DOI: 10.1006/jasc.2001.0776.

Gowland, R. L., and P. Garnsey. 2010. "Skeletal evidence for health, nutritional status and malaria in Rome and the Empire." In *Roman Diasporas: Archaeological Approaches to Mobility and Diversity in the Roman Empire*, ed. H. Eckhardt. Portsmouth, RI, Journal of Roman Archaeology, 131–156.

Gowland, R. L., and R. C. Redfern. 2010. "Childhood health in the Roman world: perspectives from the centre and margin of the Empire." *Childhood in the Past* 3: 15–42. DOI: 10.1179/cip.2010.3.1.15.

Green, S., S. Green, and G. Armelagos. 1974. "Settlement and mortality of the Christian site (1050 A.D.—1300 A.D.) of Meinarti (Sudan)." *Journal of Human Evolution* 3 (4): 297–316. DOI: 10.1016/0047-2484(74)90024-4.

Guyomarc'h, P., and J. Bruzek. 2011. "Accuracy and reliability in sex determination from skulls: a comparison of Fordisc3.0 and the discriminant function analysis." *Forensic Science International* 208: 180–e1. DOI: 10.1016/j.forsciint.2011.03.011.

Hassan, N.A-M., et al. 2014. "Ancient DNA study of the remains of putative infanticide victims from the Yewden Roman villa site at Hambleden, England." *Journal Archaeological Science* 43: 192–197. DOI: 10.1016/j.jas.2013.12.017.

Hawass, Z., et al. 2010. "Ancestry and pathology in King Tutankhamun's family." *The Journal of the American Medical Association* 303: 638–647. DOI: 10.1001/jama.2010.121.

Hedges, R.E.M. 2002. "Bone diagenesis: an overview of process." *Archaeometry* 44 (3): 319–328. DOI: 10.1111/1475-4754.00064.

Hedges, R.E.M. 2006. "Where does our protein carbon come from?" *British Journal of Nutrition* 95: 1031–1032. DOI: 10.1079/BJN20061782.

Hedges, R.E.M., and L. M. Reynard. 2007. "Nitrogen isotopes and the trophic level of humans in archaeology." *Journal of Archaeological Science* 34: 1240–1251. DOI: 10.1016/j.jas.2006.10.015.

Hedges, R.E.M., et al. 2007. "Collagen turnover in the adult femoral mid-shaft: modeled from anthropogenic radiocarbon tracer measurements." *American Journal of Physical Anthropology* 133: 808–816. DOI: 10.1002/ajpa.20598.

Hens, S. M., E. Rastelli, and G. Belcastro. 2008. "Age estimation from the human *os coxa*: a test on a documented Italian collection." *Journal of Forensic Sciences* 53: 1040–1043. DOI: 10.1111/j.1556-4029.2008.00818.x.

Hin, S. 2013. *The Demography of Roman Italy: Population Dynamics in an Ancient Conquest Society, 201 BCE–14 CE.* Cambridge: Cambridge University Press.

Holleran, C. 2011. "Migration and the urban economy of Rome." In *Demography and the Graeco-Roman World: New Insights and Approaches,* eds. C. Holleran and A. Pudsey. Cambridge: Cambridge University Press, 155–180.

Hoppa, R. D., and J. W. Vaupel. 2002. *Paleodemography: Age Distributions from Skeletal Samples.* Cambridge: Cambridge University Press.

Hough, A. J. 2001. "Pathology of osteoarthritis." In *Arthritis and Allied Conditions,* ed. W. J. Koopman. Philadelphia: Lea and Febiger, 2167–2194.

Humphrey, L. T., et al. 2007. "Unlocking evidence of early diet from tooth enamel." *Proceedings of the National Academy of Sciences of the United States of America* 105 (19): 6834–6839. DOI: 10.1073/pnas.0711513105.

İşcan, M. Y., and K. A. Kennedy. 1989. *Reconstruction of Life from the Skeleton.* New York: Alan R. Liss.

Jackes, M. 2011. "Representativeness and bias in archaeological skeletal samples." In *Social Bioarchaeology,* eds. S. C. Agarwal, and B. A. Glencross. Chichester: Blackwell, 107–146. DOI: 10.1002/9781444390537.ch5.

Jacobi, K. P., and M. E. Danforth. 2002. "Analysis of interobserver scoring patterns in porotic hyperostosis and *cribra orbitalia*." *International Journal of Osteoarchaeology* 12: 248–258. DOI: 10.1002/oa.619.

Jaeger, L. H., and A. M. Iñiguez. 2014. "Molecular paleoparasitological hybridization approach as effective tool for diagnosing human intestinal parasites from scarce archaeological remains." *PLoS ONE* 9 (8): e105910. DOI: 10.1371/journal.pone.0105910.

Johansson, S. R., and S. Horowitz. 1986. "Estimating mortality in skeletal populations: influence of the growth rate on the interpretation of levels of trends during the transition to agriculture." *American Journal of Physical Anthropology* 71: 233–250. DOI: 10.1002/ajpa.1330710211.

Jurmain, R., et al. 2012. "Bioarchaeology's Holy Grail: the reconstruction of activity." In *A Companion to Paleopathology,* ed. A. L. Grauer. New York: Wiley-Blackwell, 531–552. DOI: 10.1002/9781444345940.ch29.

Kanz, F., and K. Grosschmidt. 2006. "Head injuries of Roman gladiators." *Forensic Science International* 13: 207–216. DOI: 10.1016/j.forsciint.2005.10.010.

Katz, D., and J. M. Suchey. 1989. "Race differences in pubic synphyseal aging patterns in the male." *American Journal of Physical Anthropology* 80: 167–172. DOI: 10.1002/ajpa.1330800204.

Katzenberg, M. A. 2000. "Stable isotope analysis: a tool for studying past diet, demography and history." In *Biological Anthropology of the Human Skeleton,* eds. M. A. Katzenberg and S. R. Saunders. New York: Wiley-Liss, 305–328. DOI: 10.1002/9780470245842.ch13.

Keenan-Jones, D., J. Hellstrom, and R. Drysdale. 2012. "Lead contamination in the drinking water of Pompeii." In *Art, Industry and Infrastructure in Roman Pompeii,* eds. E. Poehler, M. Flohr, and K. Cole. Oxford: Oxford University Press, 130–181.

Keenleyside, A., et al. 2009. "Stable isotopic evidence for diet in a Roman and Late Roman population from Leptiminus, Tunisia." *Journal of Archaelogical Science* 36: 51–63. DOI: 10.1016/j.jas.2008.07.008.

Kemkes-Grottenthaler, A. 2002. "Aging through the ages: historical perspectives on age indicator methods." In *Paleodemography: Age Distributions from Skeletal Samples*, eds. R. D. Hoppa and J. W. Vaupel. Cambridge: Cambridge University Press, 48–72.

Killgrove, K. 2010a. "Response to C. Bruun's 'Water, oxygen isotopes, and immigration to Ostia-Portus.'" *Journal of Roman Archaeology* 23: 133–136. DOI: 10.1017/S1047759400002336.

Killgrove, K. 2010b. "Identifying immigrants to Imperial Rome using strontium isotope analysis." In *Roman Diasporas: Archaeological Approaches to Mobility and Diversity in the Roman Empire*, ed. H. Eckardt. Portsmouth, RI: Journal of Roman Archaeology, Supplementary volume 78, 157–174.

Killgrove, K., and J. Montgomery. 2016. "All roads lead to Rome: exploring human migration to the eternal city through biochemistry of skeletons from two Imperial-era cemeteries (1st–3rd c AD)." *PLoS ONE* 11 (2): e0147585. DOI: 10.1371/journal.pone.0147585.

Kimmerle, E. H., D. A. Prince, and G. E. Berg. 2008. "Inter-observer variation in methodologies involving the pubic symphysis, sternal ribs, and teeth." *Journal of Forensic Sciences* 53: 594–600. DOI: 10.1111/j.1556-4029.2008.00715.x.

Klepinger, L. L. 2006. *Fundamentals of Forensic Anthropology*. Wiley-Liss: New Jersey.

Konigsberg, L. W. and S. R. Frankenberg, 1994. "Paleodemography: 'Not quite dead.'" *Evolutionary Anthropology* 3: 92–105. DOI: 10.1002/evan.1360030306.

Konigsberg, L. W., and S. R. Frankenberg. 2002. "Deconstructing death in paleodemography." *American Journal of Physical Anthropology* 117: 297–309. DOI: 10.1002/ajpa.10039.

Krogman, W. M., and M. Y. İşcan. 1986. *The Human Skeleton in Forensic Medicine*. Springfield, IL: Charles C. Thomas.

Larsen, C. S. 2006. "The changing face of bioarchaeology: an interdisciplinary science." In *Bioarchaeology: The Contextual Analysis of Human Remains*, eds. J. E. Buikstra and L. A. Beck. Burlington, MA: Academic Press, 359–374.

Lazer E. 2011. *Resurrecting Pompeii*. Abingdon, Oxon, and New York: Routledge. DOI: 10.3764/ajaonline1161.

Lee, U. Y., I. B. Kim, and D. S. Kwak. 2015. "Sex determination using discriminant analysis of upper and lower extremity bones: new approach using the volume and surface area of digital model." *Forensic Science International* 253: 135.e1–4. DOI: 10.1016/j.forsciint.2015.05.017.

Lewis, M. E. 2010. "Life and death in a *civitas* capital: metabolic disease and trauma in the children from Late Roman Dorchester, Dorset." *American Journal of Physical Anthropology* 143: 405–416. DOI: 10.1002/ajpa.21239.

Lewis, M. E. 2011. "The osteology of infancy and childhood: misconceptions and potential." In *Rethinking The Little Ancestor: New Perspectives on the Archaeology of Infancy and Childhood*, eds. M. Lally and A. Moore. Oxford: Archaeopress, 1–13.

Lightfoot, E., M. Slaus, and T. C. O'Connell. 2014. "Water consumption in Iron Age, Roman and early medieval Croatia." *American Journal of Physical Anthropology* 154: 535–543. DOI: 10.1002/ajpa.22544.

Ligt, L. de. 2012. *Peasants, Citizens and Soldiers: Studies in the Demographic History of Roman Italy 225BC–AD 100*. Cambridge: Cambridge University Press.

Ligt, L. de, and S. Northwood, eds. 2008. *People, Land and Politics: Demographic Developments and the Transformation of Roman Italy, 300 BC–AD 14*. Leiden: Brill.

Ligt, L. de, and L. E. Tacoma, eds. 2016. *Migration and Mobility in the Early Roman Empire*. Leiden: Brill.

Lorkiewicz, W. 2011. "Nonalimentary tooth use in the Neolithic population of the Lengyel culture in central Poland (4600–4000 BC)." *American Journal of Physical Anthropology* 144: 538–551. DOI: 10.1002/ajpa.21435.

Lovejoy C. O., et al. 1985. "Multifactorial determination of skeletal age at death: a method and blind test of its accuracy." *American Journal of Physical Anthropology* 68: 1–14. DOI: 10.1002/ajpa.1330680102.

Lunardini, A., et al. 2005. "A severe case of rickets in the Roman Imperial Age (I–II century A.D.)." *Journal of Paleopathology* 17: 137–143.

Lunardini, A., et al. 2008. "A severe case of ankylosing spondylitis from the Roman Imperial Age (I–II century A.D.)." *Journal of Paleopathology* 20: 29–35.

Luongo, G., et al. 2003. "Impact of the AD 79 explosive eruption on Pompeii. II. Causes of death of the inhabitants inferred by stratigraphic analysis and areal distribution of the human casualties." *Journal of Volcanology and Geothermal Research* 126: 169–200. DOI: 10.1016/S0377-0273(03)00147-1.

Macchiarelli, R., A. Sperduti, and L. Bondioli. 1990. "L'indagine radiografica dello scheletro nella attribuzione dell'età alla morte. II. Analisi sperimentale dei corpi vertebrali." *Rivista di Antropologia* 68: 103–127.

Macchiarelli, R., et al. 1994. "Intra- and interobserver concordance in scoring Harris lines: a test on bone sections and radiographs." *American Journal of Physical Anthropology* 95: 77–83. DOI: 10.1002/ajpa.1330950107.

Macchiarelli, R., et al. 2006. "How Neanderthal molar teeth grew." *Nature* 444 (7120): 748–751. DOI: 10.1038/nature05314.

Maggiano, I. S., et al. 2008. "Cross-sectional analysis of long bones, occupational activities and long-distance trade of the Classic Maya from Xcambó: archaeological and osteological evidence." *American Journal of Physical Anthropology* 136: 470–477. DOI: 10.1002/ajpa.20830.

Mahoney, P. 2011. "Human deciduous mandibular molar incremental enamel development." *American Journal of Physical Anthropology* 144: 204–214. DOI: 10.1002/ajpa.21386.

Manzi, G., A. Sperduti, and P. Passarello. 1991. "Behavior-induced auditory exostoses in imperial Roman society: evidence from coeval urban and rural communities near Rome." *American Journal of Physical Anthropology* 85: 253–260. DOI: 10.1002/ajpa.1330850303.

Manzi, G., et al. 1989. "Linee di Harris e ipoplasia dello smalto nei resti scheletrici delle popolazioni umane di Isola Sacra e *Lucus Feroniae* (Roma, I–III sec. d.C.)." *Rivista di Antropologia* 67: 129–148.

Manzi, G., et al. 1999. "Discontinuity of life conditions at the transition from the Roman Imperial age and early Middle ages: example from central Italy evaluated by pathological dental alveolar lesions." *American Journal of Human Biology* 11 (3): 327–341. DOI: 10.1002/(SICI)1520-6300(1999)11:3<327::AID-AJHB5>3.0.CO;2-M.

Maples, W. R. 1989. "The practical application of age-estimation techniques." In *Age Markers in the Human Skeleton*, ed. M. Y. İşcan. Springfield, IL: Charles C. Thomas, 319–324.

Marciniak, S., et al. 2016. "*Plasmodium falciparum* malaria in 1st–2nd century CE southern Italy." *Current Biology* 26: 1220–1222.

Martyn, R.E.V., et al. 2015. "Capturing Roman dietary variability in the catastrophic death assemblage at Herculaneum, AD 79." *United Kingdom Archaeological Science, Environmental and Archaeological Science Conference.* Durham, UK, April 2015: 8–11.

Mays, S. 2014. "A test of a recently devised method of estimating skeletal age at death using features of the adult acetabulum." *Journal of Forensic Sciences* 59: 184–187. DOI: 10.1111/1556-4029.12293.

Mays, S., and M. Faerman. 2001. "Sex identification in some putative infanticide victims from Roman Britain using ancient DNA." *Journal of Archaeological Science* 28: 555–559. DOI: 10.1006/jasc.2001.0616.

Mays, S., E. Fysh, and G. M. Taylor. 2002. "Investigation of the link between visceral surface rib lesions and tuberculosis in a medieval skeletal series from England using ancient DNA." *American Journal of Physical Anthropology* 119: 27–36. DOI: 10.1002/ajpa.10099.

Mays, S., et al. 2001. "Paleopathological and biomolecular study of tuberculosis in a medieval skeletal collection from England." *American Journal of Physical Anthropology* 114: 298–311. DOI: 10.1002/ajpa.1042.

McNeill, W. H. 1979. "Historical patterns of migration." *Current Anthropology* 20: 95–102. DOI: 10.1086/202206.

Merritt, C. E. 2014. "A test of Hartnett's revisions to the pubic symphysis and fourth rib methods on a modern sample." *Journal of Forensic Sciences* 59: 703–711. DOI: 10.1111/1556-4029.12380.

Merritt, C. E. 2015. "The influence of body size on adult skeletal age estimation methods." *American Journal of Physical Anthropology* 156: 35–57. DOI: 10.1002/ajpa.22626.

Minozzi, S., et al. 2012a. "Pituitary disease from the past: a rare case of gigantism in skeletal remains from the Roman Imperial Age." *The Journal of Clinical Endocrinology and Metabolism* 97: 4302–4303. DOI: 10.1210/jc.2012-2726.

Minozzi, S., et al. 2012b. "Palaeopathology of human remains from the Roman Imperial Age." *Pathobiology* 79: 268–283. DOI: 10.1159/000338097.

Minozzi, S., et al. 2013. "A case of gout from Imperial Rome (1st–2nd Century AD)." *Journal of Clinical Research & Bioethics* 4: 162. DOI: 10.4172/2155-9627.1000162.

Mirdha, B. R., et al. 1999. "Bone marrow examination for identifying malaria in fever of unknown origin." *Journal Association of Physicians of India* 47: 177–179.

Molleson, T. I. 1989. "Social implications of mortality patterns of juveniles from Poundbury Camp, Romano-British Cemetery." *Anthropologischer Anzeiger* 47: 27–38.

Molto, J. E. 1990. "Differential diagnosis of rib lesions: a case study from Middle Woodland, Southern Ontario. Circa 230 AD." *American Journal of Physical Anthropology* 83: 439–447. DOI: 10.1002/ajpa.1330830405.

Montgomery, J., et al. 2010. "'Gleaming white and deadly': the use of lead to track human exposure and geographic origins in the Roman period in Britain." In *Roman Diasporas: Archaeological Approaches to Mobility and Diversity in the Roman Empire*, ed. H. Eckardt. Portsmouth, RI: Journal of Roman Archaeology, Supplementary volume 78, 199–226.

Müldner, G., C. Chenery, and H. Eckardt. 2011. "The 'Headless Romans:': multi-isotope investigations of an unusual burial ground from Roman Britain." *Journal of Archaeological Science* 38: 280–290. DOI: 0.1016/j.jas.2010.09.003.

Musumeci, I., et al. 2015. "Osteoarthritis in the XXIst Century: risk factors and behaviours that influence disease onset and progression." *International Journal of Molecular Sciences* 16: 6093–6112. DOI: 10.3390/ijms16036093.

Nafplioti, A. 2011. "Tracing population mobility in the Aegean using isotope geochemistry: a first map of local biologically available $^{87}Sr/^{86}Sr$ signatures." *Journal of Archaeological Science* 38: 1560–1570. DOI: 10.1016/j.jas.2011.02.021.

Naito, Y. I., et al. 2010. "Quantitative evaluation of marine protein contribution in ancient diets based on nitrogen isotope ratios of individual amino acids in bone collagen: an investigation at the Kitakogane Jomon site." *American Journal of Physical Anthropology* 143: 31–40. DOI: 10.1002/ajpa.21287.

Nawrocki, S. P. 2010. "The nature and sources of error in the estimation of age at death from the skeleton." In *Age Estimation of the Human Skeleton*, eds. K. Lathamand and M. Finnegan. Springfield, IL: Charles C. Thomas, 79–101.

Noy, D. 2000. *Foreigners at Rome: Citizens and Strangers.* London: Duckworth.

O'Connell, T. C., et al. 2012. "The diet-body offset in human nitrogen isotopic values: a controlled dietary study." *American Journal of Physical Anthropology* 149: 426–434. DOI: 10.1002/ajpa.22140.

Ortner, D. J. 1991. "Theoretical and methodological issues in paleopathology." In *Human Paleopathology: Current Syntheses and Future Options*, eds. D. Ortner and A. Aufderheide. Washington, DC: Smithsonian Institution Press, 5–11.

Ortner, D. J. 2012. "Differential diagnosis and issues in disease classification." In *A Companion to Paleopathology*, ed. A. L. Grauer. Malden, MA: Wiley-Blackwell, 250–267. DOI: 10.1002/9781444345940.ch14.

Ortner, D. J., and G. J. Putschar. 1981. *Identification of Pathological Conditions in Human Skeletal Remains*. Washington, D.C.: Smithsonian Institution Press.

Ottini, L., et al. 2001. "A subject with abnormally short stature from Imperial Rome." *Journal of Endocrinological Investigation* 24: 546–548. DOI: 10.1007/BF03343890.

Paine, R. R., and H. C. Harpending. 1998. "Effect of sample bias on palaeodemographic fertility estimates." *American Journal of Physical Anthropology* 105: 231–240. DOI: 10.1002/(SICI)1096-8644(199802)105:2<231::AID-AJPA9>3.0.CO;2-X.

Paine, R. R., and G. R. Storey. 2008. "Epidemics, age at death and mortality in Ancient Rome." In *Urbanism in the Preindustrial World*, ed. G. R. Storey. Tuscaloosa: University of Alabama Press, 69–85.

Paine, R. R., et al. 2009. "A health assessment for Imperial Roman burials recovered from the necropolis of San Donato and Bivio CH, Urbino, Italy." *Journal of Anthropological Science* 87: 193–210.

Palubeckaitė, Ž., R. Jankauskas, and J. Boldsen. 2002. "Enamel hypoplasia in Danish and Lithuanian Late Medieval / Early Modern samples: a possible reflection of child morbidity and mortality patterns." *International Journal of Osteoarchaeology* 12: 189–201. DOI: 10.1002/oa.607.

Parfitt, A. 2004. "What is the normal rate of bone remodeling?" *Bone* 35: 1–3. DOI: 10.1016/j.bone.2004.03.022.

Petrone, P. P. 1993. "Schiavitù, stress da attività lavorativa, malnutrizione: condizioni socio-culturali quali principali cause di morbilità e mortalità in popolazioni di età imperiale dell'area flegrea (Napoli, Campania)." *Abstracts X Congresso Antropologi Italiani*: 18.

Petrone, P., et al. 2011. "Enduring fluoride health hazard for the Vesuvius area population: the case of AD 79 *Herculaneum*." *PLoS ONE* 6 (6): e21085. DOI: 10.1371/journal.pone.0021085.

Piccioli, A., V. Gazzaniga, and P. Catalano, eds. 2015. *Bones: Orthopaedic Pathologies in Roman Imperial Age*. New York: Springer International Publishing.

Pollard, A. M., et al. 2011a. "Technical note: some observations on the conversion of dental enamel $\delta^{18}O_p$ values to $\delta^{18}O_w$ to determine human mobility." *American Journal of Physical Anthropology* 145: 499–504. DOI: 10.1002/ajpa.21524.

Pollard, A. M., et al. 2011b. "'These boots were made for walking': the isotopic analysis of a C_4 Roman inhumation from Gravesend, Kent, UK." *American Journal of Physical Anthropology* 146: 446–456. DOI: 10.1002/ajpa.21602.

Prowse, T. L. 2016. "Isotopes and mobility in the ancient Roman world." In *Approaches to Migration in the Early Roman Empire*, eds. L. de Ligt and L. E. Tacoma. Leiden: Brill Publishers, 205–233. DOI: 10.1163/9789004307377_011.

Prowse, T. L., et al. 2004. "Isotopic paleodiet studies of skeletons from the imperial Roman-age cemetery of Isola Sacra, Rome, Italy." *Journal of Archaeological Science* 31: 259–272. DOI: 10.1016/j.jas.2003.08.008.

Prowse, T. L., et al. 2005. "Isotopic evidence for age-related variation in diet from Isola Sacra, Italy." *American Journal of Physical Anthropology* 128: 2–13. DOI: 10.1002/ajpa.20094.

Prowse, T. L., et al. 2007. "Isotopic evidence for age-related immigration to Imperial Rome." *American Journal Physical Anthropology* 132: 510–519. DOI: 10.1002/ajpa.20541.

Prowse, T. L., et al. 2008. "Isotopic and dental evidence for infant and young child feeding practices in an imperial Roman skeletal sample." *American Journal Physical Anthropology* 137: 294–308. DOI: 10.1002/ajpa.20870.

Prowse, T. L et al. 2010. "Stable isotope and ancient DNA evidence for geographic origins at the site of Vagnari (2nd–4th centuries AD), Italy." In *Roman Diasporas: Archaeological Approaches to Mobility and Diversity in the Roman Empire*, ed. H. Eckardt. Portsmouth, RI: Journal of Roman Archaeology, 175–198.

Ragsdale, B. D., and L. M. Lehmer. 2012. "A knowledge of bone at the cellular (histological) level is essential to paleopathology." In *A Companion to Paleopathology*, ed. A. L. Grauer. Malden, MA: Wiley-Blackwell, 227–249. DOI: 10.1002/9781444345940.ch13.

Redfern, R. C., and S. N. DeWitte. 2011a. "A new approach to the study of Romanization in Britain: a regional perspective of cultural change in late Iron Age and Roman Dorset using the Siler and Gompertz-Makeham models of mortality." *American Journal of Physical Anthropology* 144: 269–285. DOI: 10.1002/ajpa.21400.

Redfern, R. C., and S. N. DeWitte. 2011b. "Status and health in Roman Dorset: the effect of status on risk of mortality in post-conquest populations." *American Journal of Physical Anthropology* 146: 197–208. DOI: 10.1002/ajpa.21563.

Retief, F. P., and L. Cilliers. 2005. "Lead poisoning in ancient Rome." *Acta Theologica Supplementum* 7: 147–164. DOI: /10.4314/actat.v26i2.52570.

Ricci, R., et al. 1997. "Pattern of porotic hyperostosis and quality of life in a II century A.D. farm near Rome." *Rivista di Antropologia* 75: 117–128.

Richards, M. P., and R.E.M. Hedges. 1998. "Stable isotope analysis reveals variations in human diet at the Poundbury camp cemetery site." *Journal of Archaeological Science* 25: 1247–1252. DOI: 10.1006/jasc.1998.0307.

Richards, M. P., R. J. Schulting, and R.E.M. Hedges. 2003. "Sharp shift in diet at onset of Neolithic." *Nature* 425 (6956): 366–366. DOI: 10.1038/425366a.

Richards, M. P., et al. 2000. "Neanderthal diet at Vindija and Neanderthal predation: the evidence from stable isotopes." *Proceedings of the National Academy of Sciences of the United States of America* 97 (13): 7663–7666. DOI: 10.1073/pnas.120178997.

Richards, M. P., et al. 2005. "Isotope evidence for the intensive use of marine foods by Late Upper Palaeolithic humans." *Journal of Human Evolution* 49 (3): 390–394. DOI: 10.1016/j.jhevol.2005.05.002.

Roberts, S. B., et al. 1988. "Effect of weaning on accuracy of double labeled water method in infants." *American Journal of Physiology* 254: 622–627.

Rogers, T., and S. Saunders. 1994. "Accuracy of sex determination using morphological traits of the human pelvis." *Journal of Forensic Sciences* 39: 1047–1056. DOI: 10.1016/1353–1131(95)90091–8.

Rothman, K., and S. Greenlander. 1998. *Modern Epidemiology*. Philadelphia: Lippincott–Raven.

Rubini, M., et al. 2014. "Paleopathological and molecular study on two cases of ancient childhood leprosy from the Roman and Byzantine empires." *International Journal of Osteoarcheology* 24: 570–582. DOI: 10.1002/oa.2242.

Ruff, C. B. 1987. "Sexual dimorphism in human lower limb bone structure: relationship to subsistence strategy and sexual division of labour." *Journal of Human Evolution* 16 (5): 391–416. DOI: 10.1016/0047–2484(87)90069–8.

Ruff, C. B., C. S. Larsen, and W. C. Hayes. 1984. "Structural changes in the femur with the transition to agriculture on the Georgia coast." *American Journal Physical Anthropology* 64: 125–136. DOI: 10.1002/ajpa.1330640205.

Rutgers, L. V., et al. 2009. "Stable isotope data from the early Christian catacombs of ancient Rome: new insights into the dietary habits of Rome's early Christians." *Journal of Archaeological Science* 36: 1127–1134. DOI: 10.1016/j.jas.2008.12.015.

Sabel, N., et al. 2008. "Neonatal lines in the enamel of primary teeth: a morphological and scanning electron microscopic investigation." *Archives of Oral Biology* 53: 954–963. DOI: 10.1016/j.archoralbio.2008.05.003.

Salesse, K., et al. 2014. "Variability of bone preservation in a confined environment: the case of the catacomb of Sts Peter and Marcellinus (Rome, Italy)." *Palaeogeography Palaeoclimatology Palaeoecology* 416: 43–54. DOI: 10.1016/j.palaeo.2014.07.021.

Sallares, R. 2002. *Malaria and Rome: A History of Malaria in Ancient Italy.* Oxford: Oxford University Press. DOI: 10.1093/acprof:oso/9780199248506.001.0001.

Sallares, R., A. Bouwman, and C. Anderung. 2004. "The spread of malaria to Southern Europe in antiquity: new approaches to old problems." *Medical History* 48: 311–328.

Sallares, R., and R. Gomzi. 2001."Biomolecular archaeology of malaria." *Ancient Biomolecules* 3: 195–213.

Salvadei, L., F. Ricci, and G. Manzi. 2001. "Porotic hyperostosis as a marker of health and nutritional conditions during childhood: studies at the transition between Imperial Rome and the early Middle Ages." *American Journal of Human Biology* 13 (6): 709–717. DOI: 10.1002/ajhb.1115.

Sandberg, P. A., et al. 2014. "Intra-tooth stable isotope analysis of dentine: a step toward addressing selective mortality in the reconstruction of life history in the archaeological record." *American Journal of Physical Anthropology* 155: 281–293. DOI: 10.1002/ajpa.22600.

Santos, A. L., and C. A. Roberts. 2001. "A picture of tuberculosis in young Portuguese people in the early 20th century: a multidisciplinary study of the skeletal and historical evidence." *American Journal of Physical Anthropology* 115: 38–49. DOI: 10.1002/ajpa.1054.

Schaefer, M., S. M. Black, and L. Scheuer. 2009. *Juvenile osteology: a laboratory and field manual.* Amsterdam: Academic.

Scheidel, W. 2001a. "Progress and problems in Roman demography." In *Debating Roman Demography*, ed. W. Scheidel. Leiden: Brill, 1–81.

Scheidel, W. 2001b. "Roman age structure: evidence and models." *Journal of Roman Studies* 91: 1–26. DOI: 10.1017/S0075435800015811.

Scheidel, W. 2004. "Human mobility in Roman Italy, I: The free population." *The Journal of Roman Studies* 94: 1–26. DOI: 10.2307/4135008.

Scheidel, W. 2005. "Human mobility in Roman Italy, II: The slave population." *The Journal of Roman Studies* 95: 64–79. DOI: 0.3815/000000005784016270.

Scheidel, W. 2007. "A model of real income growth in Roman Italy." *Historia: Zeitschrift für Alte Geschichte* 56: 322–346. DOI: 10.2307/25598399.

Scheidel, W. 2012. "Physical well-being." In *The Cambridge Companion to the Roman Economy*, ed. W. Scheidel. Cambridge: Cambridge University Press, 321–333. DOI: 10.1017/CCO9781139030199.020.

Schmitt, A. 2004. "Age-at-death assessment using the os pubis and the auricular surface of the ilium: a test on an identified Asian sample." *International Journal of Osteoarchaeology* 14: 1–6. DOI: 10.1002/oa.693.

Schmitt, A., et al. 2002. "Variability of the pattern of aging on the human skeleton: evidence from bone indicators and implications on age at death estimation." *Journal of Forensic Sciences* 47: 1203–1209. DOI: 10.1520/JFS15551J.

Schuenemann, V. J., et al. 2011. "Targeted enrichment of ancient pathogens yielding the pPCP1 plasmid of *Yersinia pestis* from victims of the Black Death." *Proceedings of the National Academy of Sciences of the United States of America* 108 (38): 746–752. DOI: 10.1073/pnas.1105107108.

Schuenemann, V. J., et al. 2013. "Genome-wide comparison of medieval and modern *Mycobacterium leprae.*" Science 341 (6142): 179–183. DOI: 10.1126/science.1238286.

Sealy, J. 2001. "Body tissue chemistry and palaeodiet." In *Handbook of Archaeological Sciences*, eds. D. R. Brothwell and A. M. Pollard. Chichester: John Wiley, 269–279.

Setzer, T. J. 2014. "Malaria detection in the field of paleopathology: a meta-analysis of the state of the art." *Acta Tropica* 140: 97–104. DOI: 10.1016/j.actatropica.2014.08.010.

Shaw, C. N., and J. T. Stock. 2009a. "Habitual throwing and swimming correspond with upper limb diaphyseal strength and shape in modern athletes." *American Journal of Physical Anthropology* 140, 160–172. DOI: 10.1002/ajpa.21063.

Shaw, C. N., and J. T. Stock. 2009b. "Intensity, repetitiveness, and directionality of habitual adolescent mobility patterns influence the tibial diaphysis morphology of athletes." *American Journal of Physical Anthropology* 140: 149–159. DOI: 10.1002/ajpa.21064.

Shaw, H., et al. 2016. "Identifying migrants in Roman London using lead and strontium stable isotopes." *Journal of Archaeological Science* 66: 57–68. DOI: 10.1016/j.jas.2015.12.001.

Sládek, V., et al. 2007. "Human manipulative behavior in the central European late eneolithic and early Bronze Age: humeral bilateral asymmetry." *American Journal of Physical Anthropology* 133: 669–681. DOI: 10.1002/ajpa.20551.

Slovak, N. M., and A. Paytan. 2012. "Applications of Sr isotopes in archaeology." In *Handbook of Environmental Isotope Geochemistry, Advances in Isotope Geochemistry*, ed. M. Baskaran. Berlin: Springer-Verlag, 743–768. DOI: 10.1007/978-3-642-10637-8_35.

Small, A. M. 2011. *Vagnari: The Village, the Industries, the Imperial Property.* Bari: Edipuglia.

Smith, T. M., D. J. Reid, and J. E. Sirianni. 2006. "The accuracy of histological assessments of dental development and age at death." *Journal of Anatomy* 208: 125–138. DOI: 10.1111/j.1469-7580.2006.00500.x.

Sofeso, C., et al. 2012. "Verifying archaeological hypotheses: investigations on origin and genealogical lineages of a privileged society in Upper Bavaria from Imperial Roman times (Erding, Kletthamer Feld)." In *Population Dynamics in Prehistory and Early History: New Approaches Using Stable Isotopes and Genetics*, eds. E. Kaiser, J. Burger, and W. Schier. Berlin: De Gruyter, 113–130. DOI: 10.1515/9783110266306.113.

Soomer, H., et al. 2003. "Reliability and validity of eight dental age estimation methods for adults." *Journal of Forensic Sciences* 48: 149–152.

Sperduti, A. 1995. "I resti scheletrici umani nella necropoli di eta' romano-imperiale di Isola Sacra (1–3 sec. d.C.). Analisi paleodemografica." PhD Dissertation, Università di Roma.

Sperduti, A. 1997. "Life conditions of a Roman Imperial Age population: occupational stress markers and working activities in *Lucus Feroniae* (Rome, I–II cent. AD)." *Human Evolution* 12 (4): 253–267. DOI: 10.1007/BF02438179.

Sperduti, A., L. Bondioli, and P. Garnsey. 2012. "Skeletal evidence for occupational structure and diet at the coastal towns of Portus and Velia (Central Italy, I–III cent. AD)." In *More Than Just Numbers? The Role of Science in Roman archaeology*, ed. I. Schrüfer-Kolb. Portsmouth, RI: Journal of Roman Archaeology, Supplementary volume 91, 53–70.

Sperduti, A., and G. Manzi. 1990. "*Hyperostosis frontalis interna* in a cranial sample from the Roman population of Portus." *Rivista di Antropologia* 68: 279–286.

Sperduti, A., et al. 2011. "Non-masticatory tooth wear at Gricignano d'Aversa, Italy (2500–1750 BCE): the importance of macro- and microscopic analysis." *American Association of Physical Anthropologists 80th Annual Meeting.* Minneapolis, Minnesota: 12–16 April 2011, 281.

Sperduti, A., et al. in press. "Differential burial treatment of newborn infants from Late Roman Age: children and dogs depositions at Peltuinum." In *Archeologia e Antropologia della Morte, Atti del III Incontro Internazionale di Studi di Antropologia e Archeologia a confronto*, Vol. 1, ed. V. Nizzo. Rome: Editorial Service System.

Spigelman, M., D. H. Shin, and G.K.B. Gal. 2012. "The promise, the problems, and the future of DNA analysis in paleopathology studies." In *A Companion to Paleopathology*, ed. A. L. Grauer. Malden, MA: Wiley-Blackwell, 133–151. DOI: 10.1002/9781444345940.ch8.

Spirduso, W. W. 1995. *Physical dimensions of aging.* Champaign, IL: Human Kinetics.

Stepańczak, B., K. Szostek, and J. Pawlyta. 2014. "The human bone oxygen isotope ratio changes with aging." *Geochronometria* 41 (20): 147–159. DOI: 10.2478/s13386-013-0146-1.

Storey, R. 1997. "Individual frailty, children of privilege, and stress in Late Classic Copan." In *Bones of the Maya*, eds. S. Whittington and D. Reed. Tuscaloosa: University of Alabama Press, 116–126.

Stoyanova, D., B.F.D. Algee-Hewitt, and D. E. Slice. 2015. "An enhanced computational method for age-at-death estimation based on the pubic symphysis using 3D laser scans and thin plate splines." *American Journal of Physical Anthropology* 158: 431–40. DOI: 10.1002/ajpa.22797.

Spradley, M. K., and R. L. Jantz. 2011. "Sex estimation in forensic anthropology: skull versus postcranial elements." *Journal of Forensic Sciences* 56: 289–296. DOI: 10.1111/j.1556-4029.2010.01635.x.

Stuart-Macadam, P. 1989. "Nutritional deficiency diseases: a survey of scurvy, rickets, and iron deficiency anemia." In *Reconstruction of Life from the Skeleton*, eds. M. Y. İşcan and K.A.R. Kennedy. New York: Alan R. Liss, 201–222.

Stuart-Macadam, P. 1992. "Anemia in past human populations." In *Diet, Demography, and Disease: Changing Perspectives on Anemia*, eds. P. Stuart-Macadam and S. Kent. New York: Aldine de Gruyter, 151–170.

Sutter, R. C. 2003. "Nonmetric subadult skeletal sexing traits: I. A blind test of the accuracy of eight previously proposed methods using prehistoric known-sex mummies from Northern Chile." *Journal of Forensic Sciences* 48: 927–935.

Tafuri, M. A, et al. 2016. "Life and death in Neolithic southeastern Italy: the strontium isotope evidence." *International Journal of Osteoarcheology* DOI: 10.1002/oa.2516.

Tang, N., D. Antoine, and S. Hillson. 2014. "Application of the Bang and Ramm age at death estimation method to two known-age archaeological assemblages." *American Journal of Physical Anthropology* 155: 332–351. DOI: 10.1002/ajpa.22566.

Van der Merwe, A. E., M. Steyn, and E. N. L'Abbé. 2010. "Trauma and amputations in 19th century miners from Kimberley, South Africa." *International Journal of Osteoarcheology* 20: 291–306. DOI: 10.1002/oa.1035.

Van Gerven, D. P., and G. J. Armelagos. 1983. " 'Farewell to paleodemography?' Rumors of its death have been greatly exaggerated." *Journal of Human Evolution* 12 (4): 353–360. DOI: 10.1016/S0047-2484(83)80162-6.

van Klinken, G. J., M. P. Richards, and R.E.M. Hedges. 2000. "An overview of causes for stable isotopic variations in past human environmental, ecophysiological, and cultural effects." In *Biogeochemical Approaches to Paleodietary Analysis*, eds. S. H. Ambrose and M. A. Katzenberg New York: Kluwer Academic/Plenum, 39–58. DOI: 10.1007/0-306-47194-9_3.

Villotte, S., and C. J. Knüsel. 2013. "Understanding entheseal changes: definition and life course changes." *International Journal of Osteoarcheology* 23: 135–146. DOI: 10.1002/oa.2289.

Wagner, D., et al. 2014. "*Yersinia pestis* and the plague of Justinian 541–543 AD: a genomic analysis." *The Lancet Infectious Diseases* 14: 319–326. DOI: 10.1016/S1473-3099(13)70323-2.

Waldron, T., and J. Rogers. 1991. "Inter-observer variation in coding osteoarthritis in human skeletal remains." *International Journal of Osteoarcheology* 1: 49–56. DOI: 10.1002/oa.1390010107.

Walker, P. L. 2008. "Sexing skulls using discriminant function analysis of visually assessed traits." *American Journal of Physical Anthropology* 136: 39–50. DOI: 10.1002/ajpa.20776.

Walker, P. L., et al. 2009. "The causes of porotic hyperostosis and *cribra orbitalia*: a reappraisal of the iron-deficiency-anemia hypothesis." *American Journal of Physical Anthropology* 139: 109–125. DOI: 10.1002/ajpa.21031.

Walrath, D. E., P. Turner, and J. Bruzek. 2004. "Reliability test of the visual assessment of cranial traits for sex determination." *American Journal of Physical Anthropology* 125: 132–137. DOI: 10.1002/ajpa.10373.

Warinner, C., et al. 2014. "Pathogens and host immunity in the ancient human oral cavity." *Nature Genetics* 46 (4): 336–344. DOI: 10.1038/ng.2906.

Waters-Rist, A. et al. 2010. "Activity-induced dental modification in Holocene Siberian hunter-fisher-gatherers." *American Journal of Physical Anthropology* 143: 266–278. DOI: 10.1002/ajpa.21313.

Weaver, D. S., et al. 2000. "A surgical amputation in 2nd century Rome." *Lancet* 356 (9230), 686. DOI: 10.1016/S0140-6736(05)73840-X.

Weiss, E., and R. Jurmain. 2007. "Osteoarthritis revisited: a contemporary review of aetiology." *International Journal of Osteoarcheology* 17: 437–450. DOI: 10.1002/oa.889.

Weiss K. M. 1972. "On the systematic bias in skeletal sexing." *American Journal of Physical Anthropology* 37: 239–250. DOI: 10.1002/ajpa.1330370208.

Weyrich, L. S., K. Dobney, and A. Cooper. 2015. "Ancient DNA analysis of dental calculus." *Journal of Human Evolution* 79: 119–124. DOI: 10.1016/j.jhevol.2014.06.018.

White, T. D., and P. A. Folkens. 2005. *The Human Bone Manual.* Boston: Elsevier Academic Press.

Wood, J. W., et al. 1992. "The Osteological Paradox: Problems of inferring prehistoric health from skeletal samples." *Current Anthropology* 33: 343–358. DOI: 10.1086/204084.

Wright, L., and H. P. Schwarcz. 1998. "Stable carbon and oxygen isotopes in human tooth enamel: identifying breastfeeding and weaning in prehistory." *American Journal of Physical Anthropology* 106: 1–18. DOI: 10.1002/(SICI)1096-8644(199805)106:1<1::AID-AJPA1>3.0.CO;2-W.

Yurtsever, Y., and J. R. Gat. 1981. "Atmospheric Waters." In *Stable Isotope Hydrology: Deuterium and Oxygen-18 in the Water Cycle*, eds. J. Gat and R. Gonfiantini. Vienna: International Atomic Energy Agency, 103–142.

Zink, A. R., et al. 2006. "Molecular analysis of ancient microbial infections." *FEMS Microbiology Letters* 213: 141–147. DOI: 10.1111/j.1574-6968.2002.tb11298.x.

Zuckerman, M. K., K. N. Harper, and G. J. Armelagos. 2016. "Adapt or die: three case studies in which the failure to adopt advances from other fields has compromised paleopathology." *International Journal of Osteoarchaeology* 26: 375–383. DOI: 10.1002/oa.2426.

Zuckerman, M. K., B. L. Turner, and G. J. Armelagos. 2012. "Evolutionary thought in paleopathology and the rise of the biocultural approach." In *A Companion to Paleopathology*, ed. A. L. Grauer. Malden MA: Wiley-Blackwell, 34–57. DOI: 10.1002/9781444345940.ch3.

CHAPTER FIVE

Human Growth and Stature

Rebecca Gowland & Lauren Walther

Introduction

Stature is the most ancient form of biometric data collected from human populations and continues to be one of the most frequently recorded physiological parameters today. While skin wrinkles, hair is lost, and weight fluctuates; height is considered to be a stable identifying feature of adults. In actuality, stature does decrease slightly in older age and can even fluctuate by 1.5–2 centimeters throughout the course of a single day. Between waking and sleeping, compression of the soft tissues occurs during weight-bearing activities, thus reducing height. By contrast, astronauts enduring prolonged periods of weightlessness may 'grow' in height by as much as four to six centimeters because of abnormal expansion of the inter-vertebral discs.[1] After returning to earth, their stature undergoes a period of readjustment, during which they are at risk of back injury. For the rest of us mere earth dwellers, fluctuations in stature are relatively minor, and height is considered a robust descriptor of individuals and populations.

Despite this, final adult stature is a product of the interplay between an individual's genetics and a range of environmental and social factors impacting on the body during the growth period.[2] As a consequence of extrinsic influences, a substantial degree of intrapopulation variation can occur, even within genetically homogenous populations.[3] The sensitivity of stature to life circumstances means that it can be usefully harnessed to investigate the impact of a variety of environmental and social variables on the health of populations.[4] Steckel provides a valuable synthesis of the range of applications of stature data, including studies of colonialism, migration, slavery, infectious disease and occupation.[5]

In today's industrialized world "taller populations are generally richer populations."[6] Studies of both ancient and modern groups demonstrate that taller individuals tend to have longer average life expectancies than their shorter counterparts.[7] This is because short adult stature is associated with adverse conditions during childhood, which can prevent an individual from achieving their genetic height potential. However, growth stunting during childhood can be masked by the phenomenon of "catch-up growth": an accelerated period of growth following an earlier episode of stasis.[8] Nonetheless, in order for catch-up growth to occur, the child requires improved nutritional and environmental conditions; further, there has to be something remaining of the potential growth-period.

The 1958 British Birth Cohort Study found that the height of children at age seven was a useful predictor of future employment status because growth provides a sensitive indicator of social conditions. The chances of unemployment in the shortest fifth of children were found to be three time greater than the tallest fifth. In this study, height at seven years was viewed to be reliably correlated with poor socioeconomic and psychosocial environment.[9] In a similar vein, historical accounts from nineteenth-century England frequently attest to stunted growth and the lamentable physiological condition of child factory workers, with social commentators such as Engels warning that child labor would result in a "race of pygmies."[10] Public health reformers of the period, such as Edwin Chadwick measured the heights of children and adults, but ultimately determined that the unhygienic urban environment, rather than factory work was responsible for the observed growth deficits in children and small adult stature.[11]

The relationship between adult stature and health is contingent upon the growth *journey* rather than the *endpoint*. This is well-illustrated by Barker and colleagues' study of the Helsinki birth cohort, which demonstrated that boys who were tall when they entered school (indicative of adequate nutrition and environment) had a longer life span.[12] However, those boys who were tall as a consequence of rapid catch-up growth after a period of stunting had shorter life spans. In this example, adult stature was comparable, but the hidden consequences of earlier life stressors were expressed in terms of elevated frailty in later life.

Studies of adult stature in past populations have traditionally drawn upon historical records. Such data, however, is heavily skewed towards more recent populations, usually of eighteenth–nineteenth century date or later (for example, military recruitment records).[13] In order to examine trends in stature from earlier periods of history, researchers must turn to human skeletal remains from archaeological contexts as the primary source of evidence.[14] Many studies of past stature have utilized human skeletal remains, including the analyses of hominid remains, temporal trends within and between different regions, the effects of changing subsistence practice, urbanization, climate change, and the impact of disease on human populations.[15]

A number of authors have undertaken large-scale studies of stature in Roman-period skeletons as a means of assessing physical well-being in different regions and over the duration of the Empire. There is some debate between those who argue for favorable economic and living conditions and those who maintain that the regime was not conducive to good health. Jongman broadly falls into the former camp, asserting that mean femur length was at its longest between the mid-first to mid-second centuries whereas it declined, along with the Empire's fortunes, during the third and fourth centuries.[16] Kron likewise argues for a "healthy" Roman period, stating that male adult stature in Italy during the millennium from 500 BCE to 500 CE was a robust 168cm, which compared favorably with the average stature (163 cm) of nineteenth-century Italian males (historical data from military conscripts).[17] However, as Scheidel highlights, Kron's data-set spans an entire millennium and thus lacks any temporal resolution and nuance.[18] Gowland and Garnsey further criticized the regression formulae used in Kron's study and the potential incompatibility of these with Italian archaeological populations.[19] This critique is supported by Gianecchini and Moggi-Cecchi's study, which suggests that, based on current data, 164cm is a reasonable approximation of mean adult male height in Roman central Italy: similar to that of the nineteenth century conscripts.[20] They argue that the formulae frequently used to estimate stature in Roman Italy (Trotter and Gleser's 'white' formulae) result in overestimates, because the Italian archaeological population expresses different body proportions than the skeletal sample used to devise the formulae.[21] The lack of reliability in reported stature estimates has led some authors to recommend that stature calculation equations be bypassed altogether, in favor of a direct comparison of long bone length.[22] Jantz and Jantz likewise concluded that long bone length is a robust proxy for stature, with lower limb bones tending to show clearer secular trends than upper limb bones.[23] Goldewijk and Jacobs analysed a large data-set of c.10,000 burials in their study of stature across the Roman Empire.[24] They found that the ratio of femur length to other long bones in their sample was significantly different from those implied in most popular stature reconstruction methods. These authors support the current trend of directly comparing lower limb lengths as a proxy for adult stature.

This chapter reviews the study of stature in the archaeological record with a focus on the Roman period. A critical review of current methods of estimating stature, including an evaluation of the differing techniques and their application to the Roman period will follow. We do not advocate the comparison of long bone lengths alone for the study of stature in the Roman world. We argue that such an approach overlooks the biocultural significance of overall body proportions (including trunk height); instead more effort should be made to establish population-specific regression formulae. Finally, we argue that because adult stature and final body proportions are strongly influenced by environmental conditions during childhood, a more nuanced consideration of growth and adversity during infancy and childhood is required. First, this chapter will provide a detailed discussion of the methods currently used to calculate stature from skeletal remains.

Estimating Stature from Skeletal Remains

The stature of a living individual is a composite of articulated soft and hard tissue elements. Within an archaeological context, human remains are most often skeletonized, frequently incomplete, and unless *in situ*, are no longer articulated. Since the late nineteenth century, methods for deriving living stature from the "dry bones" of disarticulated skeletons have been produced from documented skeletal collections with known living heights. Pearson created regression formulae to calculate stature from long bone length in 1899 and remarkably this method is still in use by some anthropologists today. Broadly, there are two "types" of method for estimating stature from skeletal remains: the anatomical method (e.g., the "Fully technique") and the mathematical method.[25] A description of each of these, together with their potentials and limitations, is provided below.

THE ANATOMICAL METHOD

The anatomical method was first proposed by Fully in 1956, who used the technique on French World War II prisoners of war for whom ante-mortem records were available.[26] This method reconstructs living stature from the measurement of all contiguous skeletal elements that directly contribute to a person's height, including: cranial height, vertebral height, lengths of the femur and tibia, and the height of the articulated talus and calcaneus of the foot. It additionally incorporates soft tissue corrections to account for the absence of these components in skeletonized remains. Through the measurement of all of those bones that contribute towards height, this technique allows for individual idiosyncrasies in body proportions; for example, short lower limbs compared to long trunk length, or even pathological features impacting on stature, such as spinal curvature, to be taken into consideration.[27]

In a review of the Fully method, Raxter and colleagues found that it was strongly correlated with living stature, but could underestimate height by as much as 2.4 centimeters, particularly when applied to African American populations.[28] These results were thought to relate to inaccuracies in the soft tissue correction factor, which had been derived from European populations. There was also a lack of clarity in terms of the precise anatomical landmarks used by Fully to execute his measurements. Raxter and colleagues have provided useful revisions to Fully's method, including the creation of new soft tissue corrections.[29] In addition, Raxter and colleagues have highlighted the potential source of error introduced through age-related changes in the spinal column and have proposed the use of broad age-correction calculations.[30]

The primary limiting factor regarding this method is poor preservation and the need for so many contiguous skeletal elements to be present and

undamaged. To combat this problem, Auerbach developed methods for estimating values for those skeletal elements that are frequently missing, such as
the vertebrae.[31] While such estimates can potentially introduce error, these
should be minor compared to the much greater error incurred through the
use of regression formulae to estimate stature from a single long bone (see
below, 2.2).

Overall, the anatomical method provides an effective means of ascertaining population-specific differences in body proportions from well-preserved
skeletons. These measurements can then be used to create tailored regression
formulae that can be applied to the less complete skeletons. A number of authors have attempted such an approach.[32] Unfortunately, due to problems of
preservation and the time-consuming nature of this type of analysis, the anatomical method is only infrequently used in archaeological studies.[33]

THE MATHEMATICAL METHOD

The mathematical method involves the use of simple regression formulae to
calculate adult stature from the measurement of single long bones (though
the femur and tibia may be used in conjunction if both are present). These
formulae are derived from the long bones of documented skeletal individuals
for whom living height (or cadaveric stature) is known. Pearson (1899) was the
first to produce a "mathematical method" for calculating stature; however, the
most widely applied formulae today are those produced by Trotter and Gleser
and Trotter (Table 5.1).[34] The Trotter and Gleser formulae were created from
a sample of white and black Americans from the Terry collection of known
individuals in the United States. Formulae were created for the femur, tibia,
humerus, radius, and ulna—each with different associated error ranges. Stature estimates calculated from the lower limb bones, in particular the femur,

Table 5.1. The regression formulae developed by Trotter and Gleser for estimating stature
from the femur

Reference	Male Formula	Female Formula
Trotter and Gleser 1952,1958 ("White" Formula)	2.32 * *Max Fem Length* + 65.53	2.47 * *Max Fem Length* + 54.1
Trotter and Gleser 1952, 1958 ("Black" Formula)	2.10 * *Max Fem Length* + 72.22	2.28 * *Max Fem Length* + 59.76
Trotter 1970	2.38 * *Max Fem Length* + 61.41	2.47 * *Max Fem Length* + 54.1
Pearson 1899	1.88 * *Max Fem Length* + 81.306	1.95 * *Max Fem Length* + 72.844

are considered to be the most accurate. Before calculating stature from a skeleton, it is important to first establish the individual's sex and the (crudely defined) ancestry (i.e., white Americans / black Americans), as the regression equations will differ. In common with the "anatomical method," an age-related correction factor must also be applied to stature calculated using this method, because individuals (females in particular) display reduced height with age.[35]

Given that this method requires only a single measurement, stature can be calculated significantly more quickly than by the anatomical method and from even very incomplete skeletons.[36] However, the accuracy of reconstructed living stature from mathematical regression formulae is affected by intra- and interpopulation differences in body proportions that arise through genetics, ecogeographic variation, and life circumstances.[37] The body proportions of the archaeological population may be very different from those of the reference sample on which the formulae were derived. As Table 5.1 illustrates, each of these formulae differ, resulting in stature estimates that can vary by several centimeters. Ideally, one would choose equations based on a reference population that demonstrates similar body proportions, but this is not easily determined or accomplished, as there are only a limited number of skeletal collections available with documented living stature.[38] A variety of population-specific regression equations have been produced, including Allbrook (British and East African Males), Genovés (modern Mesoamerican and US Southwest), and de Beer (Dutch). However, in practice, the Trotter and Gleser "white" formulae tend to be applied to most archaeological populations, irrespective of either period or place.[39]

One last point regarding the Trotter and Gleser stature calculation tables involves a methodological controversy. When applying the technique to the original reference population, Jantz noted significant differences between estimated and documented stature.[40] Further study concluded that the method described in the original paper was not in fact followed in the creation of the published formulae—of particular concern was the omission of the medial malleolus in measurements of the tibia.[41] The implication of this widespread mismeasurement of the tibia for such a prolonged period of time in the forensic and archaeological communities is hard to quantify.[42] The following section will consider the applicability of these methods to skeletal samples from late Roman Britain.

APPLYING THE ANATOMICAL AND MATHEMATICAL METHODS TO ADULT SKELETONS: A CASE STUDY FROM ROMAN BRITAIN

Tests of stature calculated via the anatomical method versus the mathematical regression formulae tend to find the former to be the most accurate.[43]

However, the use of the anatomical method and mathematical method are not mutually exclusive. Vercellotti and colleagues used the anatomical method to create new population-specific regression formulae for European archaeological populations.[44] The results were found to be more accurate than using generic regression methods such as Trotter and Gleser's. We have adopted a similar approach to a skeletal sample from Roman Britain, with the aim of more accurately characterizing body proportions and of arriving at more reliable stature estimations. First, the anatomical method was applied to a total of 76 (36 males and 40 females) well-preserved Romano-British skeletons from five late Roman cemetery sites in southern and eastern England. The resulting calculations formed the basis of "known" stature and body proportions. Secondly, different regression methods were then applied to the sample of 76 individuals and the results compared to the "known" stature (i.e., calculating using the anatomical method). Table 5.2 shows the "known" stature and the deviations from this when each of the commonly used regression formulae were applied. Stature estimations using Trotter and Gleser and Trotter showed a statistically significant difference to the stature calculated using the Fully Anatomical Method (paired t-test; $p<0.001$) for both males and females.[45] When the "black" formula from Trotter and Gleser was used to calculate male stature, the femur provided a significantly different stature (paired t-test; $t=-2.1$, $p<0.01$), but there is no significant difference for the tibia (paired t-test; $t=1.9$, $p=0.06$).[46] This exercise, again, highlights the influence of differential body proportions. When the Pearson formulae were applied, a statistically significant difference was observed for female femoral length (paired t-test; $t=2.6$, $p=0.01$), but not for male femoral length (paired t-test; $t=-1.1$, $p=0.28$).[47] The tibia length equations demonstrated statistically significant differences for both females and males. Finally, when stature was estimated using the sum of the femur and tibia, there was no statistically significant difference with regard to male stature (paired t-test; $t=0.5$, $p=0.64$), however a significant difference was observed with the females (paired t-test; $t=4.2$ $p<0.01$). This analysis reveals different lower limb body proportions between, not only the Romano-British sample and the Trotter and Gleser and Pearson reference populations, but also males and females.

Finally, population-specific regression formulae were developed from this sample of 76 well-preserved Romano-British based on the "known" body proportions and statures. These regression formulae were then applied to the remaining Romano-British skeletons in the sample which had both a left femur and left tibia preserved. This sample consisted of 174 females and 213 males and mean statures of 154.6 cm (±2.19 cm) for Romano-British females and 164.3 cm (±2.46 cm) for Romano-British males were calculated. These figures are comparable with those estimated by Gianecchini and Moggi-Cecchi for Roman period Central Italy.[48]

Table 5.2. Stature from Romano–British skeletons calculated using the anatomical method, a range of commonly used regression techniques, and a newly developed population-specific regression method for Roman Britain

	Fully Method Stature (cm)		New RB regression Formulae		Trotter and Gleser 1952/58-White		Trotter and Gleser 1952/58-Black		Pearson 1899		Trotter 1970	
	Male $N=36$	Female $N=40$	Male $N=36$	Female $N=40$	Male $N=36$	Female $N=40$	Male $N=36$	Female $N=40$	Male $N=36$	Female $N=40$	Male $N=36$	Female $N=40$
Femur												
Max	173.43	161.80	170.97	159.27	178.86	164.14	174.81	161.33	173.14	159.49	177.67	164.14
Min	149.35	143.89	152.92	146.66	155.20	146.72	153.39	145.49	153.97	145.98	153.40	146.97
Average	163.18	153.78	163.19	153.77	168.66	156.66	165.57	154.43	164.87	153.60	167.20	156.66
Paired T-Test			$P=0.998$	$P=0.999$	$P=2.67E\text{-}09$	$P=9.36E\text{-}07$	$P=0.000$	$P=0.180$	$P=0.013$	$P=0.711$	$P=1.77E\text{-}06$	$P=9.36E\text{-}07$
Tibia												
Max	173.43	161.80	170.34	159.52	174.74	165.35	169.35	160.36	169.78	158.98	175.26	165.35
Min	149.35	143.89	152.04	145.79	156.22	146.65	152.59	144.56	151.61	143.81	155.98	146.65
Average	163.18	153.78	163.18	153.78	167.50	157.52	162.79	153.75	162.68	152.63	167.72	157.52
Paired T-Test			$P=0.999$	$P=0.999$	$P=2.89E\text{-}07$	$P=4.27E\text{-}09$	$P=0.557$	$P=0.954$	$P=0.460$	$P=0.019$	$P=1.10E\text{-}07$	$P=4.27E\text{-}09$
Femur + Tibia												
Max	173.43	161.80	170.72	159.80	176.71	164.83	171.80	161.07	172.21	159.60	176.39	164.89
Min	149.35	143.89	153.85	147.17	155.67	148.01	152.60	145.54	152.91	146.10	154.68	148.14
Average	163.18	153.78	163.20	153.74	167.65	156.89	163.53	153.84	163.86	153.18	167.04	159.93
Paired T-Test			$P=0.979$	$P=0.933$	$P=9.07E\text{-}08$	$P=8.07E\text{-}08$	$P=0.598$	$P=0.880$	$P=0.306$	$P=0.187$	$P=1.75E\text{-}06$	$P=6.37E\text{-}08$

VERTEBRAL MEASUREMENTS

The above has demonstrated the utility of the anatomical method for assessing the accuracy of stature estimation techniques and for producing population-specific formulae. However, in this Romano-British sample, the number of skeletons that were sufficiently well-preserved to apply the anatomical method in full was only 35 individuals. The spinal column is an important component of the Fully anatomical method, but taphonomic damage, disease processes, and recovery bias in excavations often lead to missing vertebrae. We were able to more than double this sample to 76 individuals by incorporating estimates for missing vertebral elements. This is possible because the spine exhibits relatively little variation in vertebral body heights between adjacent vertebrae. Individual vertebrae and also whole vertebral sections (e.g., cervical or thoracic sections) can be estimated using mathematical equations based on the size of those vertebrae present.

Vertebral columns with *known* body heights for all vertebrae (second cervical to fifth lumbar) were used to calculate a coefficient to estimate missing individual vertebral elements. This coefficient was then used to estimate missing vertebrae with the following formula:

$$k \times \frac{x + y}{2}$$

where k is the coefficient calculated from known body heights of the vertebra to be estimated, x is the known superior vertebra and y is the inferior vertebra. This calculation allowed the addition of six more individuals, four males and two females, to the "known" sample.

Auerbach produced equations for estimating missing vertebral regions (i.e., cervical, thoracic or lumbar) from known sums of those vertebral sections present.[49] A similar approach was trialed on this Romano-British sample, to determine if Auerbach's equations accurately predicted missing vertebral regions. Vertebral columns with known summed body heights of all vertebrae (C2 through to L5) were entered into Auerbach's regression formulae for estimating missing regions or total vertebral column length. While the technique generally performed well, there was a statistically significant deviation when estimating the total vertebral column length using Auerbach's equation.[50] New mathematical regression formulae were therefore created to more accurately estimate vertebral column length from the known sum of vertebral regions for this specific Romano-British population. Statistically significant differences were observed between males and females in regard to the summed lumbar vertebral heights. As a result, separate equations were created for males and females.

To estimate the whole vertebral column from the known sum of thoracic and lumbar vertebrae the following equations were created:

Females:

1.2216*Sum of Thoracic Sum of Thoracic + 1.0588
*Sum of Lumbar Sum of Lumbar+ 39.333

Males:

1.0801*Sum of Thoracic Sum of Thoracic Sum of Thoracic + 1.3493
*Sum of Lumbar Sum of Lumbar+ 39.921

If thoracic vertebrae were missing, estimation of the whole vertebral column could be estimated from the sum of lumbar vertebrae, though the error associated with this equation is larger than the previous equations.

Females:

2.0395*Sum of LumbarSum of Lumbar + 188.62

Males:

2.8165*Sum of LumbarSum of Lumbar + 98.872

These four equations make possible the addition of 35 individuals to the sample. While the above process may seem rather laborious, it is important to be confident that stature estimates are as accurate as possible and that differential body proportions between populations can be accommodated. The following section will undertake a comparison of long bone lengths only, ignoring the vertebrae entirely.

DIRECT COMPARISON OF LONG BONES

As discussed in the introduction, the direct comparison of long bone lengths, rather than calculated stature data, is now being advocated.[51] A direct comparison of average maximum femur length between Romano-British and Anglo-Saxon skeletons shows a significantly reduced femur length for both males and females in Romano-British skeletons when compared to the post-Roman period (Figure 5.1, Table 5.3, t-test: p=0.0001). Furthermore, the mean femur length for males in Roman Britain is very similar to Gianecchini and Moggi-Cecchi's figure of 445.5mm for males in Roman central Italy.[52]

One difficulty encountered when using femoral length alone to compare sites is that many skeletal reports simply provide final stature, rather than the raw data (i.e., long bone length). While the current exasperation expressed with the lack of standardization in stature calculation methodology is understandable, and the direct comparison of long bones certainly seems sensible to

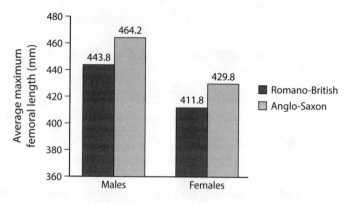

FIGURE 5.1. Comparison of femoral length between Roman
and Anglo-Saxon populations in England.

Table 5.3. Comparison of mean femur length between Romano-British and Anglo-Saxon cemeteries in Southern and Eastern England

| | Romano-British Average Femur | | | Anglo-Saxon Average Femur | | | |
| | $\overline{X}$ | | | | $\overline{X}$ | | |
	n	Mean	SD	SE	n	Mean	SD	SE
Males	297	443.8	28.2	1.6	158	464.2	26.6	2.1
Females	236	411.8	21.9	14	135	429.8	23.3	2.0

meet the needs outlined by Goldewijk and Jacobs, we would recommend an approach that considers the full body proportions of a skeletal sample, when possible.[53] For example, with the Romano-British and Anglo-Saxon skeletal samples analysed above, we noted differences in mean trunk heights, with the latter having relatively shorter trunk heights (compared to limb lengths) than the former. Therefore, while the Romano-Britons may have had substantially shorter femur lengths, in some instances, this was partially mitigated in terms of overall height by a relatively longer trunk. Goldewijk and Jacobs state that there is "no way of finding out which method renders the correct body heights," but with the more widespread use of the anatomical method, as described above, it is feasible to characterize the range and variation in body proportions in different regions.[54] This analysis of Romano-British stature will be discussed below in the context of recent large-scale studies of Roman-period skeletons.

Stature in the Roman Empire

Stature as an index of well-being has been widely utilized in studies of Roman-period skeletons. As discussed above, an estimated stature of 168cm for

Roman-period males is widely cited and in some instances used to support the suggestion that life at this time incorporated a reasonable standard of living. This estimate is substantially greater than historical data for Italian male conscripts from the nineteenth century—by comparison a measly 163cm. There are a number of criticisms that have been leveled at these estimates, including the choice of regression formulae used to calculate stature. Other criticisms can also be brought to bear; not least a consideration of the actual *ages* of the Italian conscripts at the time their heights were measured, which was approximately 20 years.[55] The adolescent growth spurt is known to be delayed in individuals exposed to adverse circumstances in early life, prolonging growth into the early twenties. Additional height attained during late adolescence is from the trunk, which will continue to grow after the long bones of the limbs have completed fusion. Living conditions for ordinary Italians were far from optimum in the nineteenth century, and the majority of European countries experienced a reduction in mean stature during this period.[56] It is therefore very likely that many of these conscripts had not yet finished growing: approximately 9% of final height in males is gained during puberty.[57] Some of this putative five-centimeter deficit between ancient Romans and nineteenth-century conscripts will be accounted for by the mismatched comparison of adult (archaeological) with adolescent/young adult (conscripts) data-sets. In addition, and as Gowland and Garnsey have argued, it is highly likely that the incompatibility of the regression formulae used to calculate stature from the Roman-period skeletons is also responsible for an overestimation of male stature.[58] A'Hearn and colleagues' study of growth and stature using the same nineteenth-century military conscript data as Kron further highlights the regional heterogeneity in height at the age of 20 years across Italy.[59] Geographically and temporally broad archaeological studies of the Roman Empire should be mindful of subregional complexities in the data and the social and environmental causes of these.

Giannecchini and Moggi-Cecchi suggest that, based on current data, 164cm is a reasonable approximation of mean adult male height in Roman central Italy: similar to that of the nineteenth-century conscripts.[60] They argue that the body proportions of the Roman period skeletal samples were more congruent with Trotter and Gleser's "black" formulae, and furthermore, the use of the "white" formulae produced an overestimate of stature.[61] Gianecchini and Moggi-Cecchi's findings are also remarkably close to the stature calculated independently from the Romano-British sample analysed here using the anatomical method (Table 5.2). Likewise, the Romano-British skeletal sample exhibited body proportions closer to the Trotter and Gleser "black" formulae and were also overestimated by the "white" formulae.

Kopeke studied a sample of Roman skeletons dating from the first to fourth centuries CE and hypothesized that stature would be greatest during the second and third centuries when the Empire was more integrated and stable.[62]

Her results, however, found that stature was largely unchanged, though it did decline very slightly during the fourth century. Kopeke and Baten argued that stature "stagnates" during the Roman period in central, western, and southern Europe and increases during the fifth and sixth centuries CE.[63] Indeed, a number of studies have identified a post-Roman increase in stature across various parts of the Empire.[64] For Britain, this tends to be interpreted in terms of an influx of Germanic migrants rather than improved local, environmental conditions. In a similar vein, Koepke and Baten have argued that Roman-period migrants to central Europe (identified as such through grave goods) were on average four cm *shorter* than the locals.[65] This difference in stature between local and nonlocals in both regions may indicate contrasting social and cultural practices, including the greater consumption of milk and meat protein in northern and eastern Europe.[66] Interestingly, increased milk consumption is thought to have played a significant part in Japan's dramatic rise in population stature during the last forty years.[67]

During the Roman period in Britain the skeletal evidence reveals a decrease in stature compared to the preceding period and an overall increase in nonspecific indicators of health stress.[68] A similar skeletal pattern has been observed in Roman-period Italy.[69] One contributing factor may be the imposition of an increasingly hierarchical structure onto the local populace with Roman occupation of Britain, resulting in greater social inequalities and increased psycho-social stresses. Scheidel likewise suggests that declining stature in Roman Gaul may be linked to an increase in population size and social inequality.[70] Data on stature from modern populations demonstrates a correlation between equality and taller adult stature.[71] The stress hormone cortisol is known to inhibit growth and this may be responsible for some of the link between psychosocial stress and reduced stature.[72]

Adult stature as a measure of well-being lacks nuance because it can mask an array of early life episodes of stress, which have significant implications for morbidity and mortality. A more fruitful approach to the study of Roman health status would be to focus attention on children (i.e., during the growth period). The skeletal remains of children are often overlooked in studies of Roman health.[73] Yet children's bodies are highly sensitive to social and environmental challenges and act as sensitive barometers for overall population health.[74] As discussed previously, poor nutrition, infection, and the synergistic interaction between these stressors, are the most influential environmental factors impacting growth.[75] Individuals suffering from prolonged episodes of health stress will exhibit a slower rate of bone growth, delayed dental eruption, a prolonged period of growth and ultimately diminished adult stature.[76] Growth stunting has been conceptualized in terms of a life-history trade-off in the face of adversities.[77] However, in addition to short-term implications, there are significant longer-term health consequences to growth stunting, such as a compromised immune system and impaired cognition.[78]

Growth, Development and Environment.

The following provides a brief discussion of growth, potential impediments to growth, and the identification of these in the archaeological record.

FETAL AND INFANT GROWTH

At birth the newborn has already had an eventful history that relates most strongly to maternal health and nutritional status.[79] Intrauterine conditions have important prolonged implications for growth trajectories in early childhood; therefore it is important to consider the degree to which maternal health can also impact growth and adult stature.[80] Women of smaller stature and pelvic dimensions will suffer an increased likelihood of obstetric risks and of giving birth to infants who are 'small for gestational age' (SGA).[81] Sibley and colleagues' study of the pelvic dimension of females from a medieval skeletal population in Sudanese Nubia observed a high proportion of females with contracted pelves.[82] The authors noted a link between evidence for growth retardation in the mother and neonatal/maternal morbidity and mortality. From a life course perspective, growth stunting of the mother as a consequence of her own poor childhood environment results in the poor health of her offspring, thus perpetuating a cycle of health inequality.[83]

Studies of growth stunting in developing countries have shown that faltering occurs early in life and is most pronounced by two years of age.[84] After birth, there is a period of adjustment in growth as the newborn makes the transition towards a regulatory system based upon its own homeostatic and genetic make-up, rather than the mother's.[85] However, the infant/mother nexus is not yet decoupled; breastfeeding regimes and the nutritional status of the mother continue to exert a strong influence on infant well-being. Poor nutrition profoundly influences growth during fetal, infant, and early childhood development phases—the period of life during which food is predominantly sourced from the maternal body, via the placenta or breast milk.[86] A longitudinal study by Chávez and colleagues showed a clear decline in breast milk consumption by the infants of poorly nourished mothers by two to three months of age, contributing to early-childhood malnutrition and poorer growth.[87] Height deficits occurring by two years tend to be maintained into adulthood. For example, the four-centimeter secular increase in the adult height of Japanese populations from 1950 to 1990 was already present at the age of two years.[88] Thus, Kuzawa and Quinn argue that adult size is most closely linked to matrilineal nutritional well-being and history.[89]

Postnatal growth velocities are highest during the first year after birth (approximately 30 cm/year in the first two months), but drop dramatically after 12 months.[90] Such rapid growth necessitates high dietary requirements during the postnatal period and leaves the infant at greater risk of malnutrition, infection, and death.[91] Infant mortality, specifically post-neonatal mortality (death occurring between one month and one year after birth), is strongly correlated with average final adult stature.[92] Ideally, therefore, it would be useful to examine infant mortality in conjunction with adult stature as complementary indices of health in the Roman world. While Lewis and Gowland were able to successfully comment on neonatal versus post-neonatal infant mortality in Medieval England, a similar study is not feasible for the Roman period, due to the differential burial treatment of infants.[93]

The analysis of different skeletal parameters of growth and stress during childhood potentially allows for the production of *biographical* data. While limb length has been shown to be more 'plastic' than trunk height in relation to environmental stresses, differences in vertebral dimensions between archaeological populations have been identified and correlated with increased frailty.[94] In other words, growth stunting during infancy, observed in vertebral dimensions, results in increased frailty during adulthood. The transverse and anteroposterior diameters of the neural canal of the vertebrae are "locked-in" by five years of age, due to fusion of this skeletal region, providing an indicator of early post-natal growth, while vertebral body height may continue growing into early adulthood.

The complementary analysis of these parameters, along with long bone length, can be used to construct an osteobiography of growth stunting in early and later childhood, with implications for adult morbidity.[95] Intra- and interpopulation comparisons of episodes of growth retardation can be identified and analysed in relation to differing environmental or social variables. Newman and Gowland illustrate the potential of this approach in their study of vertebral dimensions in children from postmedieval London.[96] Here, patterns of growth stress were correlated with socially constructed life course norms, including status-driven child-care practices and child labor.

PUBERTY

A mid-childhood growth spurt at around the ages of eight or nine years has been proposed, though this has been contested by some,[97] followed by an adolescent growth spurt around puberty. Upon reaching puberty there is an increasing divergence between individuals and sexes (boys experiencing a growth spurt at a later age than girls). Puberty is strongly affected by socioenvironmental conditions, with adverse circumstances leading to delays in growth, pubertal onset, and an extended period of maturation into early adulthood.[98] Recently developed osteological techniques now allow age-at-puberty to be assessed from skeletal remains.[99] These methods have been applied to studies of Medieval adolescence in England and have important implications for the study of past social age trajectories and fertility. Shapland

and colleagues' study of a large Medieval sample found that the onset of menstruation was delayed until approximately 15 years and over (on the basis of osteological criteria), compared to an average of 13 years today.[100] Likewise, male puberty was found to be prolonged, with maturation continuing into the early twenties.[101] The first independent application of these new skeletal techniques to Roman-period adolescent skeletons from Britain similarly indicated that puberty was delayed compared to modern norms. A later age of menarche in Romano-British females from rural and urban sites suggests that they were unlikely to have been able to reproduce until their late teens, whilst males were still developing into their early twenties.[102] These skeletal data correlated with burial rites accorded some Romano-British females, which indicate a marked change in social status from 18 to 24 years, possibly associated with age-at-marriage, or motherhood.[103] The skeletal data is also in line with Galen's assertions that male development continued until the early to mid-twenties.[104] For higher-status groups, with better nutrition, including more dietary protein, age of onset of menarche may well have been much lower, potentially contributing to much younger ages of marriage amongst Roman elites.

CONSTRUCTING GROWTH PROFILES

When analyzing skeletal growth in archaeological populations of children, it is necessary to use dental development age as a proxy for known age. Dental development (not to be confused with dental eruption) is closely correlated with chronological age:[105] if a child is unwell for a period of time, or malnourished, his or her teeth will continue to develop, despite stasis in postcranial growth. This observation was established as early as the nineteenth century, when dental age was used to enforce minimum working ages for children in factories.[106] Growth parameters such as long bone diaphyseal length can be plotted against dental age to produce growth profiles, thus allowing interpopulation comparisons (Figure 5.2). As always, there are a number of caveats that apply: firstly, dental age, while a close approximation is *not* a known age and therefore such profiles incorporate any associated error; secondly, these profiles are based on nonsurvivors (i.e., children who died) who may not be representative of the living population. The significance of this latter point has been debated,[107] with Saunders and Hoppa arguing that the likely effects are minimal.[108] Either way, the comparison of growth profiles between archaeological sites eliminates this as a source of bias because the datasets are comparable.

An example of growth profiles constructed from Romano-British and Anglo-Saxon children is provided in Figure 5.2. Children in both periods fall below the modern values.[109] This is true for the majority of growth profiles

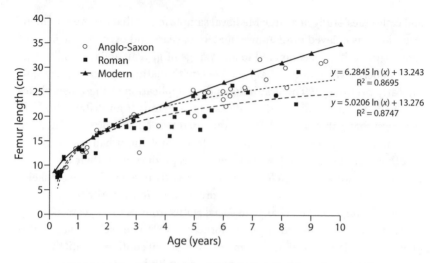

FIGURE 5.2. Long bone length plotted against dental age for a sample
of Romano-British and Anglo-Saxon skeletons. The trend-line
for the Romano-British skeletons (dashed) falls below
that of the Anglo-Saxons (dotted line).

produced from archaeological populations and is to be expected due to the
better nutritional status and reduced infectious disease burden of modern
children. The Anglo-Saxon children exhibit growth that is slightly closer to
the modern values than the Romano-British sample and this is to be expected
given the increased stature in post-Roman Britain. Rohnbogner compared
skeletal growth profiles produced from a large sample of Romano-British children with a middle-class eighteenth-to-nineteenth-century population from
London.[110] Interestingly, she found that the Romano-British children exhibited stronger growth up until approximately five years of age, when the situation reverses. Given the discussion above, this could indicate better maternal
health and infant feeding strategies in Roman-Britain compared to industrialized London, but more detrimental conditions in later childhood. Isotopic
studies suggest that infants in Roman Britain were generally breastfed until
approximately three years of age, with a gradual program of weaning.[111] This
is much longer than in postmedieval London, and the Roman-British feeding
regime may have buffered children against the worst of the environmental circumstances, providing passive immunity, as well as a hygienic and nutritious
food source. Interestingly, the isotope evidence shows some differences in infant feeding practices as between Roman Britain and Portus Romae (Isola
Sacra), where a shorter period of transitional feeding and earlier cessation of
breastfeeding was the norm.[112]

CORTICAL THICKNESS, VERTEBRAL BODY HEIGHT
AND SKELETAL INDICATORS OF 'STRESS'

In a study of the skeletal remains of children from nineteenth-century Birmingham, England, Mays and colleagues found that the cortical thickness of bones was adversely affected in individuals of lower socioeconomic status.[113] Indeed, the authors suggest that appositional growth may be a more sensitive indicator of environmental stressors than longitudinal bone growth. These results were confirmed in a study of postmedieval children by Newman and Gowland, which showed a correlation between poor appositional growth and pathological indicators of poor health, including cribra orbitalia, enamel hypoplasia, rickets, and scurvy (see also Sperduti and colleagues, this volume).[114] (Growth patterns and palaeopathological indicators can profitably be considered together in any assessment of the health of individuals.) Figure 5.3 shows a growth profile based on the height of the cervical bodies of the vertebrae. Three individuals exhibit particularly low values relative to their age (skeletons 208, 338, and 262), and it is of note that all three have severe and active pathological indicators of poor health.[115]

BODY PROPORTIONS

Skeletal growth is heterochronic, meaning that different elements grow at different rates and developmental stages. For example, gains in sitting height (i.e., trunk length) are generally made during infancy and puberty, while

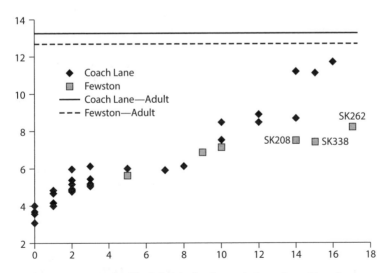

FIGURE 5.3. Vertebral body height for the cervical vertebrae. Dental age
in years is plotted on the x-axis and height in millimeters along the
y-axis. The Coach Lane and Fewston sites are both postmedieval.

gains in leg length occur during childhood. As a consequence, in young children, in respect of growth, legs are ahead of torsos.[116] Thus, at one year of age, leg length represents 35% of adult length, but by 10 years this has risen to 77%.[117] Therefore poor nutrition or adversity during this period is likely to disproportionately inhibit leg length. A study by Bogin and Baker shows that absolute and relative leg lengths (RLL) provide a summary of postnatal growth between birth and puberty and that RLL is established by eight years of age.[118] In general, the timing and duration of a growth insult during childhood can differentially impact upon adult body proportions, resulting in intrapopulation heterogeneity in limb and trunk length.[119] A study of living children from highland and lowland Peru is a pertinent illustration of this truth: Pomeroy and colleagues found the ulna and tibia to be most sensitive to poor nutrition.[120] Their study would translate well to archaeological datasets, in the context of studies of the effects of environmental stressors on relative limb segment length, stature and age-at-death in the past. For example, we may hypothesize that the relatively long trunk height identified above in Romano-British individuals compared to the post-Roman period could represent prolonged growth during later adolescence—when most stature increase is derived from the vertebrae. This longer period of growth could be the body's attempt to mitigate earlier life deficits as a consequence of the higher levels of environmental stressors in Roman compared to post-Roman Britain.

A Holistic Approach to Adult Stature

Adult stature in past societies has tended to be studied cross-sectionally, that is, as a moment in time. However, skeletal growth is diachronic in character, in that particular bones and teeth form at different times and rates. The study of children in the past has the potential to provide insights into how stature was *achieved* through the analysis of a variety of skeletal parameters relevant to different life course stages. For example, analysis of the body proportions of adults and children, which gives special attention to the proportional lengths of the distal limb segments, may be suggestive of environmental stressors.[121] Whilst the effects of growth stunting on individual long bone lengths may be masked by catch-up growth, the analysis of skeletal parameters that fuse earlier in life can provide additional information regarding early childhood.[122] Evidence for compensatory growth can then be compiled (e.g., small transverse diameter / distal limb segment / average adult stature) and correlated with mortality risk. This information can be assessed in relation to the presence of childhood indicators of poor health (e.g., cribra orbitalia), with a particular focus on the age of onset of these lesions and the presence of both active and healed lesions within the skeletal sample.[123] These data should then be integrated with longitudinal dietary data obtained from incremental isotope analysis of dentine. High resolution isotope analysis of teeth can reveal periods of severe health stress (i.e., elevated nitrogen values) in both the adult survivors and the nonsurvivors who died in childhood.[124] Growth deficits in survivors, whose isotope evidence also indicated a period of deprivation, would be of particular interest for correlating with other skeletal parameters of growth stunting. The age at which growth deficits and paleopatholgoical lesions begin to occur is of crucial importance to interpretations. Currently, there is a tendency to interpret such lesions with reference to breastfeeding practices or child care alone. However, if, within a cemetery population the growth of females in the sample was compromised, then growth deficits and paleopathological lesions observed in infants and children could represent an intergenerational legacy, including compromised immune response.[125] Social and economic factors such as poverty may carry a heritable biological legacy, resulting in physiological disadvantage that interacts synergistically with social environment to become mutually reinforcing.[126]

Within the skeletal samples analysed here it is clear that there are not just differences in stature, but also in body proportions between the Romano-British and Anglo-Saxon skeletal samples. It is also apparent that these differences are not only the product of genetics, but higher levels of childhood stress in the earlier period. The timings of these periods of stress in terms of the life course are crucial in terms of those bones that are most affected. Body proportion data can therefore contribute to a biographical understanding of adversity during the period in question. By comparing femur length alone between different cemetery populations, a wealth of additional and pertinent data for examining population well-being is simply being overlooked.

Conclusion

The study of stature as an index of well-being in the Roman world has been beset by problems relating to the lack of standardization in osteological analyses and resulting incompatibility of datasets. Furthermore, there is a tendency within bioarchaeology to adopt a position of parsimonious uniformitarianism, in other words, to apply techniques developed on specific skeletal samples in a universal manner, irrespective of period or place. This approach stems in part from the notion that skeletons are inert biological objects, rather than living tissue, which interacts with social as well as physical environments in a dynamic way.[127] Whilst the practicalities and limitations of osteological analysis must be acknowledged, much more can be done to individualize techniques and to strive to construct more informative osteobiographies. This can only be achieved through an approach that incorporates multiple techniques and integrates them within a life course perspective. A life course approach explicitly acknowledges the cumulative and inter-related nature of individual biographies. Therefore, it is recommended that the anatomical method be used when possible to estimate stature from human skeletal remains and to create population-specific formulae. In addition, a variety of skeletal indices should be calculated to establish the difference in limb segment proportions as well as a comparison of trunk and limb lengths. Differences between sexes within the same population may suggest gendered exposure or response to stress, while differences between populations could point to specific biocultural factors. Integration of these data with palaeopathological information and, where possible, isotopic evidence should aid interpretations.

As the above discussion has highlighted, adversity affecting a young girl can impact on her obstetric future and the stature of her offspring and grandchildren. We need to consider the inter-related fortunes of individuals and the range of variables that effect these, otherwise quite basic, physiological parameters. The above discussion has also highlighted the role of nutrition, including infant feeding strategies, in the determination of adult stature. The Roman Empire was a time of high mobility. The calculation of stature in respect of cemetery populations is therefore a complex matter, given that individuals within the sample are likely to have had childhood origins which were diverse. The way ahead, in our view, is to give close attention to the skeletal remains of infants and children, which have tended to be marginalized. As to the mobility of the population in general, the aim should be to arrive at as concrete an analysis of its nature and extent as is possible, making full use of isotopic evidence (via the stable isotopes of strontium, lead, and oxygen), and DNA analysis.

We are sympathetic to the motivations behind the current trend of directly comparing long bones lengths between skeletal samples as a proxy for stature. While such studies have value, we argue that they also have the potential to mislead because they fail to fully account for differential body proportions, including trunk height. Implicit in such studies is the view that different body proportions are a confounding factor—a source of error. We advocate an approach that instead harnesses this complexity with the aim of attaining additional insights into a variety of biocultural factors affecting the growth of individuals at different stages of the life course.

Acknowledgements

We are grateful to Walter Scheidel for inviting us to contribute to this volume and to Tim Thompson for comments on an earlier draft. I am indebted to Peter Garnsey for his encouragement, patience, and wisdom. Finally we are grateful to an anonymous reviewer for helpful comments on an earlier draft.

Notes

1. Sayson and Hargens 2008.
2. Steckel et al. 2002.
3. Vercelloti et al. 2011.

4. Bogin 2001; Steckel et al. 2002.

5. Steckel 2009.

6. Bozzoli et al. 2009, 647.

7. Kemkes-Grottenhaler 2005.

8. Tanner 1974.

9. Blane 2006.

10. Engels 1850, 158.

11. Chadwick 1965.

12. Barker and et al. 2011.

13. Steckel 2004; Gustafsson et al. 2007; Cardoso and Gomes 2009.

14. E.g., Steckel 2007; Steckel and Floud 1997; Roberts and Cox 2010.

15. Frayer 1980; Feldesman et al. 1990 (hominid remains); Eveleth and Tanner 1990; Ruff 1994; de Beer 2004; Steckel 2004; Gustafsson et al. 2007.

16. Jongman 2007.

17. Kron 2005.

18. Scheidel 2012.

19. Gowland and Garnsey 2010. See also Giannecchini and Moggi-Cecchi 2008.

20. Giannecchini and Moggi-Cecchi 2008.

21. Trotter and Gleser 952.

22. Goldewijk and Jacobs 2013.

23. Jantz and Jantz 1999.

24. Goldewijk and Jacobs 2013.

25. E.g., Trotter and Gleser 1952.

26. Fully 1956; Raxter et al. 2006.

27. Raxter et al. 2006; Maijanen 2009; Maijanen and Niskanen 2010; Shin et al. 2012.

28. Raxter, et al. 2006.

29. Raxter et al. 2006; Fully 1956.

30. Raxter et al. 2007.

31. Auerbach 2011.

32. E.g., Scuilli et al. 1990; Scuilli and Giesen 1993; Formicola and Franceschi 1996; Sciulli and Hetland 2007; Raxter et al. 2008.

33. De Mendonça 2000; Raxter et al. 2006; Vercellotti et al. 2009; Auerbach 2011.

34. Pearson 1899; Trotter and Gleser 1952, 1958; and Trotter 1970.

35. Raxter et al. 2006.

36. E.g., Trotter and Gleser 1952, 1958.

37. Vercellotti et al. 2009.

38. Feldesman et al. 1990; Konigsberg et al. 1998; Holliday 1999; Raxter et al. 2006; Sciulli and Hetland 2007; Auerbach and Ruff 2010.

39. Allbrook 1961; Genovés 1967; de Beer 2004; and Trotter and Gleser 1952.

40. Jantz 1992.

41. Jantz 1992; Jantz et al. 1994.

42. Gowland and Thompson 2013.

43. Raxter et al. 2006.

44. Vercellotti et al. 2009.

45. Trotter and Gleser 1952, 1958; and Trotter 1970.

46. Trotter and Gleser 1952, 1958.

47. Pearson 1899.

48. Giannecchini and Moggi-Cecchi 2008.

49. Auerbach 2011.

50. Auerbach 2011.

51. Goldewijk and Jakobs 2013.

52. Gianecchini and Moggi-Cecchi's 2008.

53. Goldewijk and Jacobs 2013.

54. Goldewijk and Jacobs 2013, 5.

55. A'Hearn et al. 2009.

56. Cole 2003.

57. Karlberg 1998.

58. Gowland and Garnsey 2010.

59. A'Hearn, et al. 2009.

60. Giannecchini and Moggi-Cecchi 2008.

61. Trotter and Gleser 1952, 1958.

62. Kopeke 2002.

63. Kopeke and Baten 2005.

64. E.g., Roberts and Cox 2003; Giannecchini and Moggi-Cecchi 2008. See above, Figure 5.1.

65. Koepke and Baten 2005.

66. King 1999; Scheidel 2012.

67. Takahashi 1984.

68. E.g., Roberts and Cox 2003 Gowland and Redfern 2010; Redfern and DeWitte 2011.

69. E.g., Gianecchini and Moggi-Cecchi 2008.

70. Scheidel 2012.

71. Bozzoli et al. 2009.

72. Walsh 2015.

73. Redfern and Gowland 2012.

74. Lewis 2007.

75. King and Ulijaszek 1999, 161; Humphrey 2000.

76. Humphrey 2000.

77. Bogin et al. 2007.

78. Pelletier 2000; McDade 2003, 2012 (immune system); Chávez 2000; Uauy et al. 2011 (cognition).

79. Tanner 1974; Kuzawa and Bragg 2012.

80. Barker et al. 2001; Barker et al. 2002.

81. Kemkes-Grottenhaler 2005.

82. Sibley et al. 1992.

83. Uauy et al. 2011.

84. Kuzawa and Quinn 2009.

85. Johnston 1986.

86. Kuzawa and Quinn 2009.

87. Chávez and colleagues 2000.

88. Cole 2003.

89. Kuzawa and Quinn 2009.

90. Johnston 1986.

91. Saunders and Barrans 1999, 184.

92. Bozzoli et al. 2009.

93. Lewis and Gowland 2007 (medieval); Gowland et al. 2014 (Roman).

94. Limbs: e.g., Wadsworth et al. 2002. Vertebrae: Watts 2013.

95. Newman and Gowland 2015.

96. Newman and Gowland 2015.

97. E.g., Smith and Buschang 2004.

98. Bogin 1999.

99. Shapland and Lewis 2013, 2014.

100. Shapland, et al. forthcoming.

101. Lewis et al. Forthcoming.

102. Arthur et al. 2016.

103. Gowland 2001.

104. Harlow and Laurence 2002.

105. Saunders 2000.

106. Kirby 2013.

107. Humphrey 2000.

108. Saunders and Hoppa 1993.

109. Maresh 1955.

110. Rohnbogner 2015.

111. Fuller et al. 2006; Powell et al. 2014.

112. Prowse et al. 2008; Powell et al. 2014.

113. Mays et al. 2009.

114. Newman and Gowland 2016.

115. Gowland. Forthcoming.

116. Karlberg 1998.

117. Tardieu 2010, 174.

118. Bogin and Baker 2012.

119. Vercellotti et al. 2011.

120. Pomeroy and colleagues 2012.

121. Pomeroy et al. 2012; Chung and Kuzawa 2014.

122. E.g., transverse diameter of the neural arch; Watts 2013; Newman and Gowland 2015.

123. Walker et al. 2009; DeWitte et al. 2014.

124. Beaumont et al. 2013, 2015; Montgomery et al. 2013.

125. Chung and Kuzawa 2014; Gowland 2015.

126. Chávez et al. 2000.

127. Gowland 2006.

References

A'Hearn, B., F. Peracchi, and G. Vecchi. 2009. "Height and the normal distribution: evidence from Italian military data." *Demography* 46: 1–25. DOI: 10.1353/dem.0.0049.

Allbrook, D. 1961. "The estimation of stature in British and East African males." *Journal of Forensic Medicine* 8: 15–28.

Arthur, N. A., R. L. Gowland, and R. C. Redfern. 2016. "Coming of age in Roman Britain: osteological evidence for pubertal timing." *American Journal of Physical Anthropology* 59: 698–713. DOI: 10.1002/ajpa.22929.

Auerbach, B. M. 2011. "Methods for estimating missing human skeletal element osteometric dimensions employed in the revised Fully technique for estimating stature." *American Journal of Physical Anthropology* 145: 67–80. DOI: 10.1002/ajpa.21469.

Auerbach, B. M., and C. B. Ruff. 2010. "Stature estimation formulae for indigenous North American populations." *American Journal of Physical Anthropology* 141: 190–207. DOI: 10.1002/ajpa.21131.

Barker, D.J.P., et al. 2001. "Size at birth and resilience to effects of poor living conditions in adult life: longitudinal study." *British Medical Journal* 323 (7324): 1273–1276. DOI: 10.1136/bmj.323.7324.1273.

Barker, D.J.P., et al. 2002. "Fetal origins of adult disease: strength of effects and biological basis." *International Journal of Epidemiology* 31: 1235–1239.

Barker, D.J.P, et al. 2011. "How boys grow determines how long they live." *American Journal of Human Biology* 23 (3): 412–416. DOI: 10.1002/ajhb.21165.

Beaumont, J., et al. 2013. "Victims and survivors: identifying survivors of the Great Famine in 19th century London using carbon and nitrogen isotope ratios." *American Journal of Physical Anthropology* 150: 87–98. DOI: 10.1002/ajpa.22179.

Beaumont, J., et al. 2015. "Infant mortality and isotopic complexity: new approaches to stress, maternal health, and weaning." *American Journal of Physical Anthropology* 157: 441–457. DOI: 10.1002/ajpa.22736.

Béguelin, M. 2011. "Stature estimation in a Central Patagonian prehispanic population: development of new models considering specific body proportions." *International Journal of Osteoarchaeology* 21: 150–158. DOI: 10.1002/oa.1117.

Blane, D. 2006. "The life course, the social gradients, and health." In *Social Determinants of Health* (second edition), eds. M. Marmot and R. G. Wilkinson. Oxford: Oxford University Press, 54–77. DOI: 10.1093/acprof:oso/9780198565895.003.04.

Bogin, B. 1999. *Patterns of Human Growth*. Cambridge: Cambridge University Press.

Bogin, B. 2001. *The Growth of Humanity*. New York: Wiley-Liss

Bogin, B., and J. Baker. 2012. "Low birth weight does not predict the ontogeny of relative leg length of infants and children: an allometric analysis of the NHANES III sample." *American Journal of Physical Anthropology* 148: 487–494. DOI: 10.1002/ajpa.22064.

Bogin, B., M. Inês Varela Silva, and L. Rios. 2007. "Life history trade-offs in human growth: adaptation or pathology?" *American Journal of Human Biology* 19 (5): 631–642. DOI: 10.1002/ajhb.20666.

Bozzoli, C., A. Deaton, and C. Quintana-Domeque. 2009. "Adult height and childhood disease." *Demography* 46: 647–669. DOI: 10.1353/dem.0.0079.

Cardoso, H.F.V., and J.E.A. Gomes. 2009. "Trends in adult stature of peoples who inhabited the modern Portuguese territory from the mesolithic to the late 20th century." *International Journal of Osteoarchaeology* 19: 711–725. DOI: 10.1002/oa.991.

Chadwick, E. 1965. *The Sanitary Conditions of the Labouring Population of Great Britain: Report 1842*. Edinburgh: Edinburgh University Press

Chávez, A., C. Martinez, and B. Soberanes. 2000. "The effect of malnutrition on human development: a 24-year study of well-nourished and malnourished children living in a poor Mexican village." In *Nutritional Anthropology: Biocultural Perspectives on Food and Nutrition*, eds. A. H. Goodman, D. L. Dufour, and GH. Pelto. Berkeley: University of California Press, 234–252.

Chung, G .C., and C. W. Kuzawa. 2014. "Intergenerational effects of early life nutrition: maternal leg length predicts offspring placental weight and birth weight among women in rural Luzon, Philippines." *American Journal of Human Biology* 26 (5): 652–659. DOI: 10.1002/ajhb.22579.

Cole, T. J. 2003. "The secular trend in human physical growth: a biological view." *Economics and Human Biology* 1: 161–168. DOI: 10.1016/S1570-677X(02)00033-3.

de Beer, H. 2004. "Observations on the history of the Dutch physical stature from the late-Middle Ages to the present." *Economics and Human Biology* 2: 45–55. DOI: 10.1016/j.ehb.2003.11.001.

de Mendonça, M. C. 2000. "Estimation of height from the length of long bones in a Portuguese adult population." *American Journal of Physical Anthropology* 112: 39–48. DOI: 10.1002/(SICI)1096-8644(200005)112:1<39::AID-AJPA5>3.0.CO;2-#.

DeWitte, S. et al., 2014. "Differential survival among individuals with active and healed periosteal new bone formation." *International Journal of Palaeopathology* 7: 38–44. DOI: 10.1016/j.ijpp.2014.06.001.

Engels, F. 1950. *The Condition of the Working-Class in England in 1844*. With a preface written in 1892. London: George Allen and Unwin Ltd.

Eveleth, P. B., and J. M. Tanner. 1990. *Worldwide Variation in Human Growth*. Second edition. Cambridge: Cambridge University Press.

Feldesman, M. R., J. G. Kleckner, and J. K. Lundy. 1990. "Femur/stature ratio and estimates of stature in mid-and late-Pleistocene fossil hominids." *American Journal of Physical Anthropology* 83: 359–372. DOI: 10.1002/ajpa.1330830309.

Feldesman, M. R., and L. K. Lundy. 1988. "Stature estimates for some African Plio-Pleistocene fossil hominids." *Journal of Human Evolution* 17 (6): 583–596. DOI: 10.1016/0047-2484(88)90086-3.

Formicola, V. 1993. "Stature reconstruction from long bones in ancient population samples: an approach to the problem of its reliability." *American Journal of Physical Anthropology* 90: 351–358. DOI: 10.1002/ajpa.1330900309.

Formicola, V., and M. Franceschi. 1996. "Regression equations for estimating stature from long bones of early Holocene European samples." *American Journal of Physical Anthropology* 100: 83–88. DOI: 10.1002/(SICI)1096-8644(199605)100:1<83::AID-AJPA8>3.0.CO;2-E.

Frayer, D. W. 1980. "Sexual dimorphism and cultural evolution in the Late Pleistocene and Holocene of Europe." *Journal of Human Evolution* 9: 399–415. DOI: 0.1016/0047-2484(80)90050-0.

Fully, G. 1956. "Une nouvelle méthode de détermination de lataille." *Annales de Medecine Legale* 35: 266–273.

Fuller, B. T., et al. 2006. "Isotopic evidence for breastfeeding and possible adult dietary differences from late/sub Roman Britain." *American Journal of Physical Anthropology* 129: 45–54. DOI: 10.1002/ajpa.20244.

Gaskell, P. 1833. *The Manufacturing Population of England*. London: Baldwin and Cradock.

Genoves, S. 1967. "Proportionality of the long bones and their relation to stature among Mesoamericans." *American Journal of Physical Anthropology* 26: 67–78. DOI: 10.1002/ajpa.1330260109.

Giannecchini, M., and J. Moggi-Cecchi. 2008. "Stature in archaeological samples from Central Italy: method issues and diachronic changes." *American Journal of Physical Anthropology* 135: 284–292. DOI: 10.1002/ajpa.20742.

Goldewijk, G. K., and J. Jacobs. 2013. *The Relation between Stature and Long Bone Length in the Roman Empire*. Groningen: University of Groningen.

Gowland, R. L. 2001. "Playing dead: implications of mortuary evidence for the social construction of childhood in Roman Britain." In *Proceedings of the Tenth Annual Theoretical Roman Archaeology Conference*, eds. G. Davies, A. Gardner, and K. Lockyear Oxford: Oxbow, 152–168.

Gowland, R. L. 2006. "Age as an aspect of social identity: the archaeological funerary evidence." In *Social Archaeology of Funerary Remains*, eds. R. L. Gowland and C. Knüsel. Oxford: Oxbow, 143–154.

Gowland, R. L. 2015. "Entangled lives: implications of the developmental origins of health and disease (DOHaD) hypothesis for bioarchaeology and the life course." *American Journal of Physical Anthropology* 158: 530–40. DOI: 10.1002/ajpa.22820.

Gowland, R. L., A. T. Chamberlain, and R. C. Redfern. 2014. "On the brink of being: re-evaluating infant death and infanticide in Roman Britain." In *Infant Health and Death in Roman Italy and Beyond*, eds. M. Carroll, and E.-J. Graham. Portsmouth, RI: Journal of Roman Archaeology, 69–88.

Gowland, R. L. and P. Garnsey. 2010. "Skeletal evidence for health, nutritional status and malaria in Rome and the empire." *Roman Diasporas. Archaeological Approaches to Mobility and Diversity in the Roman Empire*, ed. H. Eckardt. Portsmouth, RI: Journal of Roman Archaeology, 131–156.

Gowland, R. L., and S. L. Newman. in press. "Children of the revolution: childhood health inequalities and the life course during industrialisation of the 18th to 19th centuries." In *Children and Childhood in the Past,* eds. P. Beauchesne and S. Agarwal. Gainesville: University of Florida Press.

Gowland, R. L., and R. C. Redfern. 2010. "Childhood health at the core and periphery of the Roman Empire." *Childhood in the Past* 3: 15–42

Gowland, R. L., and T.J.U. Thompson. 2013. *Human Identity and Identification.* Cambridge: Cambridge University Press.

Gustafsson, A., et al. 2007. "Stature and sexual dimorphism in Sweden, from the 10th to the end of the 20th century." *American Journal of Human Biology* 19 (6): 861–870. DOI: 10.1002/ajhb.20657.

Harlow, M., and R. Laurence. 2002. *Growing Up and Growing Old in Ancient Rome: A Life Course Approach.* London: Routledge.

Holliday, T. W. 1999. "Brachial and crural indices of European Late Upper Paleolithic and Mesolithic humans." *Journal of Human Evolution* 36: 549–566. DOI: 10.1006/jhev.1998.0289.

Humphrey, L. T. 2000. "Growth studies of past populations: an overview and an example." In *Human Osteology in Archaeology and Forensic Science,* eds. M. Cox and S. Mays. London: Greenwich Medical Media, 23–38.

Jantz, R. L. 1992. "Modification of the Trotter and Gleser female stature estimation formulae." *Journal of Forensic Sciences* 37: 1230–1235. DOI: 10.1520/JFS13310J.

Jantz, R. L., D. R. Hunt, and L. Meadows. 1994. "Maximum length of the tibia: how did Trotter measure it?" *American Journal of Physical Anthropology* 93: 525–528. DOI: 10.1002/ajpa.1330930410.

Jantz, L. M., and R. L. Jantz. 1999. "Secular change in long bone length and proportion in the United States, 1800–1970." *American Journal of Physical Anthropology* 111: 57–67. DOI: 10.1002/(SICI)1096-8644(199909)110:1<57::AID-AJPA5>3.0.CO;2-1.

Johnston, F. E. 1986. "Somatic growth of the infant and preschool child." In *Human Growth,* eds. F. Falkner, and J. M. Tanner. New York: Plenum Press, 3–24.

Jongman, W. 2007. "Gibbon was right: the decline and fall of the Roman economy." In *Crises and the Roman Empire,* eds. O. Hekster, G. de Kleijn, and D. Slootjes. Brill: Leiden, 183–199. DOI: 10.1163/ej.9789004160507.i-448.38.

Jongman, W. M. 2009. "Archaeology, demography and Roman economic growth." In *Quantifying the Roman Economy: Problems and Methods,* eds. A. Bowman, and A. Wilson. Oxford: Oxford University Press, 115–26. DOI: 10.1093/acprof:oso/9780199562596.003.0004.

Karlberg, J. 1998. "The human growth curve." In *The Cambridge Encyclopaedia of Human Growth and Development,* eds. S. Ulijaszek, F. Johnston, and M. Preece. Cambridge: Cambridge University Press, 108–113.

Kemkes-Grottenthaler, A. 2005. "The short die young: the inter-relationship between stature and longevity—evidence from skeletal remains." *American Journal of Physical Anthropology* 128: 340–347. DOI: 10.1002/ajpa.20146.

King, A. 1999. "Meat diet in the Roman world: a regional inter-site comparison of the mammal bones." *Journal of Roman Archaeology* 12: 168–202. DOI: 10.1017/S1047759400017979.

King, S. E., and S. J. Ulijaszek. 1999. "Invisible insults during growth and development." In *Human Growth in the Past,* eds. R. D. Hoppa and C. M. Fitzgerald. Cambridge: Cambridge University Press, 161–182.

Kirby, P. 2013. *Child Workers and Industrial Health in Britain, 1780–1850.* London: Boydell Press.

Koepke, N. 2002. "Regional differences and temporal development of the quality of nutrition in the Roman provinces of Germania and Raetia from the first century to the fourth century AD." *Proceedings of the XIII International Economic History Association Congress.* Buenos Aires.

Koepke, N., and J. Baten. 2005. "The biological standard of living in Europe during the last two millennia." *European Review of Economic History* 9: 61–95. DOI: 10.1017 /S1361491604001388.

Konigsberg, L. W., et al. 1998. "Stature estimation and calibration: Bayesian and maximum likelihood perspectives in physical anthropology." *Yearbook of Physical Anthropology* 41: 65–92. DOI: 10.1002/(SICI)1096-8644(1998)107:27+<65::AID-AJPA4>3.0.CO;2-6.

Kron, G., 2005. "Anthropometry, physical anthropology, and the reconstruction of ancient health, nutrition, and living standards." *Historia* 54: 68–83. DOI: 10.2307/4436756.

Kuzawa, C. W., and J. M. Bragg. 2012. "Plasticity in human life history strategy: implications for contemporary human variation and the end of genus *Homo*." *Current Anthropology* 53: 369–382. DOI: 10.1086/667410.

Kuzawa, C. W., and W. A. Quinn. 2009. "Developmental origins of adult function and health: evolutionary hypotheses." *Annual Review of Anthropology* 38: 131–147. DOI: 10.1146 /annurev-anthro-091908-164350.

Lewis, M. E. 2007. *The Bioarchaeology of Children.* Cambridge: Cambridge University Press.

Lewis, M. E. Forthcoming. "Work and the adolescent in Medieval England (AD 900–1550): the osteological evidence." *Medieval Archaeology.*

Lewis, M. E., and R. L. Gowland. 2007. "Brief and precarious lives: infant mortality in contrasting sites from medieval and post-medieval England (AD850–1859)." *American Journal of Physical Anthropology* 134: 117–129. DOI: 10.1002/ajpa.20643.

Maijanen, H. 2009. "Testing anatomical methods for stature estimation on individuals from the wm bass donated skeletal collection." *Journal of Forensic Science* 54:746–752.

Maijanen, H., and M. Niskanen. 2010. "New regression equations for stature estimation for Medieval Scandinavians." *International Journal of Osteoarchaeology* 20: 472–480. DOI: 10.1002/oa.1071.

Maresh, M. M. 1955. "Linear growth of long bones of extremities from infancy through adolescence: continuing studies." *American Journal of Diseases in Children* 89: 725–742. DOI: 10.1001/archpedi.1955.02050110865010.

Mays, S., R. Ives, and M. Brickley. 2009. "The effects of socioeconomic status on endochondral and appositional bone growth, and acquisition of cortical bone in children from 19th century Birmingham, England." *American Journal of Physical Anthropology* 140: 410–416. DOI: 10.1002/ajpa.21076.

McDade, T. W. 2003. "Life history theory and the immune system steps toward a human ecological immunology." *American Journal of Physical Anthropology* 46: 100–125. DOI: 10.1002/ajpa.10398.

McDade, T. W. 2012. "Early environments and the ecology of inflammation." *Proceedings of the National Academy of Sciences of the United States of America* 109 (2): 17281–17288. DOI: 10.1073/pnas.1202244109.

Minozzie, S., et al. 2013. "'The Roman Giant': overgrowth syndrome in skeletal remains from the Imperial Age." *International Journal of Osteoarchaeology* 25: 574–584. DOI: 10.1002/oa.2322.

Molleson, T. 1995. "Rates of ageing in the eighteenth century." In *Grave Reflections: Portraying the Past through Cemetery Studies*, eds. S. R. Saunders and A. Herring. Toronto: Canadian Scholars Press, 199–222.

Montgomery, J., et al. 2013. "Strategic and sporadic marine consumption at the onset of the Neolithic: increasing temporal resolution in the isotope evidence." *Antiquity* 87: 1060–1072. DOI: 10.1017/S0003598X00049863.

Newman, S. L., and R. L. Gowland. 2015. "The use of non-adult vertebral dimensions as indicators of growth disruption and non-specific health stress in skeletal populations." *American Journal of Physical Anthropology* 158: 155–164. DOI: 10.1002/ajpa.22770.

Newman, S. L., and R. L. Gowland. Forthcoming. "Dedicated followers of fashion? Bio-archaeological perspectives on socio-economic status, inequality, and health in urban children from the Industrial Revolution." *International Journal of Osteoarchaeology*.

Pearson, K. 1899. "Mathematical contribution to the theory of evolution: on the reconstruction of the stature of prehistoric races." *Philosophical Transactions of the Royal Society London* 192: 169–244. DOI: 10.1098/rsta.1899.0004.

Pelletier, D. L. 2000. "The potentiating effects of malnutrition on child mortality: epidemiologic evidence and policy implications." In *Nutritional Anthropology: Biocultural Perspectives on Food and Nutrition*, eds. A. H. Goodman, D. L. Dufour, and G. H. Pelto. Berkeley: University of California, 227–234.

Pomeroy, E., et al. 2012. "Trade-offs in relative limb length among Peruvian children: extending the Thrifty Phenotypes hypothesis to limb proportions." *PLoS ONE* 7 (12): e51795. DOI: 10.1371/journal.pone.0051795

Powell, L. A., et al. 2014. "Infant feeding practices in Roman London: evidence from isotope analysis." In *Infant Health and Death in Roman Italy and Beyond*, eds. M. Carroll and E.-J. Graham. Portsmouth, RI: Journal of Roman Archaeology, 89–110.

Prowse, T. L., et al. 2008. "Isotopic and dental evidence for infant and young child feeding practices in an imperial Roman skeletal sample." *American Journal of Physical Anthropology* 137: 294–308. DOI: 10.1002/ajpa.20870.

Raxter, M.H., B.M. Auerbach, and C.B. Ruff. 2006. "Revision of the Fully technique for estimating stature." *American Journal of Physical Anthropology* 130: 374–384. DOI: 10.1002/ajpa.20361.

Raxter, M. H., C. B. Ruff, and B. M. Auerbach. 2007. "Technical note: Revised Fully stature estimation technique." *American Journal of Physical Anthropology* 133: 817–818. DOI: 10.1002/ajpa.20588.

Raxter, M. H., et al. 2008. "Stature estimation in Ancient Egyptians: a new technique based on anatomical reconstruction of stature." *American Journal of Physical Anthropology* 136: 147–155. DOI: 10.1002/ajpa.20790.

Redfern, R. C., and S. N. Dewitte. 2011. "Status and health in Roman Dorset: the effect of status on risk of mortality in post-conquest populations." *American Journal of Physical Anthropology* 146: 197–208. DOI: 10.1002/ajpa.21563.

Redfern, R. C., and R. L. Gowland. 2012. "A bioarchaeological perspective on the pre-adult stages of the life course: implications for the care and health of children in the Roman Empire." In *Families in the Roman and Late Antique World*, eds. M. Harlow and L. L. Loven. New York: Continuum, 111–140.

Roberts, C., and M. Cox. 2007. "The impact of economic intensification and social complexity on human health in Britain from 6000BP (Neolithic) and the introduction of farming to the mid-nineteenth century AD." In *Ancient Health: Skeletal Indicators of Agricultural and Economic Intensification*, eds. M. N. Cohen and G.M.M. Crank-Kramer. Gainesville: University of Florida Press, 149–163.

Roberts, C.A., and M. Cox. 2003. *Health and Disease in Britain: From Prehistory to the Present Day*. Gloucester: Sutton.

Rohnbogner, A. 2015. "Dying young: a palaeopathological analysis of child health in Roman Britain." PhD Dissertation, University of Reading, England.

Roseboom, T. J., J.H.P. van der Meulen, and A.C.J. Ravelli. 2001. "Effects of prenatal exposure to the Dutch famine on adult disease in later life: an overview." *Molecular and Cellular Endocrinology* 185: 93–98.

Ross, M. G., and M. H. Beall. 2008. "Adult sequelae of intrauterine growth restriction." *Seminars in Perinatology* 32: 213–218. DOI: 10.1053/j.semperi.2007.11.005.

Ruff, C. B. 1994. "Morphological adaptation to climate in modern and fossil hominids." *American Journal of Physical Anthropology* 37: 65–107. DOI: 10.1002/ajpa.1330370605.

Saunders, S. R. 2000. "Subadult skeletons and growth related studies." In *Skeletal Biology of Past Peoples: Research Methods*, eds. M. A. Katzenberg and S. R. Saunders. New York: Wiley-Liss, 135–161.

Saunders, S. R., and L. Barrans. 1999. "What can be done about the infant category in skeletal samples?" In *Human Growth in the Past*, eds. R. D. Hoppa and C. M. Fitzgerald. Cambridge: Cambridge University Press, 183–209.

Saunders, S. R. and R. D. Hoppa. 1993. "Growth deficit in survivors and non-survivors: biological correlates of mortality bias in subadult skeletal samples." *Yearbook of Physical Anthropology* 36: 127–151.

Sayson, J. V., and A. R. Hargens. 2008. "Pathophysiology of lower back pain during exposure to microgravity." *Aviation, Space and Environmental Medicine* 79: 365–373. DOI: 10.3357/ASEM.1994.2008.

Scheidel, W. 2012. "Physical well-being." In *The Cambridge Companion to the Roman Economy*, ed. W. Scheidel. Cambridge: Cambridge University Press, 321–333. DOI: 10.1017/CCO9781139030199.020.

Sciulli, P. W., and M. J. Giesen. 1993. "Brief Communication: an update on stature estimation in Prehistoric Native Americans of Ohio." *American Journal of Physical Anthropology* 92: 395–399. DOI: 10.1002/ajpa.1330920309.

Sciulli, P. W., and B. M. Hetland. 2007. "Stature estimation for Prehistoric Ohio Valley Native American populations based on revisions of the Fully Technique." *Archaeology of Eastern North America* 35: 105–113.

Sciulli, P. W., K. N. Schneider, and M. C. Mahaney. 1990. "Stature estimation in Prehistoric Native Americans of Ohio." *American Journal of Physical Anthropology* 83: 275–280. DOI: 10.1002/ajpa.1330830302.

Shapland F., M. E. Lewis. 2013. "Brief Communication: a proposed osteological method for the estimation of pubertal stage in human skeletal remains." *American Journal of Physical Anthropology* 151: 302–310.

Shapland, F., and M. E. Lewis. 2014. "Brief Communication: a proposed method for the assessment of pubertal stage in human skeletal remains using cervical vertebrae maturation." *American Journal of Physical Anthropology* 153: 144–153. DOI: 10.1002/ajpa.22416.

Shapland, F., M. Lewis, and R. Watts. Forthcoming. "The lives and deaths of young medieval women: the osteological evidence." *Medieval Archaeology*.

Shin, D. H, et al. 2012. "Ancient-to-modern secular changes in Korean stature." *American Journal of Physical Anthropology* 147: 433–442.

Sibley, L. M., G. J. Armelagos, and D. P. Van Gerven. 1992. "Obstetric dimensions of the true pelvic in a medieval population from Sudanese Nubia." *American Journal of Physical Anthropology* 60: 279–317. DOI: 10.1002/ajpa.1330890403.

Sletner, L., et al. 2014. "Maternal life course socio-economic position and offspring body composition at birth in a multi-ethnic population." *Paediatric and Perinatal Epidemiology* 28: 445–454. DOI: 10.1111/ppe.12137.

Smith, S. L., and P. H. Buschang. 2004. "Variation in longitudinal diaphyseal long bone growth in children three to ten years of age." *American Journal of Human Biology* 16 (6): 648–657. DOI: 10.1002/ajhb.20077.

Steckel, R. H. 2004. "New light on the 'Dark Ages': the remarkably tall stature of Northern European Men during the Medieval Era." *Social Science History* 28: 211–228. DOI: 10.1215/01455532-28-2-211.

Steckel, R. H. 2005. "Health and nutrition in the Preindustrial Era: insights from a millennium of average heights in northern Europe." In *Living Standards in the Past*, eds.

R.C. Allen, T. Bengtsson, and M. Dribe. Oxford: Oxford University Press, 227–254. DOI: 10.1093/0199280681.003.0010.

Steckel, R. H. 2007. *A Pernicious Side of Capitalism: The Care and Feeding Slave Children.* Available at http://web.econ.ohio-state.edu/rsteckel/vita.pdf.

Steckel, R. H. 2009. "Heights and human welfare: recent developments and new directions." *Explorations in Economic History* 46: 1–23. DOI: 10.3386/w14536.

Steckel, R. H., and R. Floud. 1997. *Health and Welfare during Industrialization.* Chicago: University of Chicago Press.

Steckel, R. H., et al. 2002. "Skeletal health in the Western Hemisphere from 4000 B.C. to the present." *Evolutionary Anthropology* 11: 142–155. DOI: 10.1002/evan.10030.

Takahashi, E. 1984. "Secular trend in milk consumption and growth in Japan." *Human Biology* 56 (3): 427–437.

Tardieu, C. 2010. "Development of the human hind limb and its importance for the evolution of bipedalism." *Evolutionary Anthropology* 19: 174–186. DOI: 10.1002/evan.20276.

Tanner, J. M. 1974. "Variation in growth and maturity of newborns." In *The Effect of the Infant on its Caregiver*, eds. M. Lewis and L. A. Rosenblum. New York; London: Wiley-Interscience, 77–103.

Trotter, M. 1970. "Estimation of stature from intact long limb bones." In *Personal Identification in Mass Disasters*, ed. T. D. Stewart. Washington, DC: Smithsonian Press, 71–83.

Trotter, M., and G. C. Gleser. 1952. "Estimation of stature from long bones of American whites and Afroamericans." *American Journal of Physical Anthropology* 10: 463–512.

Trotter, M., and G. C. Gleser. 1958. "A re-evaluation of estimation of stature based on measurements of stature taken during life and of long bones after death." *American Journal of Physical Anthropology* 16: 79–123. DOI: 10.1002/ajpa.1330160106.

Uauy, R., J. Kain, and C. Corvalan. 2011. "How can the developmental origin of health and disease (DOHaD) hypothesis contribute to improving health in developing countries. "*American Journal of Clinical Nutrition* 96: 1759S–1764S. DOI: 10.3945/ajcn.110.000562.

Vercellotti, G., et al. 2009. "Stature estimation in an early Medieval (XI–XII c.) Polish population: testing the accuracy of regression equations in a bioarchaeological sample." *American Journal of Physical Anthropology* 140: 135–142. DOI: 10.1002/ajpa.21055.

Vercellotti, G., et al. 2011. "Intrapopulation variation in stature and body proportions: social status and sex differences in an Italian medieval population (Trino Vercellese, VC)." *American Journal of Physical Anthropology* 145: 203–214. DOI: 10.1002/ajpa.21486.

Wadsworth, M. E., et al. 2002. "Leg and trunk length at 43 years in relation to childhood health, diet and family circumstances: evidence from the 1946 national birth cohort." *International Journal of Epidemiology* 31: 383–390. DOI: 10.1093/ije/31.2.383.

Walker, P. L., et al. 2009. "The causes of porotic hyperostosis and cribra orbitalia: a reappraisal of the iron deficiencyanemia hypothesis." *American Journal of Physical Anthropology* 139: 109–125.

Walsh, J. 2015. "Normal bone physiology, remodeling and its hormonal regulation." *Surgery* 33: 1–6. DOI: 10.1016/j.mpsur.2014.10.010.

Watts, R. 2011. "Non-specific indicators of stress and their relationship to age-at-death in medieval York: using stature and vertebral canal neural size to examine the effects of stress occurring during different stages of development." *International Journal of Osteoarchaeology* 21: 568–576. DOI: 10.1002/oa.1158.

Watts, R. 2013. "Childhood development and adult longevity in an archaeological population from Barton-upon-Humber, Lincolnshire, England." *International Journal of Palaeopathology* 3: 95–104. DOI: 10.1016/j.ijpp.2013.05.001.

Werdelin, L. 1985. "The stature of some medieval Swedish populations." *Fornvännen* 80: 133–141.

CHAPTER SIX

Ancient DNA

Noreen Tuross & Michael G. Campana

Introduction

History leaves traces in the human genome as well as those of pathogens and domesticates. While much can be gleaned from the genetic fossils preserved in extant genomes (see Chapter 7), genomes are palimpsests, with more recent events overwriting previous ones in part. The study of ancient DNA (aDNA)—DNA preserved in archaeological, paleontological, and museum sources—permits us to investigate the genome before and after historic events and observe how it evolves in real time. The field of aDNA also has a palimpsestic nature in which older results are not only extended and revised, but totally discarded due to rapid technological advances.

In this chapter, we examine how aDNA analysis has helped reconstruct ancient history, with a particular focus on cases investigating Roman history. We briefly describe the biochemistry of ancient DNA and the history of its research. Through several key case studies, we show the potential for aDNA research to clarify the course of ancient history, and also highlight some of its weaknesses and limitations.

Biochemistry of Ancient DNA

In living organisms, DNA (deoxyribonucleic acid) is the biomolecule that encodes the instructions to produce polypeptides and functional RNAs (ribonucleic acids). It is the primary molecule of heredity that determines biological traits passed from parents to offspring. DNA contains both coding and noncoding regions. In coding regions, DNA sequences encode genes, the instructions to produce polypeptides and functional RNAs. Conversely, noncoding regions do not contain genes, but may serve other roles such as gene regulation. In its primary structure, DNA is a directional, linear, polymeric biomolecule comprised of nucleotides. Each nucleotide consists of a deoxyribose sugar, a phosphate, and one of four nitrogenous bases: adenine (A), thymine (T), guanine (G), and cytosine (C). DNA strands are directional: strand synthesis proceeds from the 5'- to 3'-terminus of the molecule. Typically, DNA in cells is double-stranded: each strand is paired with a complementary DNA strand running in the opposite orientation via hydrogen bonds between corresponding nucleotides. A and T are complementary, as are G and C.

In eukaryotes, the majority of DNA (and thus genes) is contained within the cell nucleus (nuclear DNA). Nuclear DNA is bound to proteins (histones) and packaged into chromosomes. Each nucleus contains two copies of each non-sex chromosome, one from each of the organism's parents. Sex chromosomes are also inherited biparentally, but the number of each chromosome varies by the sex-determination system. In mammals, females have two copies of the X chromosome (one from each parent), while males have a maternally inherited X and a paternally inherited Y chromosome. Conversely, in birds, males have two Z chromosomes, while females have a paternally inherited Z and a maternally inherited W chromosome. Additionally, eukaryotic cells contain mitochondrial DNA (mtDNA), a circular piece of DNA found in the mitochondria (cellular compartments responsible for energy generation): mtDNA are almost exclusively maternally inherited and can exist in hundreds to thousands of copies per cell. Many organisms have additional extranuclear DNA. For instance, plants have chloroplastid DNA within their chloroplasts (organelles responsible for photosynthesis).

In living cells, DNA is under constant attack (e.g., from ultraviolet light exposure, hydrolysis, reactive oxygen species, etc.). DNA repair mechanisms prevent DNA from accumulating damage. Upon death, cellular repair mechanisms stop. Cell membranes break down, releasing digestive enzymes into the cytoplasm. These enzymes rapidly destroy DNA unless they are inactivated via mechanisms including freezing and desiccation. Even if enzymatic digestion is arrested, DNA decays over time due to hydrolytic and oxidative damage.[1] The rates of these reactions are environmentally dependent. DNA is most likely to survive in stable, cold, dry environments (e.g., permafrost) with neutral to slightly basic pH and high salt concentrations.[2] Protection from ultraviolet light (which promotes oxidation) and microbial and fungal attack also increases the probability of DNA preservation.[3]

Surviving endogenous ancient DNA is fragmented into short pieces (< 500 base pairs [bp][4]). The majority of aDNA research has focused on DNA fragments between 100 and 200 bp.[5] Recent research using high-throughput sequencing methodologies, however, has shown that the majority of endogenous aDNA molecules in most samples are less than 100 bp in length.[6] These molecules typically bear significant damage including single-strand nicks, miscoding and blocking lesions, and crosslinks with other biomolecules.[7] Although some lesions are reparable (e.g., via phenacylthiazolium bromide treatment to remove protein crosslinks),[8] most prevent the affected DNA from being analyzed.

Due to the small quantities of analyzable target DNA molecules in most specimens, contamination with nonendogenous DNA poses the greatest risk to aDNA research.[9] Endogenous aDNA rarely exceeds 10% of the total DNA content in ancient samples.[10] To combat contamination, laboratories studying ancient DNA developed "criteria of authenticity" to ensure reliable results.[11] Although criteria vary between research groups, almost all laboratories employ control reactions to weed out contaminants, replication to ensure that results are trustworthy, and separation of ancient DNA preparations from modern DNA laboratory space to limit cross-contamination.[12]

The predominant source of sequence error deriving from authentic aDNA is cytosine-to-thymine (C→T) transitions.[13] C→T transitions occur when endogenous cytosine residues are deaminated to uracil. During PCR, uracil is read and replaced as thymine by DNA polymerase.[14] These lesions occur in a predictable pattern in aDNA strands: C→T transitions are more frequent at the 5′ end of the DNA molecule, with the complementary guanine-to-adenine (G→A) transition appearing more frequently at the 3′ end.[15] The presence of this pattern is now used as a criterion of ancient DNA authenticity in paleogenomic analyses.[16] Additionally, treatment of DNA extracts with uracil *N*-glycosylase before PCR can reduce the presence of C→T transitions by cleaving DNA molecules at uracil residues.[17] Recently, some C→T transitions that survive uracil *N*-glycosylase treatment have been shown to be due to ancient DNA methylation.[18] This discovery opens the door to investigations of ancient epigenomics, which could provide information on development, nutritional status, and overall health in the past.[19]

History of Ancient DNA Research

In 1984, Russell Higuchi and colleagues demonstrated that small fragments of endogenous DNA were preserved in a 140-year-old museum skin of the extinct quagga (*Equus quagga quagga*), a subspecies of zebra. DNA extracted from the skin was both of low molecular weight (< 500 bp) and consisted primarily of contaminants. The team had to screen 25,000 bacterial clones of randomly inserted quagga-skin DNA to find two containing recognizable mitochondrial DNA sequences!

Although Higuchi and co-workers had proved the survival of ancient DNA, it was not until the invention of the polymerase chain reaction (PCR) in the mid-1980s by Kary Mullis and his colleagues at Cetus Corporation that systematic study of this biomolecule became possible.[20] Using synthesized

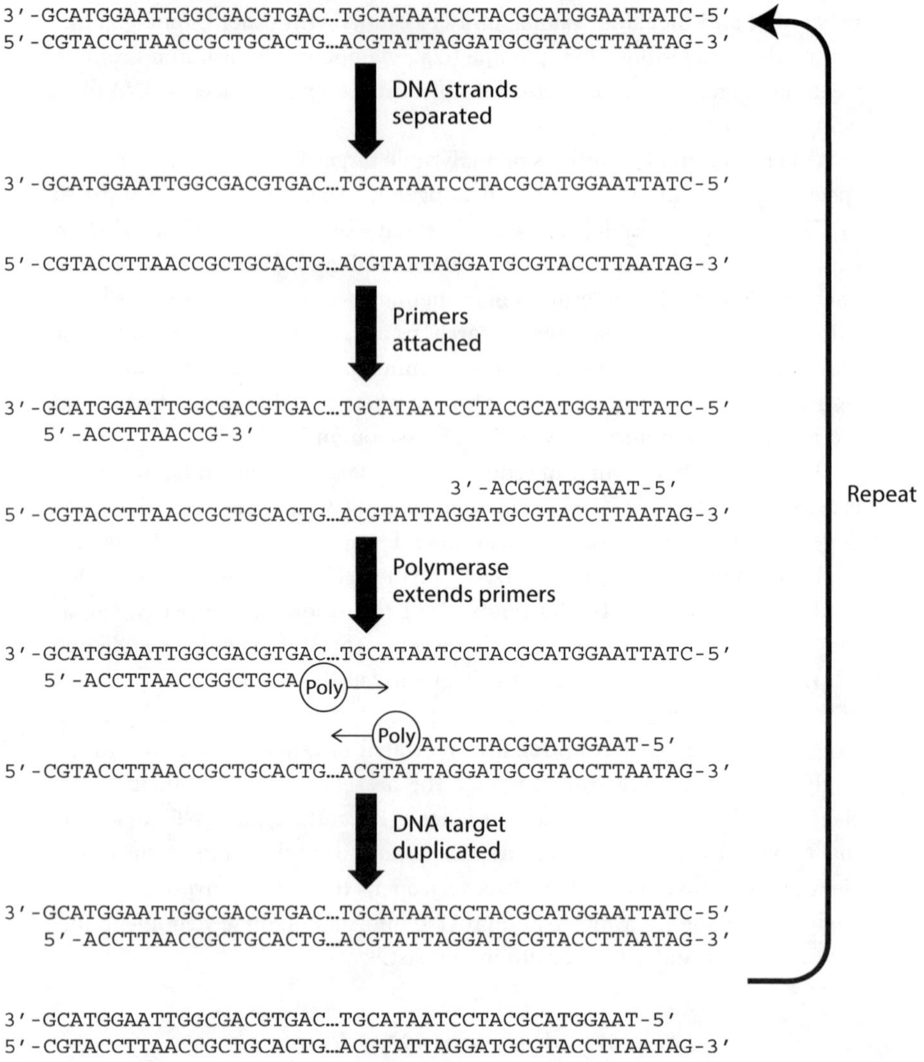

```
3'-GCATGGAATTGGCGACGTGAC...TGCATAATCCTACGCATGGAATTATC-5'
5'-CGTACCTTAACCGCTGCACTG...ACGTATTAGGATGCGTACCTTAATAG-3'
```

                    DNA strands
                    separated

```
3'-GCATGGAATTGGCGACGTGAC...TGCATAATCCTACGCATGGAATTATC-5'
```

```
5'-CGTACCTTAACCGCTGCACTG...ACGTATTAGGATGCGTACCTTAATAG-3'
```

                    Primers
                    attached

```
3'-GCATGGAATTGGCGACGTGAC...TGCATAATCCTACGCATGGAATTATC-5'
   5'-ACCTTAACCG-3'
```

```
                                    3'-ACGCATGGAAT-5'
5'-CGTACCTTAACCGCTGCACTG...ACGTATTAGGATGCGTACCTTAATAG-3'
```

                    Polymerase   Repeat
                    extends primers

```
3'-GCATGGAATTGGCGACGTGAC...TGCATAATCCTACGCATGGAATTATC-5'
   5'-ACCTTAACCGGCTGCA(Poly)⟶
```

```
                         ⟵(Poly)ATCCTACGCATGGAAT-5'
5'-CGTACCTTAACCGCTGCACTG...ACGTATTAGGATGCGTACCTTAATAG-3'
```

                    DNA target
                    duplicated

```
3'-GCATGGAATTGGCGACGTGAC...TGCATAATCCTACGCATGGAATTATC-5'
   5'-ACCTTAACCGCTGCACTG...ACGTATTAGGATGCGTACCTTAATAG-3'
```

```
3'-GCATGGAATTGGCGACGTGAC...TGCATAATCCTACGCATGGAAT-5'
5'-CGTACCTTAACCGCTGCACTG...ACGTATTAGGATGCGTACCTTAATAG-3'
```

FIGURE 6.1. Diagram of the polymerase chain reaction. DNA molecules are duplicated
using repeated cycles of heat denaturation, annealing of short oligonucleotide
primers, and extension of the primers using thermostable DNA polymerase.

single-stranded DNA molecules of known sequence ("primers") and a thermostable DNA polymerase, PCR copies specific DNA regions of interest up to
billions of times, permitting the analysis of even single molecules of DNA (Figure 6.1). Repeated cycles of heating and cooling render DNA single-stranded,
attach the primers to the DNA target and then synthesize complementary
DNA strands using the primers as starting points for DNA polymerase. The
newly synthesized double-stranded DNA copies can then be used as templates
for the next heating and cooling cycle.

The application of PCR techniques to archaeological, historical and pa-leontological questions triggered an explosion of ancient DNA research in the early 1990s. By the middle of the decade, the fervor was tempered by the discovery that many of the most spectacular discoveries—notably, DNA from Cretaceous dinosaurs, insects and bacteria preserved in amber, and Miocene plant fossils—were the results of laboratory contamination.[21] These cases led to the development of the strict "criteria of authenticity" that ensured reliability and replicability of aDNA analyses.

Despite the advent of PCR, aDNA research was hampered by its inability to generate large quantities of reliable data. The vast majority of aDNA research focused on mitochondrial DNA since its high copy number per cell facilitated PCR amplification and its maternal inheritance simplified phylogenetic analysis. This situation changed in 2005 when high-throughput sequencing-by-synthesis sequencers were put on the market.[22] Instead of targeting short pieces of known DNA regions, these approaches could potentially sequence all the molecules in a sample by attaching specific DNA adapters to the ends of the original DNA strands, immobilizing the adapted molecules on a bead or plate using these adapters, and then sequencing millions to billions of molecules in simultaneous sequencing reactions.[23] These technologies could sequence entire ancient genomes,[24] but at the cost of shorter and less-accurate sequence reads and sequencing large quantities of unwanted contaminants present in the samples. These downsides have since been mitigated by the improvement of the sequencers such that their sequence length and quality rival those of traditional PCR-based techniques and the development of capture technologies that help exclude unwanted contaminants from sequencing pools.[25]

Population Relationships:
Ancient Etruscans and Modern Tuscans

Documentary and archaeological evidence often shed little light on the genetic origins of ancient peoples and their relationships to historical and modern populations. Ancient DNA research can contribute to our understanding of ancient history by determining the genetic affiliation of ancient peoples and the extent to which extant human populations are their direct descendants.

A notable and debated case has been that of the ancient Etruscans, a non-Indo-European population of preclassical central Italy (Etruria). Archaeological, paleoanthropological, and textual records provide little evidence as to the Etruscans' genetic affinities, although (at least) cultural exchange has been documented with eastern Mediterranean societies.[26] Furthermore, the Etruscans' relationship to modern-day Tuscans (the current inhabitants of the area) is unclear. To resolve these questions, Vernesi and colleagues analyzed mitochondrial control region (a rapidly evolving, noncoding locus frequently used to infer intraspecific phylogenies) sequences from 30 Etruscan individuals.[27] The specimens were dated archaeologically to seventh to third century B.C. and derived from various towns within the former Etruria. Comparison of the ancient Etruscan haplotypes (unique variants of a genetic locus) to those of modern European and Mediterranean populations showed that the ancient Etruscans were more closely related to eastern Mediterranean peoples than to extant Tuscans and other modern Italians. Based on these data, Vernesi and colleagues argue that the Etruscans derived at least partially from eastern Mediterranean populations and that they underwent population replacement after incorporation into the Roman Empire. Computer simulations exploring a variety of population parameters (including mutation rate, population size, migration, social stratification, etc.) also supported population replacement.[28] Sequencing of 27 medieval Tuscans' mitochondrial control regions (tenth to fifteenth century A.D.) further supported population replacement before 1000 A.D. by showing that modern and medieval Tuscans are related to each other, but do not descend from ancient Etruscans.[29] Notably, many of the Etruscan haplotypes are not found in modern databases, which may be the result of lineage extinction after assimilation,[30] but also has been interpreted as evidence of DNA sequence infidelity due to DNA damage or contamination.[31]

The possibility of DNA sequence inaccuracy and the small sample sizes employed in the study of the ancient Etruscans has caused debate about some of the conclusions of these studies. While Barbujani and colleagues and Guimaraes and colleagues deny that DNA sequence error affected their analyses, Achilli and colleagues argue that the discontinuity between the Etruscans and modern Tuscans is an artifact of sequencing errors and technical issues.[32] Achilli and colleagues surveyed 322 extant Tuscan mitochondrial control regions. They found that Tuscan populations, especially the inhabitants of the Etruscan-founded town of Murlo, were more closely related to Near Eastern populations than were other modern Italians. The frequencies of Near Eastern haplotypes were relatively low (~5%), a pattern that could be missed without sufficient sampling. Achilli and colleagues argue that their data shows continuity between modern Tuscans and ancient Etruscans, but agree that the Etruscans derived from Near Eastern populations.

The discrepancy between these two interpretations highlights many of the issues confounding current human population genetic studies. While the ancient DNA studies have time depth, they lack sufficient sample size and can overlook low-level patterns in the genetic data. Meanwhile, Achilli and colleagues attribute the Near Eastern haplotypes in Tuscany to the Etruscans, but, since their data lack time depth, alternate explanations are possible (e.g., later gene flow from the eastern Mediterranean). In both cases, mitochondrial DNA only represents noncoding maternal lineages. Paternal and coding genetic history remains unexplored. Genomic analyses would provide a more complete picture of Etruscan genetic affiliations.

Ancient Historic Genealogy:
Egyptian Royal Mummies

One of the more media-attention-grabbing aspects of aDNA research has been its ability to clarify ancient family trees. The most noted (and controversial) example has been that of the Egyptian royal mummies. The identities and relationships of many Egyptian mummies are uncertain due to the looting of tombs, possible reuse or mislabeling of sarcophagi during ancient history, and fragmentary archaeological and textual evidence.[33] In 2010, Zahi Hawass and colleagues published a 5-generation reconstructed pedigree of 11 mummies belonging to the royal lineage of Tutankhamun. In 2012, the researchers followed up their initial study with a second analysis using computed tomography that confirmed that Ramesses III had been assassinated during the historically documented harem conspiracy. Using genetic evidence, the authors putatively identified an unknown mummy ("E") found in the same royal cache as Ramsesses III as the body of Pentawere, one of the chief architects of the conspiracy.[34] The unknown mummy shared Y chromosome haplotypes with Ramsesses III and had been covered in a ritually impure goat skin, suggestive of punishment.

On the surface, these studies provided a great leap forward in our understanding of Ancient Egyptian history. Among the aDNA research community, however, they were instantly controversial. Due to the warm climate and unknown effects of mummification, the long-term survival of authentic ancient DNA in Egypt has been questioned.[35] Moreover, the two analyses were based on the analysis of ancient microsatellites. Microsatellites are a type of genetic marker consisting of repeat motifs 1–6 bp in length (e.g., CACACA . . .) whose alleles vary by number of repeat units. Due to their rapid evolution and high population specificity, microsatellites are frequently analyzed in forensics for missing persons identification as well as in maternity/paternity cases. Microsatellites' use in ancient DNA research, however, is typically denigrated due to difficulties in sequencing the molecules and their propensity to generate artifact alleles during PCR.[36] Additionally, microsatellite allelic frequencies are unknown in ancient Egyptian populations. The genetic results are therefore statistically difficult to evaluate, especially in a high-status population suspected of consanguinity.[37] Further research using more reliable genetic markers such as single nucleotide polymorphisms (SNPs), a class of genetic marker that varies at a single nucleotide position, is required to verify the mummy genealogies.

The Human Paleogenomics Revolution

Genomic sequencing of thousands of extant human individuals and hundreds of ancient human samples has revolutionized our understanding of human population history.[38] For instance, the application of whole-genomic sequencing and improved biogeographic modeling algorithms has documented the relative contributions of Iron Age and Anglo-Saxon populations to the extant British Islands population and shown that ancient population structure is still observable in the extant British.[39] In addition to studying population histories and evolution, we can now even analyze the ancient epigenome and its effects on gene expression.[40]

In many cases, the analysis of complete genomes has rewritten the history books. Karmin and colleagues documented a previously unknown Y chromosome bottleneck that coincides with the onset of the Neolithic in Europe and probably correlates with changes in societal organization at this time.[41] Both Rhagavan and colleagues and Skoglund and colleagues identified a genetic signature of Australasian ancestry in extant and ancient Native American populations.[42] Although morphological data suggested that some Native American groups have Australasian ancestry, previous genetic research had only documented Asian ancestry for these peoples.[43] The application of whole-genome sequencing to the Tuscans/Etruscans and to the ancient Egyptians may resolve the questions remaining from the previous PCR-based research.

The Peopling of the Roman Empire

Despite their prominence in European and world history, the peoples of the Roman Empire have been relatively understudied using ancient DNA technologies. Most published analyses have been small in scope or preliminary. Three PCR-based studies found that both male and female infants were subjected to infanticide practiced in Roman Britain, suggesting that infanticide was not used to manipulate sex ratios.[44] Similarly, at a bathhouse in Roman Ashkelon, Israel, infanticide was practiced on both males and females, suggesting its use as a brothel.[45] Mitochondrial haplogroup variation corresponding to common European lineages was found at the Roman Estate of Vagnari, Italy.[46] PCR-based analyses of human remains at Casti Amanti and the Caius Iulius Polybius house (Pompeii, Italy) found that sufficient DNA was preserved to identify sex using the amelogenin gene and to identify individuals through microsatellites.[47] Later mitochondrial haplogroup analyses confirmed that 6 of the 13 individuals at the Caius Iulius Polybius house were matrilineally related.[48] Based on mitochondrial haplotypes obtained from four Syrian skeletons dating from the early Bronze Age to the late Roman Period, Witas and colleagues suggested a genetic link between the Indian subcontinent and Mesopotamia.[49] More recent research conducted on four Roman Age skeletons from London using DNA capture also found mitochondrial haplogroup, eye color and hair color variation.[50] The most comprehensive analysis yet consists of seven low coverage (~1× coverage) genomes of Romano-British from York, England.[51] Six of the genomes show close affinities with modern British Celtic populations, while the seventh shows affinities with populations from the Middle East, attesting to the cosmopolitan nature of the Roman Empire.

Health and Disease in Ancient History

Human history has been shaped by infectious disease. Due to limited historical descriptions and archaeological evidence, the pathogens behind many of the great plagues of ancient history are unknown. Ancient DNA evidence greatly increases our understanding of ancient pathogen virulence and prevalence. Not only can aDNA analysis provide definitive pathogen identifications, it may also permit the detection of disease in victims without diagnostic anatomical pathology or where other biochemical assays have proven ineffective;[52] aDNA analysis can also document ancient mixed-strain infections and infections with strains that are no longer extant, a level of resolution other paleomicrobiological techniques cannot provide.[53]

Early paleopathological aDNA studies utilized PCR-based approaches to identify ancient infections. Using these methods, mycobacterial infections have been documented in Egyptian mummies.[54] Several groups have reported the PCR amplification and sequencing of DNA matching *Trypanosoma cruzi* (Chagas' Disease) from ancient Andean mummies.[55] PCR products diagnostic of *Plasmodium falciparum* (malaria) infection have been amplified from Ancient Egyptian mummies and an ancient Roman skeleton from Lugnano, Umbria, Italy.[56] Using current DNA capture techniques, Marciniak and colleagues confirmed the presence of *Plasmodium falciparum* malaria at the Imperial Roman cemeteries of Velia and Vagnari (first–second century CE, southern Italy).[57] Papagrigorakis and colleagues identified the pathological agent of the historical Plague of Athens as *Salmonella enterica* serovar *typhi* (typhoid), although this diagnosis is disputed.[58] *Escherichia coli* was identified in the bodies of Lindow Man and an Egyptian mummy.[59] *Corynebacterium* DNA was found in the head of an Egyptian mummy.[60] Finally, Luciani and colleagues found *Haemophilus parainfluenzae* DNA in the skeleton of a pre-Columbian Native American.[61]

Technological advancements recently permitted the greatest ancient DNA advance in our understanding of disease in ancient history: the isolation and sequencing of a *Yersinia pestis* strain responsible for the Plague of Justinian.[62] Based on medical descriptions, historians had postulated that *Yersinia pestis* was responsible for the episodic pandemic Plague of Justinian (541–750 CE), the Black Death (1330–1351 CE), and the ongoing modern pandemic (1855 CE–present).[63] Without confirmatory scientific evidence, attribution of the both the Plague of Justinian and the Black Death to *Yersinia pestis* was debated.[64]

Ancient DNA genomic evidence proved that *Yersinia pestis* was at least one of the causes of both the Plague of Justinian and the Black Death.[65] It also revised our understanding of *Yersinia* phylogeny. Extant *Yersinia pestis* strains are commonly divided into three biovars (*Antiqua*, *Medievalis*, and *Orientalis*) based on their biological properties, although these have little phylogenetic meaning.[66] Each of the strains had been posited as causative of the three pandemics: *Antiqua* was thought to be responsible for the Plague of Justinian, *Medievalis* for the Black Death, and *Orientalis* for modern pandemic.[67] Ancient DNA evidence quickly disproved this hypothesis. Drancourt and colleagues showed that both the Plague of Justinian and the Black Death bore haplotypes similar to extant *Orientalis* strains.[68] Using a revised nomenclature following phylogeny, Harbeck and colleagues and Wagner and colleagues showed that the strain responsible for the Plague of Justinian fell on its own branch of the *Yersinia pestis* tree, near the root with *Yersinia pseudotuberculosis*, and distant from *Medievalis* and *Orientalis* biovars.[69] Based on its location in the phylogenetic tree, the Plague of Justinian may have originated in Asia.[70] This strain has yet to be identified in extant rodent reservoirs and may be extinct.[71]

Pathogenic strain persistence has also been demonstrated in *Yersinia pestis*. Victims of one of the last eighteenth-century European outbreaks of this plague in Marseille, France, contained strains of *Yersinia pestis* that is related to those recovered from fourteenth-century Black Death skeletons.[72]

Animal and Plant Breeding Practices in Ancient History

While numerous ancient DNA analyses address prehistoric animal and plant domestication,[73] relatively few have addressed the known changes in cultivation and breeding practices during ancient history. For instance, during the Roman Imperial period, horses and cattle sizes increased across Europe, although whether this is the result of selective breeding, improved feeding practices, or importation of larger animals is unclear.[74] Colominas and colleagues attempted to answer this question using mtDNA data from Iberian cattle, but were unable to secure a large enough sample size (n = 6) to resolve the debate.[75] Other aDNA analyses have been proofs-of-method or simple species and lineage identifications. Mitochondrial and nuclear DNA was amplified from Roman remains of equids found at Pompeii and Herculaneum, Italy.[76] Nevertheless, the authenticity of at least one of these mitochondrial sequences has been questioned as a hybrid between a domestic horse and a Somalian ass sequence.[77] Schlumbaum and colleagues obtained sequences from Roman-period apple seeds from the site of Oedenberg / Biesheim-Kunheim, France.[78] Typically Near Eastern mitochondrial haplotypes were observed in Roman-period cattle from August Raurica, Switzerland.[79] Breeding and cultivation practices in ancient history merit further analysis using ancient DNA, especially the application of modern genomic techniques.

Future Directions

While ancient DNA research has made great strides since its inception in the 1980s, its application to ancient history remains relatively limited. The majority of studies utilizing current paleogenomic methodologies have examined prehistoric events.[80] Most aDNA analyses of ancient historical remains have used outdated PCR technologies with little resolving power or ability to answer vital questions. Nevertheless, as demonstrated by the complete genomic sequence of the Plague of Justinian, ancient DNA data have the capability of clarifying and revising our understanding of ancient history. Future analyses employing these technologies promise a much more in-depth understanding of the past than is possible utilizing solely historical documentation and archaeological discoveries.

Notes

1. Lindahl 1993.
2. Pääbo et al. 2004.
3. Bollongino et al. 2008.
4. Pääbo 1989.
5. E.g., Dabney et al. 2013; Pääbo et al. 2004; Römpler et al. 2006; Shapiro and Cooper 2003.
6. E.g., Dabney et al. 2013.
7. E.g., Pääbo et al. 2004; Willerslev and Cooper 2005.
8. Vasan et al. 1996; Poinar et al. 1998.
9. E.g., Kolman and Tuross 2000; Pääbo et al. 2004; Willerslev and Cooper 2005.
10. Knapp and Hofreiter 2010.
11. E.g., Cooper and Poinar 2000; Gilbert et al. 2005.
12. Cooper and Poinar 2000; Gilbert et al. 2005.
13. Briggs et al. 2007; Brotherton et al. 2007.
14. Pääbo et al. 2004.
15. Briggs et al. 2007.
16. E.g., Schuenemann et al. 2011.
17. Hofreiter et al. 2001.
18. Briggs et al. 2010; Llamas et al. 2012; Pedersen et al. 2013.
19. Pedersen et al. 2013.
20. Saiki et al. 1985; Mullis and Faloona 1987.
21. An et al. 1995; Woodward et al. 1994 (dinosaurs); Cano and Borucki 1995; Cano et al. 1994, 1995; DeSalle et al. 1992, 1993 (insects and bacteria); Golenberg et al. 1990; Soltis et al. 1992 (plant fossils).
22. Margulies et al. 2005.
23. Bentley et al. 2008; Margulies et al. 2005.
24. E.g., Miller et al. 2008.
25. E.g., Maricic et al. 2010; Carpenter et al. 2013; Hodges et al. 2009.
26. Vernesi et al. 2004.
27. Vernesi et al. 2004.
28. Belle et al. 2006.
29. Guimaraes et al. 2009.
30. Vernesi et al. 2004.
31. Bandelt 2004; Malyarchuk and Rogozin 2004.
32. Barbujani et al. 2004; Guimaraes et al. 2009; Achilli et al. 2007.
33. E.g., Hawass et al. 2010, 2012.

34. Hawass, et al. 2012.

35. Gilbert et al. 2005; Krings et al. 1999; Marota et al. 2002.

36. See Campana 2008 for a review.

37. E.g., Bixler 1982; Hawass et al. 2010.

38. E.g., Stoneking and Krause 2011; Veeramah and Hammer 2014.

39. Schiffels et al. 2016; Leslie et al. 2015.

40. Orlando et al. 2015.

41. Karmin et al. 2015.

42. Rhagavan et al. 2015; Skoglund et al. 2015.

43. Australasian: Rhagavan et al. 2015; Skoglund et al. 2015. Previous: E.g., Reich et al. 2012.

44. Hassan et al. 2014; Mays and Faerman 2001; Waldron et al. 1999

45. Faerman et al. 1997, 1998

46. Prowse et al. 2010

47. Cipollaro et al. 1998; Cipollaro et al. 1999

48. Di Bernardo et al. 2009

49. Witas et al. 2013.

50. Eaton et al. 2015

51. Martiniano et al. 2016

52. E.g., Bos et al. 2015; Guhl et al. 1999; Sallares and Gomzi 2001.

53. E.g., Bos et al. 2014; Kay et al. 2015.

54. E.g., Nerlich et al. 1997; Zink et al. 2001a; Zink et al. 2003.

55. Aufderheide et al. 2004; Ferreira et al. 2000; Guhl et al. 1997, 1999; Madden et al. 2001.

56. Mummies: Taylor et al. 1997; Hawass et al. 2010; Nerlich et al. 2008. Roman: Sallares and Gomzi 2001.

57. Marciniak and colleagues 2016.

58. Papagrigorakis et al. 2006; Shapiro et al. 2006.

59. Fricker et al. 1997 (Lindow Man); Zink et al. 2000 (mummy).

60. Zink et al. 2001b.

61. Luciani and colleagues 2006.

62. Wagner et al. 2014.

63. Drancourt and Raoult 2002.

64. E.g., Gilbert et al. 2004; McCormick 2007.

65. Bos et al. 2011; Wagner et al. 2014.

66. Harbeck et al. 2013.

67. Drancourt et al. 2004.

68. Drancourt et al. 2004, 2007.

69. Harbeck et al. 2013; Wagner and et al. 2014.

70. Harbeck et al. 2013; Wagner et al. 2014.

71. Wagner et al. 2014.

72. Bos et al. 2016; Seifert et al. 2016.

73. E.g., Bailey et al. 1996; Jaenicke-Després et al. 2003; Larson et al. 2007; Ludwig et al. 2009.

74. Albarella et al. 2008; Colominas et al. 2014.

75. Colominas et al. 2014.

76. Cipollaro and Di Bernardo 2004; Di Bernardo et al. 2004a; Di Bernardo et al. 2004b.

77. Cipollaro 2011; Gurney 2010.

78. Schlumbaum and colleagues 2012.

79. Schlumbaum et al. 2006.

80. E.g., the evolution of the genus *Homo*: Meyer et al. 2014; Reich et al. 2010.

References

Achilli, A., et al. 2007. "Mitochondrial DNA variation of modern Tuscans supports the Near Eastern origin of Etruscans." *American Journal of Human Genetics* 80 (4): 759–768. DOI: 10.1086/512822.

Albarella, U., C. Johnstone, and K. Vickers. 2008. "The development of animal husbandry from the Late Iron Age to the end of the Roman period: a case study from South-East Britain." *Journal of Archaeological Science* 35: 1828–1848. DOI: 10.1016/j.jas.2007 .11.016.

An, C., et al. 1995. "Molecular cloning and sequencing the 18S rDNA from specialized dinosaur egg fossil found in Xixia Henan, China." *Acta Scientiarum Naturalium Universitatis Pekinensis* 31: 140–147.

Aufderheide, A. C., et al. 2004. "A 9000-year record of Chagas' disease." *Proceedings of the National Academy of Sciences of the United States of America* 101 (7): 2034–2039. DOI: 10.1073/pnas.0307312101.

Bailey, J. F., et al. 1996. "Ancient DNA suggests a recent expansion of European cattle from a diverse wild progenitor species." *Proceedings of the Royal Society B, Biological Sciences* 263: 1467–1473. DOI: 10.1098/rspb.1996.0214.

Bandelt, H.-J. 2004. "Etruscan artifacts." *American Journal of Human Genetics* 75 (5): 919–920. DOI: 10.1086/425180.

Barbujani, G., et al. 2004. "Etruscan artifacts: much ado about nothing." *American Journal of Human Genetics* 75 (5): 923–927. DOI: 10.1086/425283.

Belle, E.M.S., et al. 2006. "Serial coalescent simulations suggest a weak genealogical relationship between Etruscans and modern Tuscans." *Proceedings of the National Academy of Sciences of the United States of America* 103 (21): 8012–8017. DOI: 10.1073/ pnas.0509718103.

Bentley, D. R., et al. 2008. "Accurate whole human genome sequencing using reversible terminator chemistry." *Nature* 456 (7218): 53–59. DOI: 10.1038/nature07517.

Bixler, R. H. 1982. "Sibling incest in the royal families of Egypt, Peru, and Hawaii." *The Journal of Sex Research* 18: 264–281. DOI: 10.1080/00224498209551152.

Bollongino, R., A. Tresset, and J.-D. Vigne. 2008. "Environment and excavation: pre-lab impacts on ancient DNA analyses." *Comptes Rendus Palevol* 7: 91–98. DOI: 10.1016/j .crpv.2008.02.002.

Bos, K. I., et al. 2011. "A draft genome of *Yersinia pestis* from victims of the Black Death." *Nature* 478 (7376): 506–510. DOI: 10.1038/nature10549.

Bos, K. I., et al. 2014. "Pre-Columbian mycobacterial genomes reveal seals as a source of New World human tuberculosis." *Nature* 514 (7523): 494–497. DOI: 10.1038/nature13591.

Bos, K., et al. 2015. "Parallel detection of ancient pathogens via array-based DNA capture." *Philosophical Transactions of the Royal Society B, Biological Sciences* 370: 20130375. DOI: 10.1098/rstb.2013.0375.

Bos, K. I., et al. 2016 "Eighteenth century Yersinia pestis genomes reveal the long-term persistence of an historical plague focus." *eLife* 5: e12994. DOI: 10.7554/eLife.12994.001.

Briggs A. W., et al. 2007. "Patterns of damage in genomic DNA sequences from a Neandertal." *Proceedings of the National Academy of Sciences of the United States of America* 104 (37): 14616–14621. DOI: 10.1073/pnas.0704665104.

Briggs, A. W., et al. 2010. "Removal of deaminated cytosines and detection of *in vivo* methylation in ancient DNA." *Nucleic Acids Research* 38: e87. DOI: 10.1093/nar/gkp1163.

Brotherton, P., et al. 2007. "Novel high-resolution characterization of ancient DNA reveals C > U-type base modification events as the sole cause of *post mortem* miscoding lesions." *Nucleic Acids Research* 35: 5717–5728. DOI: 10.1093/nar/gkm588.

Campana, M. G. 2008. "The use of ancient microsatellites to detect past migrations." *Archaeological Review from Cambridge* 23.2: 147–160.

Cano, R. J., and M. K. Borucki. 1995. "Revival and identification of bacterial spores in 25- to 40-million-year-old Dominican amber." *Science* 268 (5213): 1060–1064. DOI: 10.1126/science.7538699.

Cano, R. J., et al. 1993. "Amplification and sequencing of DNA from a 120–135-million-year-old weevil." *Nature* 363 (6429): 536–538. DOI: 10.1038/363536a0.

Cano, R. J., et al. 1994. "*Bacillus* DNA in fossil bees: an ancient symbiosis?" *Applied and Environmental Microbiology* 60: 2164–2167.

Carpenter, M. L., et al. 2013. "Pulling out the 1%: whole-genome capture for the targeted enrichment of ancient DNA sequencing libraries." *The American Journal of Human Genetics* 93 (5): 852–864. DOI: 10.1016/j.ajhg.2013.10.002.

Cipollaro, M. 2011. "Strengthening ancient mtDNA equid sequences from Pompeii." *Journal of Cellular Biochemistry* 112: 363–364.

Cipollaro, M., and G. Di Bernardo. 2004. "DNA from equine remains buried by the 79 A.D. Vesuvius eruption." *Rendiconti Lincei* 15, 151–157.

Cipollaro, M., et al. 1998. "Ancient DNA in human bone remains from Pompeii archaeological site." *Biochemical and Biophysical Research Communications* 247: 901–904.

Cipollaro, M., et al. 1999. "Histological analysis and ancient DNA amplification of human bone remains found in Caius Iulius Polybius house in Pompeii." *Croatian Medical Journal* 40 (3): 392–397.

Colominas, L., A. Schlumbaum, and M. Saña. 2014. "The impact of the Roman Empire on animal husbandry practices: study of the changes in cattle morphology in the north-east of the Iberian Peninsula through osteometric and ancient DNA analyses." *Archaeological and Anthropological Sciences* 6: 1–16. DOI: 10.1007/s12520-013-0116-9.

Cooper, A., and H. N. Poinar. 2000. "Ancient DNA: do it right or not at all." *Science* 289 (5482): 1139. DOI: 10.1126/science.289.5482.1139b.

Dabney, J., et al. 2013. "Complete mitochondrial genome sequence of a Middle Pleistocene cave bear reconstructed from ultrashort DNA fragments." *Proceedings of the National Academy of Sciences of the United States of America* 110 (39): 15758–15763. DOI: 10.1073/pnas.1314445110.

DeSalle, R., M. Barcia, and C. Wray. 1993. "PCR jumping in clones of 30-million-year-old DNA fragments from amber preserved termites (*Mastotermes electrodominicus*)." *Experientia* 49: 906–909. DOI: 10.1007/BF01952607.

DeSalle, R., et al. 1992. "DNA sequences from a fossil termite in Oligo-Miocene Amber and their phylogenetic implications." *Science* 257 (5078): 1933–1936. DOI: 10.1126/science.1411508.

Di Bernardo, G., et al. 2004a. "Genetic characterization of Pompeii and Herculaneum *Equidae* buried by Vesuvius in 79 AD." *Journal of Cellular Physiology* 199: 200–205. DOI: 10.1002/jcp.10461.

Di Bernardo, G., et al. 2004b. "2000 year-old ancient equids: an ancient-DNA lesson from Pompeii remains." *Journal of Experimental Zoology B: Molecular and Developmental Evolution* 302B: 550–556. DOI: 10.1002/jez.b.21017.

Di Bernardo, G., et al. 2009. "Ancient DNA and family relationships in a Pompeian house." *Annals of Human Genetics* 73, 429–437.

Drancourt, M., et al. 2004. "Genotyping, Orientalis-like *Yersinia pestis*, and plague pandemics." *Emerging Infectious Diseases* 10 (9): 1585–1592. DOI: 10.3201/eid1009.030933.

Drancourt, M., et al. 2007. "*Yersinia pestis* Orientalis in remains of ancient plague patients." *Emerging Infectious Diseases* 13: 332–333. DOI: 10.3201/eid1302.060197.

Drancourt, M., and D. Raoult. 2002. "Molecular insights into the history of plague." *Microbes and Infection* 4: 105–109. DOI: 10.1016/S1286–4579(01)01515–5.

Eaton, K., et al. 2015. *Museum of London Report on the DNA Analyses of Four Roman Individuals*. London: Museum of London.

Faerman, M., et al. 1997. "DNA analysis reveals the sex of infanticide victims." *Nature* 385: 212–213.

Faerman, M., et al. 1998. "Determining the sex of infanticide victims from the late Roman Era through ancient DNA analysis." *Journal of Archaeological Science* 25: 861–865.

Ferreira, L. F., et al. 2000. "Paleoparasitology of Chagas disease revealed by infected tissues of Andean mummies." *Acta Tropica* 75: 79–84.

Fricker, E. J., M. Spigelman, and C. R. Fricker. 1997. "The detection of *Escherichia coli* DNA in the ancient remains of Lindow Man using the polymerase chain reaction." *Letters in Applied Microbiology* 24: 351–354.

Gilbert, M.T.P., et al. 2003. "Long-term survival of ancient DNA in Egypt: response to Zink and Nerlich (2003)." *American Journal of Physical Anthropology* 128: 110–114. DOI: 10.1002/ajpa.20045.

Gilbert, M.T.P., et al. 2004. "Absence of *Yersinia pestis*–specific DNA in human teeth from five European excavations of putative plague victims." *Microbiology* 150: 341–354. DOI: 10.1099/mic.0.26594–0.

Gilbert, M.T.P., et al. 2005. "Assessing ancient DNA studies." *Trends in Ecology and Evolution* 20: 541–544. DOI: 10.1016/j.tree.2005.07.005.

Golenberg E. M., et al. 1990. "Chloroplast DNA sequence from a Miocene *Magnolia* species." *Nature* 344 (6267): 656–658. DOI: 10.1038/344656a0.

Guhl, F., et al. 1997. "*Trypanosoma cruzi* DNA in human mummies." *The Lancet* 349: 1370. DOI: 10.1016/S0140–6736(05)63207–2.

Guhl, F., et al. 1999. "Isolation of *Trypanosoma cruzi* DNA in 4,000-year-old mummified tissue from northern Chile." *American Journal of Physical Anthropology* 108: 401–407. DOI: 10.1002/(SICI)1096–8644(199904)108:4<401::AID-AJPA2>3 .0.CO;2-P.

Guimaraes, S., et al. 2009. "Genealogical discontinuities among Etruscan, Medieval, and contemporary Tuscans." *Molecular Biology and Evolution* 26 (9): 2157–2166. DOI: 10.1093/molbev/msp126.

Gurney, S.M.R. 2010. "Revisiting ancient mtDNA equid sequences from Pompeii." *Journal of Cellular Biochemistry* 111: 1080–1081.

Harbeck, M., et al. 2013. "*Yersinia pestis* DNA from skeletal remains from the 6th century AD reveals insights into Justinianic Plague." *PLoS Pathogens* 9 (5): e1003349. DOI: 10.1371/journal.ppat.1003349.

Hassan, N.A.-M., et al. 2014. "Ancient DNA study of the remains of putative infanticide victims from the Yewden Roman villa site at Hambleden, England." *Journal of Archaeological Science* 43: 192–197.

Hawass, Z., et al. 2010. "Ancestry and pathology in King Tutankhamun's family." *Journal of the American Medical Association* 303: 638–647. DOI: 10.1001/jama.2010.121.

Hawass, Z., et al. 2012. "Revisiting the harem conspiracy and death of Ramesses III: anthropological, forensic, radiological, and genetic study." *British Medical Journal* 345: e8268. DOI: 10.1136/bmj.e8268.

Higuchi, R., et al. 1984. "DNA sequences from the quagga, an extinct member of the horse family." *Nature* 312 (5991): 282–284. DOI: 10.1038/312282a0.

Hodges, E., et al. 2009. "Hybrid selection of discrete genomic intervals on custom-designed microarrays for massively parallel sequencing." *Nature Protocols* 4: 960–974. DOI: 10.1038/nprot.2009.68.

Hofreiter, M., et al. 2001. "DNA sequences from multiple amplifications reveal artifacts induced by cytosine deamination in ancient DNA." *Nucleic Acids Research* 29: 4793–4799. DOI: 10.1093/nar/29.23.4793.

Jaenicke-Després, V., et al. 2003. "Early allelic selection in maize as revealed by ancient DNA." *Science* 302 (5648): 1206–1208. DOI: 10.1126/science.1089056.

Karmin, M., et al. 2015. "A recent bottleneck of Y chromosome diversity coincides with a global change in culture." *Genome Research* 25 (4): 459–466. DOI: 10.1101/gr.186684.114.

Kay, G. L., et al. 2015. "Eighteenth-century genomes show that mixed infections were common at time of peak tuberculosis in Europe." *Nature Communications* 6: 6717. DOI: 10.1038/ncomms7717.

Knapp, M., and M. Hofreiter. 2010. "Next generation sequencing of ancient DNA: requirements, strategies and perspectives." *Genes* 1: 227–243. DOI: 10.3390/genes1020227.

Kolman, C. J., and N. Tuross. 2000. "Ancient DNA analysis of human populations." *American Journal of Physical Anthropology* 111: 5–23. DOI: 10.1002/(SICI)1096-8644(200001)111:1<5::AID-AJPA2>3.0.CO;2-3.

Krings, M., et al. 1999. "mtDNA analysis of Nile River Valley populations: a genetic corridor or a barrier to migration?" *American Journal of Human Genetics* 64 (4): 1166–1176.

Larson, G., et al. 2007. "Ancient DNA, pig domestication, and the spread of the Neolithic into Europe." *Proceedings of the National Academy of Sciences of the United States of America* 104 (39): 15276–15281. DOI: 10.1073/pnas.0703411104.

Leslie, S., et al. 2015. "The fine-scale genetic structure of the British population." *Nature* 519 (7543): 309–314. DOI: 10.1038/nature14230.

Lindahl, T. 1993. "Instability and decay of the primary structure of DNA." *Nature* 362 (6422): 709–715. DOI: 10.1038/362709a0.

Llamas, B., et al. 2012. "High-resolution analysis of cytosine methylation in ancient DNA." *PLoS One* 7 (1): e30226. DOI: 10.1371/journal.pone.0030226.

Luciani, S., et al. 2006. "Molecular characterization of a pre-Columbian mummy and in situ coprolite." *American Journal of Physical Anthropology* 129: 620–629. DOI: 10.1002/ajpa.20314.

Ludwig, A., et al. 2009. "Coat color variation at the beginning of horse domestication." *Science* 324 (5926): 485. DOI: 10.1126/science.1172750.

Madden, M., et al. 2001. "Hybridization screening of very short PCR products for paleoepidemiological studies of Chagas' disease." *BioTechniques* 30: 102–109.

Malyarchuk, B. A., and I. B. Rogozin. 2004. "On the Etruscan mitochondrial DNA contribution to modern humans." *American Journal of Human Genetics* 75 (5): 920–923. DOI: 10.1086/425220.

Marciniak, S., et al. 2016. "*Plasmodium falciparum* malaria in 1st–2nd century CE southern Italy." *Current Biology* 26: R1205–R1225.

Margulies, M., et al. 2005. "Genome sequencing in microfabricated high-density picolitre reactors." *Nature* 437 (7057): 376–380. DOI: 10.1038/nature03959.

Maricic, T., M. Whitten, and S. Pääbo. 2010. "Multiplexed DNA sequence capture of mitochondrial genomes using PCR products." *PLoS One* 5 (11): e14004. DOI: 10.1371/journal.pone.0014004.

Marota, I., et al. 2002. "DNA decay rate in papyri and human remains from Egyptian archaeological sites." *American Journal of Physical Anthropology* 117: 310–318. DOI: 10.1002/ajpa.10045.

Martiniano, R., et al. 2016. "Genomic signals of migrations and continuity in Britain before the Anglo-Saxons." *Nature Communications* 7: 10326. DOI: 10.1038/ncomms1 0326.

Mays, S. and M. Faerman. 2001. "Sex identification in some putative infanticide victims from Roman Britain using ancient DNA." *Journal of Archaeological Science* 28: 555-559. DOI: 10.1006/jasc.2001.0616.

McCormick, M. 2007. "Toward a molecular history of the Justinianic pandemic." In *Plague and the End of Antiquity: The Pandemic of 541-750*, ed. L. K. Little. Cambridge: Cambridge University Press, 290-312.

Meyer, M., et al. 2014. "A mitochondrial genome sequence of a hominin from Sima de los Huesos." *Nature* 505 (7483): 403-406. DOI: 10.1038/nature12788.

Miller, W., et al. 2008. "Sequencing the nuclear genome of the extinct woolly mammoth." *Nature* 456 (7220): 387-390. DOI: 10.1038/nature07446.

Mullis, K. B., and F. A. Faloona. 1987. "Specific synthesis of DNA *in vitro* via a polymerase-catalyzed chain reaction." *Methods in Enzymology* 155: 335-350.

Nerlich, A. G., et al. 1997. "Molecular evidence for tuberculosis in an ancient Egyptian mummy." *The Lancet* 350: 1404. DOI: 10.1016/S0140-6736(05)65185-9.

Nerlich, A. G., et al. 2008. "*Plasmodium falciparum* in Ancient Egypt." *Emerging Infectious Diseases* 14: 1317-1318. DOI: 10.3201/eid1408.080235.

Orlando, L., M.T.P. Gilbert, and E. Willerslev. 2015. "Reconstructing ancient genomes and epigenomes." *Nature Reviews Genetics* 16 (7): 395-408.

Pääbo, S. 1989. "Ancient DNA: extraction, characterization, molecular cloning and enzymatic amplification." *Proceedings of the National Academy of Sciences of the United States of America* 86 (6): 1939-1943.

Pääbo, S., et al. 2004. "Genetic analyses from ancient DNA." *Annual Review of Genetics* 38: 645-679. DOI: 10.1146/annurev.genet.37.110801.143214.

Papagrigorakis, M. J., et al. 2006. "DNA examination of ancient dental pulp incriminates typhoid fever as a probable cause of the Plague of Athens." *International Journal of Infectious Diseases* 10: 206-214. DOI: 10.1016/j.ijid.2005.09.001.

Pedersen, J. S., et al. 2013. "Genome-wide nucleosome map and cytosine methylation levels of an ancient human genome." *Genome Research* 24 (3): 454-466. DOI: 10.1101/gr.163592.113.

Poinar, H. N., et al. 1998. "Molecular coproscopy: dung and diet of the extinct ground sloth *Nothrotheriops shastensis*." *Science* 281 (5375): 402-406. DOI: 10.1126/science.281.5375.402.

Prowse, T. L., et al. 2010. "Stable isotope and mitochondrial DNA evidence for geographic origins on a Roman estate at Vagnari (Italy)." In *Roman Diasporas: Archaeological Approaches to Mobility and Diversity in the Roman Empire*, ed. H. Eckard *Journal of Roman Empire Supplement* 78: 175-197.

Raghavan, M., et al. 2015. "Genomic evidence for the Pleistocene and recent population history of Native Americans." *Science* 349 (6250): DOI: 10.1126/science.aab3884.

Reich, D., et al. 2010. "Genetic history of an archaic hominin group from Denisova Cave in Siberia." *Nature* 468 (7327): 1053-1060. DOI: 10.1038/nature09710.

Reich, D. E., et al. 2012. "Reconstructing Native American population history." *Nature* 488 (7411): 370-4. DOI: 10.1038/nature11258.

Römpler, H., et al. 2006. "Multiplex amplification of ancient DNA." *Nature Protocols* 1: 720-728. DOI: 10.1038/nprot.2006.84.

Saiki, R. K., et al. 1985. "Enzymatic amplification of β-globin genomic sequences and restriction site analysis for diagnosis of sickle cell anemia." *Science* 230 (4732): 1350-1354. DOI: 10.1126/science.2999980.

Sallares, R., and S. Gomzi. 2001. "Biomolecular archaeology of malaria." *Ancient Biomolecules* 3:195–213.

Schiffels, S., et al. 2016. "Iron Age and Anglo-Saxon genomes from East England reveal British migration history." *Nature Communications* 7: 10408. DOI: 10.1038/ncomms 10408.

Schlumbaum, A., M. Turgay, and J. Schibler. 2006. "Near east mtDNA haplotype variants in Roman cattle from Augusta Raurica, Switzerland, and in the Swiss Evolène breed." *Animal Genetics* 37: 373–375. DOI: 10.1111/j.1365-2052.2006.01435.x.

Schlumbaum, A., S. van Glabeke, and I. Roldan-Ruiz. 2012. "Toward the onset of fruit tree growing north of the Alps: ancient DNA from waterlogged apple (*Malus* sp.) seed fragments." *Annals of Anatomy* 194: 157–162. DOI: 10.1016/j.aanat.2011.03.004.

Schuenemann, V. J., et al. 2011. "Targeted enrichment of ancient pathogens yielding the pPCP1 plasmid of *Yersinia pestis* from victims of the Black Death." *Proceedings of the National Academy of Sciences of the United States of America* 108 (38): 746–752. DOI: 10.1073/pnas.1105107108.

Seifert, L., et al. 2016 "Genotyping *Yersinia pestis* in historical Plague: evidence for long-term persistence of *Y. pestis* in Europe from the 14th to the 17th century." *PloS One* 11 (1): e0145194. DOI: 10.1371/journal.pone.0145194.

Shapiro, B., and A. Cooper. 2003. "Beringia as an Ice Age genetic museum." *Quaternary Research* 60 (1): 94–100. DOI: 10.1016/S0033-5894(03)00009-7.

Shapiro, B., A. Rambaut, and M.T.P. Gilbert. 2006. "No proof that typhoid caused the Plague of Athens (a reply to Papagrigorakis et al.)." *International Journal of Infectious Diseases* 10: 334–335. DOI: 10.1016/j.ijid.2006.02.006.

Skoglund, P., et al. 2015. "Genetic evidence for two founding populations of the Americas." *Nature* 525 (7565): 104–108. DOI: 10.1038/nature14895.

Soltis, P. S., D. E. Soltis, and C. J. Smiley. 1992. "An *rbcL* sequence from a Miocene *Taxodium* (bald cypress)." *Proceedings of the National Academy of Sciences of the United States of America* 89 (1): 449–451.

Stoneking, M., and J. Krause. 2011. "Learning about human population history from ancient and modern genomes." *Nature Reviews Genetics* 12 (9): 603–614. DOI: 10.1038/nrg3029.

Taylor, G., P. Rutland, and T. Molleson. 1997. "A sensitive polymerase chain reaction method for the detection of *Plasmodium* species DNA in ancient human remains." *Ancient Biomolecules* 1: 193–203.

Vasan, S., et al. 1996. "An agent cleaving glucose-derived protein crosslinks *in vitro* and *in vivo*." *Nature* 382 (6588): 275–278. DOI: 10.1038/382275a0.

Veeramah, K. R., and M. F. Hammer. 2014. "The impact of whole-genome sequencing on the reconstruction of human population history." *Nature Reviews Genetics* 15 (3): 149–162. DOI: 10.1038/nature12736.

Vernesi, C., et al. 2004. "The Etruscans: a population-genetic study." *American Journal of Human Genetics* 74 (4): 694–704. DOI: 10.1086/383284.

Wagner, D. M., et al. 2014. "*Yersinia pestis* and the Plague of Justinian 541–543 AD: a genomic analysis." *The Lancet Infectious Diseases* 14: 319–326. DOI: 10.1016/S1473-3099(13)70323-2.

Waldron, T., et al. 1999. "Sexing of Romano-British baby burials from the Beddingham and Bignor villas." *Sussex Archaeological Collections* 137: 71–79.

Willerslev, E., and A. Cooper. 2005. "Ancient DNA." *Proceedings of the Royal Society B: Biological Sciences* 272: 3–16. DOI: 10.1098/rspb.2004.2813.

Witas, H. W., et al. 2013. "mtDNA from the early Bronze Age to the Roman Period suggests a genetic link between the Indian subcontinent and Mesopotamian cradle of civilization." *Plos One* 8 (9): e73682.

Woodward, S. R., N. J. Weyand, and M. Bunnell. 1994. "DNA sequence from Cretaceous Period bone fragments." *Science* 266 (5188): 1229–1232. DOI: 10.1126/science.7973705.

Zink, A., et al. 2000. "Molecular evidence of bacteremia by gastrointestinal pathogenic bacteria in an infant mummy from Ancient Egypt." *Archives of Pathology and Laboratory Medicine* 124: 1614–1618. DOI: 0.1043/0003–9985(2000)124<1614:MEOBBG>2.0.CO;2.

Zink, A., et al. 2001a. "Molecular analysis of skeletal tuberculosis in an ancient Egyptian population." *Journal of Medical Microbiology* 50: 355–366. DOI: 10.1099/0022–1317–50–4–355.

Zink, A., et al. 2001b. "Corynebacterium in Ancient Egypt." *Medical History* 45: 267–272. DOI:DOI: 10.1017/S0025727300067740.

Zink A. R., et al. 2003. "Characterization of *Mycobacterium tuberculosis* complex DNAs from Egyptian mummies by spoligotyping." *Journal of Clinical Microbiology* 41: 359–367. DOI: 10.1128/JCM.41.1.359–367.2003.

CHAPTER SEVEN

Modern DNA and the Ancient Mediterranean

Roy J. King & Peter A. Underhill

Foreword

The patterns of DNA variation in living human populations are an archive of our species' evolutionary history, and one goal of population genetics is to extract and understand this inherited memory. Using genetic principles tractable to the curious but uninitiated reader, this chapter illustrates to what extent modern Y chromosome sequence diversification can reflect possible prehistoric scenarios that align with evidence from other disciplines, including ancient DNA. While the mitochondrial genome traces maternal history, here we focus on recent advances in sequencing paternally transmitted Y chromosomes. These male-specific chromosomes preserve an uninterrupted record of mutations that have persisted and accumulated sequentially during the entirety of anatomically modern human history. Numerous modern Y chromosome DNA sequences can be assembled into a single time-calibrated phylogeny or gene tree that displays geographically distinctive branches. A molecular clock makes it possible to estimate when branching events (splits) in the tree began to emerge. These Y chromosome phylogenic attributes can help inform hypotheses and computational modeling as to where, when, and how populations formed and expanded. Thus, this conceptually simple genetic system can eloquently record, in time and space, the last vestiges of shared common ancestry distributed by rapidly expanding populations, as well as the onset of subsequent regional diversification. Using knowledge of the structure of Y chromosome diversification in modern populations, seasoned with penetrating clues from ancient DNA, we explore some episodes of human habitation in the Mediterranean, including relationships to the Black Sea (Pontic) and adjoining regions of Europe.

Introduction

Genetics is the study of inheritance, and DNA variation is the essence of heredity. DNA sequence differences underpin genetics overall and population genetics is the study of such diversity in populations and how it changes through time. Reconstructing human history using modern DNA has been a long-standing endeavor rooted in sampling practicality.[1] If a mutational change does not negatively affect the individual's ability to reproduce, it may be passed down to each succeeding generation, eventually becoming established in a population. Such mutations, whether beneficial, harmful, or neutral, can serve as genetic markers.

In humans, genetic information is preserved in a genome comprised of the 46 chromosomes of a cell's nucleus as well as the mitochondrial genome (mtDNA) found in the surrounding cytoplasm. The pertinent elemental units of DNA are four molecules called nucleotide bases, represented by the letters A (adenine), C (cytosine), G (guanine), and T (thymine). The most common DNA sequence variants are single nucleotide variations (SNVs), so named to be apart from the 1% frequency threshold required for the traditional single nucleotide polymorphism (SNP) designation. Thus, SNV lists contain "singletons" while SNP lists do not. While the most of the 3 billion bases in the human genome are identical, reflecting the recent ancestry of our species, every individual can have as many as 10 million nucleotide base differences (0.33%) amongst other humans. Since the base substitution mutation rate is low in nuclear DNA, the occurrence of a SNV on any chromosome typically represents a unique event in human history and thus has distinctive time and place properties potentially traceable to a common ancestor. However, since pairs of chromosomes (diploid state) occur in the genomes of sexually reproducing species, the process of recombination fractionates the original relationships amongst other SNVs that arose on the same chromosome. Thus, SNVs originating on chromosomes 1 through 22 and the X are often analyzed as independent loci. During the time that a SNV remains in the gene pool, it may spread to other regions where it gets established at different frequencies. Typically, frequencies measured systematically in large numbers of SNVs are used to estimate the degree of affinity amongst populations.

Genetic Distance

Genetic distance is any type of numerical measure of the evolutionary relatedness of two or more populations or two or more chromosomes. Genetic diversity in populations is a result of admixture and local differentiation. Spatial distributions of human genetic variation typically appear as nonuniform patches or clines.[2] Clines can manifest either because of continuous gene flow between populations that initially have different genetic constitutions or as a result of local population growth followed by range expansion.[3] When people mate with their neighbors (i.e., distant relatives) they tend to become more inbred and thus a population will trend towards having greater genetic similarity than they would if they mated equally with other people despite being separated by geographic and/or cultural barriers. This predisposition with respect to the lack of random mating creates subdivision such that metapopulations comprised of various subpopulations often get established. Subtle differences in marker frequency amongst metapopulations can be used to parse such substructure. Currently hundreds of thousands of base substitutions scattered amongst both the nuclear and mtDNA genomes have been identified, some of which are selected for genotyping in populations using instrumentation that can simultaneously determine the nucleotide status for each chosen variant. Typically, such genotype frequency–based datasets are presented either as population trees [4] or Principal Components (PCs), that is, graphical summaries of independent plot patterns in which the location of a point represents an individual's resemblance to other individuals similarly analyzed.[5] Essentially, individuals that cluster closer together in a PC analysis have more genetic relatedness (affinity) to one another than individuals who cluster further apart. By genotyping representative individuals from known populations one can also assess the biogeographic ancestry of any unknown sample(s) relative to other previously defined populations.

Caveats

However, extrapolating prehistoric events based on proxy modern DNA data alone can be limited or distorted because of issues such as 1) different population histories could generate the same genetic landscape; 2) marker frequency patterns changing over time; 3) earlier demographic episodes may be hidden or replaced by more recent events; 4) differential forces of selection at specific genes vs. random forces uniformly acting on all the genes in the population, e.g., population-size fluctuations; 5) migration/hybridization reducing genetic distance between populations; and 6) minimal and imprecise information on the timing of population splits, mergers, or migrations. While complex computational methods exist to analyze genome-wide variation to infer demographic features such as divergence times and ancestral population size,[6] these simulation models often assume an absence of recombination under a situation of constant migration without growth.[7] Although popular algorithms like ADMIXTURE can process large amounts of SNV data and reveal signals of population structure and admixture,[8] such analyses, solely on their own merits alone, are unable to directly estimate the time frames for such events and also the directionality of gene flow.

Gene Trees and Substructure

Knowledge of the time and duration of an event is obviously of considerable interest to prehistorians.[9] The challenge is to untangle the course of prehistoric events comprising the assemblage of extant genetic diversity. Ultimately additional and better quality sequences of full genomes coupled with sophisticated computational analyses will render a more accurate account. In the interim an alternative to multilocus computer simulation population approaches exists that can directly address the dimension of time. Here, time refers to a molecular clock that assumes that DNA sequence evolution occurs at a constant rate such that the divergence between two lineages accurately relates to the time both split from a common ancestor, aka the coalescent time. It is based on the haplotype analysis of gene trees using individuals instead of multiple independent marker frequencies co-analyzed in populations. The term *haplotype* refers to a combination of allelic states for a collection of two or more SNVs occurring on the same chromosome. By convention, allelic state is simply designated 0 or 1, where the 0 allele is the nascent ancestral form and 1 the more innovative, derived state. While the accumulation of mutations over time and space occurs throughout the genome, most diploid autosomal haplotype records are subject to the process of recombination. Consequently, much of the ancestry tract information content largely decays within a few thousands of years at most.[10] In addition, significant loss of genetic information also occurs nonsystematically by independent assortment during gamete formation, that is, meiosis.[11] However, since Y chromosome and mtDNA are single unpaired haploid molecules that escape recombination, they are exceptionally suitable for gene tree–based phylogenetic analysis. The pattern and length of branches proportional to the total number of sequentially accumulated mutations preserves the entire evolutionary course as one traverses the path from prehistoric root to contemporary tip of any particular lineage present in the phylogeny.

Mitochondrial DNA vs. Y Chromosome

While both haploid systems trace sex-specific histories and have time-calibrated branches with geographic structure, the mtDNA genome has both strengths and weaknesses relative to the Y chromosome. Virtues include: both sexes have mtDNA genomes making sampling more inclusive, including ancient DNA specimens, and the high number of mtDNA molecules per cell contrasts to just one copy of the Y chromosome per cell. This higher amount of mtDNA molecules vs. nuclear DNA favors a longer preservation period and

improves the chance of successful retrieval for mtDNA in ancient DNA specimens. An extensive collection of mtDNA data exists already, in part because
many complete Sanger sequences have already been obtained for living global
populations prior to availability of next-generation sequencing technologies
(discussed later). Corresponding mtDNA shortcomings include: a small genome size relative to nuclear chromosomes that limits the innate level of lineage resolution possible in the phylogeny. The higher mtDNA mutation rate
increases levels of recurrent mutations that add uncertainty to some of the
branching relationships in the phylogeny. Since most of the mtDNA genome
codes for genes, occasionally some show evidence of natural selection that reduces the observed diversity for some haplogroups.[12] In contrast, Y chromosome diversity seems to be shaped only by population-level forces that affect
the entire genome simultaneously, such as population growth, founder effect,
migration, and so on.[13] While we emphasize the Y chromosome in this chapter, we acknowledge that the mtDNA genome continues to provide valuable
insights into population history and substructure.[14] Despite the resolving
power of haploid systems, they have limits. The haploid maternally transmitted mtDNA and paternally transmitted Y chromosome gene trees only provide
respective female and male narratives of human migratory history that may,
in some cases differ, due to sex-biased reproductive success.[15] Furthermore
haploid systems may exaggerate the magnitude of population replacement[16].
Ultimately it will take phased genome sequencing that distinguishes linkage
between alleles on segments of homologous autosomes resulting in the collection of genome-wide haplotypes present in an individual, together with ex-
haustive computer simulations to evaluate scenarios capable of generating the
observed modern patterns of autosomal variation that will ultimately yield a
more comprehensive reconstruction of human history.[17]

Haplogroups

A haplogroup is a collection of genetically similar haploid lineages with common molecular ancestry. The relationships amongst haplogroups form a
phylogeny. As most of the content of this chapter will involve Y chromosome
SNVs, we will use the term *haplogroup* such that, by convention for haploid
systems, the 20 most basic phylogenetic branches are assigned alphabetic
labels. Typically, Y chromosome SNV-based trees display nested sets of bi-
furcating haplogroups that fork into descendant branches. The branches in
such phylogenies are occupied by specific single nucleotide variations (SNVs).
The SNV mutation rate in human Y chromosome nuclear DNA is low enough
such that it can be assumed that most (~98%) base substitutions arise just
once in human history. Thus, individuals who share the derived allele (i.e.,
the mutation) share a common male ancestor, a condition often referred to as
"identity by descent." Although an infrequent event, a base substitution might

recur independently creating an "identity by state" situation. These rare events are obvious since the derived allele often occurs on different branches and at different time horizons. As such these renegade markers are still readily inter-pretable since the Y phylogeny provides clear haplogroup context.

Even contemporary provincial populations generally reflect the product of multiple inputs (i.e., founder lineages) with distinctive histories. Thus, stable haploid gene trees reflect a population not as a unique unit budding from a population tree, but rather as a composite of diverse lineages, the result of several layers of prehistorical events/interactions having occurred over time and space. The shape of the tree at various nodes provides information about the dynamics of demographic change, not just what lineages are present. [18]

Molecular Game of Chance

Every generation creates a potential opportunity for the introduction of new genetic variation anywhere within the genome. When manufacturing the unique genetic heritage of each child, both biological parents contribute, as total raw material, 4 copies of each of the bi-parentally inherited autosomes to the occasion. The autosomes are the chromosomes numbered from 1 to 22, each of which is present in two copies each in both males and females, except in their egg or sperm cells, which each contain only 1 of the 4 copies available. Unlike the haploid mtDNA genome and Y chromosome, the autosomes are a mosaic of inputs from multiple ancestors. During sexual reproduction, each offspring will randomly inherit 50% of each parent's complex heritage during the formation of gametes by meiosis. Consequently, except for identical twins, each sibling will inherit a singular montage of segments, with each segment reflecting an arbitrary donation by any one of the parent's various ancestors in the family pedigree, for instance, maternal grandfather; paternal great-great grandfather, and so on. There are also 3 copies of the X chromosome involved with the female [XX] contributing 2 copies and the male [XY] donating ei-ther a X or a Y chromosome at fertilization making males solely responsible for the sex of the offspring. While men and women have multiple ancestors, who have contributed to their autosome and X chromosome heritage, a man inherits his Y chromosome from only one male ancestor.

Genetic Drift

To distinguish between two populations, the ideal marker is one in which one allele is fixed in one group but is absent in another group and has recover-able time and space characteristics. The character of mutations present in the genome is a consequence of 1) locus-specific forces, gene by gene whereby the futures of mutations are influenced by the agency of natural selection, or conversely by 2) population level forces such as size fluctuation, migration, admixture and so on that simultaneously influence all the genes in a population. Under such conditions of neutrality, it is simple chance alone (e.g., founder effect) that determines what diversity is either lost or maintained. This random wavering in allele frequencies is referred to as genetic drift.

Size does matter regarding inheritance. In a contracting population, under conditions of neutrality, a new variant can approach fixation sooner than in a large population. Within the assemblage of chromosomes in a mating couple, the consequences of chance (i.e., genetic drift) are magnified more on the Y chromosome than on the others simply because the proportion of autosomes to X chromosomes to Y chromosomes is 4:3:1 respectively. Consequently, a new Y chromosome mutation appearing in the gene pool for the first time has at least a four times greater chance of getting established, by chance alone, in a population relative to any new mutation that arises on one of the 4 autosomes in play during reproduction.[19] Furthermore, differential reproductive success will occur whenever influencing factors exclude some males from mating. This additional happenstance will further increase beyond a factor of four, the likelihood that a new Y chromosome–derived allele can be established in a population. Thus, Y chromosome distributions most often show the highest geographic relief worldwide of any genetic "identity by descent" system. Such distinguishing Y chromosome mutations are more likely to coincide with secondary bottlenecks associated with the progression of subpopulation origins such as a Siberian founder related to the colonization of North America.[20]

In addition to being quite sensitive to genetic drift, the human Y chromosome is ~50 million bases in size while the mtDNA genome totals only ~16,000 bases. Thus, theoretically the maximum number of possible SNV markers available to catalog a mtDNA lineage is 16,000. However, after ignoring a hypervariable region of ~1000 bases, phylogenetic analyses of 15,000 nucleotides of comparable mtDNA sequence from global population samples showed that the number of derived sites within any individual lineage averages just ~50 mutations ranging from the root to tip of the mtDNA gene tree.[21] In contrast the larger Y chromosome provides a greater inventory of informative stable binary mutations. This higher capacity increases possibilities that some Y chromosome lineages will, by chance, correspond more closely with specific demographic episodes, which otherwise appear less distinct as far as signals from other genetic loci are concerned.

Thousands of Y chromosome mutations have been found in nonsystematic fashion during the past ~15 years.[22] Genotyping some of these markers in sample collections confirmed the robustness of their worldwide phylogenetic relationships and the strong geographic patterning of major haplogroups and sub-haplogroups.[23] Prior to the genomic era of whole Y chromosome sequences, the hierarchy of the 20 major haplogroups, as well as the order of some of their related sub-haplogroups, was understood from the perspective of only ca. 600 canonical SNVs. Insufficient detail as to the number and length of branches connecting structural nodes only formed a topology impoverished with regards to temporal and population dynamics information. Although, we still cannot rank the order in which two or more SNVs accumulated on the same branch, we can now estimate the time interval between sequential nodes, and by doing so for every branch connection, create a time-calibrated and demographic-aware phylogeny. Thus, the Y chromosome is arguably an illuminating marker system capable of providing frameworks regarding bio-geographic ancestry and the changing structure of populations over time. In the following section, we briefly summarize recent advances leading to the era of a time-calibrated human Y chromosome phylogeny.[24] Thus, the Y chromo-some is probably best suited to approach the criteria mentioned earlier in this section for an ideal population-specific marker.

Human Genomics

The near completion of a highly accurate modern human Reference genome by an international team of researchers in 2003 was a landmark accomplish-ment, taking more than a decade of effort. Not only did it provide detailed information concerning the location, composition, and function of a complete set of genes, this unprecedented resource also created new opportunities lead-ing to major advances in high-capacity parallel short read DNA sequencing technologies. Along with advances in computational data analyses, these devel-opments have thrust human genetics into the realm of "big data" and human genomics. Typically, a consenting adult voluntarily donates saliva, blood, or a tissue specimen from which DNA is isolated. Currently the most commonly used sequencing methodology requires that a "library" of DNA molecules rep-resenting a donor's entire genome be created by first breaking up the sampled DNA into millions of pieces. A synthetic molecular tag that uniquely identifies the sample is then attached to all the ends of the fragments and the fraction of tagged molecules *ca.* 250–500 nucleotides in length is gathered afterwards for sequencing. Notably, sample-specific tagging allows the efficient simulta-neous sequencing of a mixture of different sample libraries and the subsequent computer-based retrieval of each sample's own set of sequenced fragments. Next, the bits of DNA sequence recovered for each individual are computation-ally assembled (i.e., mapped) by aligning them to the nonrepetitive regions of the Reference genome. Later an algorithm compares the sample and Reference DNA sequences, distinguishing likely nucleotide differences. The higher the number of short sequence reads that overlap the known Reference bases, the greater the sample sequencing coverage and confidence in the nucleotide base calling. Nowadays an entire human genome can be sequenced to 30X coverage in just a few days for a few thousand dollars or less. Most of the human genome information justifiably focuses on medical genetic issues. Since sequencing the entirety of the 3 billion base pair nucleotide sequence of the human genome is often not necessary, additional methods have been developed in which only selective regions of the genome are targeted and captured by hybridization. An example is the human exome, the specific chromosomal regions that contain just the ~3 million nucleotides of sequence that codes for all the (*ca.* 20,000–25,000) known protein genes. The 1000 Genomes Project Consortium was a pioneering test bed for technologies and methods with the goal of sequencing numerous genomes at low coverage and creating a global reference of genetic variation.[25] While many thousands of human genomes have now been sequenced, many global populations remain under-represented. Implementing the same capture technologies used for medical research has provided new opportunities aimed at reconstructing population histories using the genomes of individuals, be they either modern or ancient.[26]

Recent Strides

Landmark progress during the past ~4 years involved the first iteration of a calibrated Y chromosome gene tree based on single base substitution mutations observed in 3.2 million base pairs of sequence compared in 36 publically available sequences.[27] The calibration was based on a human chimp 6.5-million-year divergence time.

The next significant development regarding "whole" Y chromosome sequencing involved using a total of 69 samples from Africa, Eurasia, and the Americas.[28] The first step involved using stringent criteria to identify all the single copy regions on the Y chromosome where the current short read sequence data align to the Reference sequence reliably.[29] Next a standardized pipeline was developed to call SNVs along a tract of *ca.* 10 million base pairs of sequence in each of the samples, construct a phylogeny and estimate unbiased coalescent times (discussed below) for various branching events in the phylogeny. In contrast to the aforementioned ~50 markers defining individual mtDNA lineages, each of the 69 Y chromosome lineages averaged *ca.* 1100 substitutions from root to tip. This ~twentyfold ratio [1100/50] in marker availability translates to noticeably more granular phylogenetic and geographic resolution with the Y chromosome. Additional subsequent important Y chromosome studies,[30] using independent calibration approaches including ancient DNA, corroborated the mutation rate such that a reliable consensus was established.[31] Successful use of the capture methodology regarding modern "whole" Y chromosome sequencing has been demonstrated.[32]

Time to Most Recent Common Ancestor

Determining the time dimension requires construction of a gene tree involving SNV variant data from resequencing a known length of Y chromosome sequence (e.g., 10 million nucleotides) in tens or more (e.g., 36–1244) samples.

We now have knowledge of the average rate at which the SNVs are randomly added as the tree continues to evolve. When the dataset is based on resequencing 10 million bases of sequence, the average mutation period is such that 1 SNV accumulates every 132 years.[33] Note that this empirical value is proportional to the length of sequence under consideration. For example, if 5 million bases were sequenced and a phylogeny constructed, the average mutation period would be one SNV being added every 264 years. Therefore, phylogenies based on different sequence lengths will yield similar estimates of "wait times" between successive branch splits making different datasets comparable. The coalescent time is the time taken (going backwards in time) for two or more lineages to coalesce (i.e., join) at the nodal point of their closest shared molecular ancestral branch or precursor lineage.[34] Thus, the lengths of the branches in such a phylogeny are now meaningful since they are proportional to time. Branches connecting successive splits reflect the average period during which the reproductively relevant members of a population remained at constant size. However, it is possible, especially in the case of a long interior branch, that multiple but now invisible (i.e., lost) past episodes of population size fluctuation occurred. Thus, the connection visible today may reflect the last in a series of population size contractions. On the other hand, when one observes the feature of many lineages radiating from a single ancestral branch, this bushlike structure marks the beginning of a rapid population expansion out of a prior bottleneck. This pattern is seen most often nearest the tips, that is, the canopy of the tree. It is essential to appreciate that the time to the most recent common ancestor (TMRCA) shown as a branching event in a calibrated gene tree and the time of the fastest population growth do not need to coincide. We reiterate that the TMRCA (i.e., split time of two or more lineages) does not necessarily correspond to any significant event in a population's history. While demographic events (e.g., noticeable changes in population size) can sometimes begin very soon after the emergence of a new short-stemmed branch from which many branches emerge, in other instances growth often occurs at some considerable time after the onset of branch formation exemplified by a long stem. Thus, the TMRCA sets an upper bound for the subsequent population dynamics. Conversely Y chromosome data from securely dated ancient DNA specimens provides insight about the potential lower bound for the age of a marker in the gene pool.

Features of the Time-Calibrated Y Chromosome Phylogeny

Except for the stand-alone haplogroup A00, the modern Y chromosome phylogeny (Figure 7.1) reflects the sequential diversification of paternal lineages that have survived from the earliest beginnings, *ca.* 190,000 years ago to the present[35]. Within our species some lineages present in the past appear to no

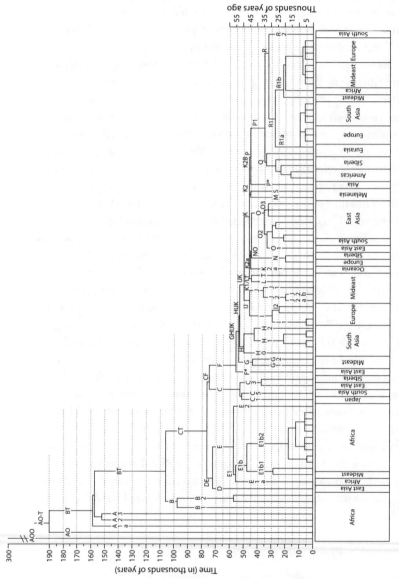

FIGURE 7.1. Y chromosome gene tree of major haplogroup relationships and estimated ages in thousands of calendar years based on single nucleotide substitutions detected while resequencing ca. 10 million nucleotide bases in each of the globally representative individuals. Branch lengths are proportional to the number of single base substitution mutations found on each branch. Figure is a simplified adaption of information published in Karmin et al. 2015 and Poznik et al. 2016. We alone are responsible for any omissions; errors, or inadvertent phylogenetic misinterpretations we might have made when merging these two publications.

longer exist, as exemplified by ancient DNA results from a well-preserved *ca.* 45,000-year-old Siberian specimen (Ust'-Ishim).[36] The consequences of drift (i.e., survival of the luckiest) are magnified when population size is low as would be expected for situations when the number of reproductively fit males are few. In this regard, the discovery of the independent Aoo lineage provides evidence that introgression (i.e., transfer of genetic information from one species to another by hybridization) occurred between a now-extinct archaic hominoid male and an anatomically modern female. This Aoo lineage has persisted, albeit at very low frequency in Africa from ~300 thousand years ago to the present day.

Figure 7.1 shows the main haplogroups and their geographic affiliations and coalescent times based on modern DNA. This tree is a simplified representation of information adapted from two pivotal studies. For the sake of simplicity, only a fraction of the known branch diversification reported in these two studies is shown. Each haplogroup drawn in the figure contains considerably more granular sub-haplogroup differentiation than shown. Figure 7.1 illustrates how the A, B, C through T core haplogroup structure is defined by early founders whose descendants have accumulated additional mutations down the line of descent into the interior branching levels of the tree, often displaying intermediate geographic distributions. Typically, populations from a geographic region contain a composite of lineages from many different portions (i.e., haplogroups) of the tree.

Signals of Climate Change Modulating Migrations

The end of the African megadrought between 135,000 and 75,000 years ago ushered in more humid conditions that may have stimulated early modern human expansions and migrations.[37] While we cannot determine where the source population, personified as the CT branch, persisted during a time span of *ca.* 30,000 years, the calibrated phylogeny and modern non-African phylogeographic distributions of D, C, and F raise the possibility that the DE and CF ancestors may have already been in Eurasia by 75,000 years ago. This approximates the model of a small bottlenecked Asian subpopulation from which offspring populations later expanded during the wetter interglacial climates (*ca.* 59,000–24,000 years ago) of Oxygen Isotope Stage 3.[38] An exceptional feature of the calibrated phylogeny is the swift formation of the scaffold of the major non-African founder haplogroups as small numbers of intrepid pioneers rapidly disperse across Eurasia and the Australian and Papuan landmasses.[39] Specifically the differentiation of C and the F related core of F*, G, H, IJ, K, K2, M, P and S haplogroups during the narrow 50,000–55,000-year time interval.[40] The Last Glacial Maximum *ca.* 27,000–19,000 years ago depopulated the northern latitudes. Increased aridity elsewhere isolated populations as they contracted into refugia where endemic variation arose, only to disperse as climatic conditions improved. Later more recently evolved variants form the tips of the tree, often displaying restricted geographic distributions. These newer clusters manifest as rakelike branching structures with very short stems that emerge around the 5,000-year coalescent time horizon for haplogroups such as E1b2, H1, I1, R1a, and much of R1b. Such branches radiating from a common short root near the canopy are consistent with population census data indicating that rapid growth in the numbers of males occurred across several continents nearly coincident with the TMRCA of the short stem, with the most explosive phase visible in just the last 2000 years.[41] These recent bursts of growth have been interpreted to reflect the consequences of economic and cultural changes.[42] Also, one can now conceptually prune back the modern Y chromosome tree to any past coalescent time horizon of interest and get a perspective as to what level of extant Y chromosome haplogroup diversity was likely available at the time. Assuming that the phylogeny is calibrated correctly, then no Y markers present in the modern phylogeny that are younger than a securely dated specimen should be anticipated. For example, results from sequencing archaic genomes from a Denisovan and European Neanderthals suggested that introgression among archaic and early modern humans was not uncommon.[43] A recent report, using sophisticated modeling of chromosome 21 sequence data from a >50,000 years ago female Neanderthal from the Altai region of Asia, claimed to have found evidence of a past modern human introgression ca. 100,000 years ago.[44] Although any single Y chromosome can be lost in a single generation, the survival of archaic haplogroup A00 suggests that someday evidence of anatomically modern human introgression into Neanderthals could materialize should an Asian male Neanderthal ever be found that displays any of human Y chromosome variants consistent with branch temporalities of the modern phylogeny (Figure 7.1) such as the BT or CT branches that coalesce *ca.* 155,000 and 105,000 years ago respectively.

Impact of Credible Ancient DNA

Haploid systems reflect a narrow slice of the genome. While ancient DNA results are arguably revealing hidden complexities previously not recognized by geneticists attempting to reconstruct human prehistory using modern DNA,[45] usually only a few ancient specimens are available, and these usually yield only a low residuum of authentic mutations. When compared to modern samples, ancient Y chromosome DNA results are now beginning to reveal past branching structures including the detection of extinct branches.[46] What is exciting, however, about the achievement of credible ancient DNA studies[47] is that, even if only a few remnants of Y chromosome markers are detected, they can be cross-checked with the large database of SNVs identified in modern DNA. The intersection of Y chromosome results retrieved from ancient DNA,[48] with temporal and geographic knowledge from the modern Y chromosome phylogeny are providing a reliable guide in charting the way into the past.

Prehistoric Europe

The cast of characters in the story of the peopling of prehistoric Europe, prior to the agricultural transition, are often framed in both terms of climate and culture based on point and blade technologies and symbolic art. Pertaining to the Aurignacians, a ~37,000-year-old Russian specimen (Kostenki 14) and a 35,000-year-old Belgium specimen (Goyet 116.1) both share genome-wide affinities to some modern Siberians and Europeans.[49] Both of these Paleolithic males also possessed a related Y chromosome sub-haplogroup C variety. This same haplogroup was also detected in a 7,000-year-old Mesolithic Iberian (La Braña 1), but genome-wide analyses indicated that different demographic events were involved rather than representing a long static period of persistence.[50] This particular haplogroup C variety is now nearly absent in modern Europeans. The Late Glacial presence of ancient mtDNA haplogroup M in Europe[51] together with mtDNA evidence of a pre-Neolithic population replacement in Europe provides further coherence to the hunter-gatherer stages of European occupation.[52]

The Neolithic Transition

Long-standing discussions regarding the roles of cultural diffusion and demographic expansion endure.[53] Both ancient and modern Y chromosome data showed for the first time that contemporary Sardinians most prominently retain Neolithic genetic heritage reflective of a demic expansion of agriculturalists into Europe.[54] Ancient Y chromosome evidence from Ötzi, the 5,300-year-old Tyrolean Iceman, revealed that he carried Y chromosome haplogroup G2a-L91, a subclade of predecessor haplogroup G2a-PF3147. Both closely related haplogroups show their highest frequencies in present day Sardinia and southern Corsica compared to low levels (~1%) elsewhere including Anatolia and Cyprus.[55] The presence of PF3147-related lineages in ancient DNA from Neolithic specimens from west Anatolia[56] as well as less resolved data compatible with its presence in Neolithic specimens from southern France[57] is consistent with Mediterranean-region-associated gene flow during the Neolithic transition. A corroborating autosomal SNV frequency-based PC analyses also showed that Iceman's affinity was closest to modern Sardinians rather than to other modern continental Europeans.[58] Subsequent rigorous autosomal analyses revealed that the modern Sardinians were also similar to other ancient farmer DNA specimens from continental Europe,[59] as well as the Aegean.[60] There is also an apparent absence of PF3147-related lineages in postglacial Mesolithic era ancient DNA specimens.[61] This aggregate of both ancient and modern evidence solidifies the likelihood that G2a-PF3147 and its descendant sub-haplogroup G2a-L91 Y chromosomes, despite both lacking frequency clines, quite specifically trace an otherwise now largely hidden demic expansion of Neolithic peoples to Europe. Their present-day low frequency indicates that these specific G2a lineages have been supplanted by rapidly expanding Eurasian steppe populations associated with the Bronze Age,[62] the composition of which approximates the modern present-day observed genetic structure with some elements recognized in Iron Age cultures.[63]

State of Affairs

Critics of the use particular markers in haploid gene trees to study population history, often point to misinterpretations resulting from the failure to recognize that coalescent times do not always closely correlate with demographic events of interest.[64] While the recent progress made in understanding the Neolithic transition in Europe starting about 7000 BCE as well as Bronze Age demography in the steppes north of the Black and Caspian Seas by leveraging the hierarchy of time-calibrated nested modern haplogroups and glimpses of ancient uniparental and autosomal DNA is undeniable, the discussion in the following sections associating DNA patterns with more recent complex post-Neolithic transitions including the seafaring Minoan, Greek, Phoenician, and Roman cultures, while plausible, is still not as yet clear-cut. Is it possible to identify genetic patterns capable of differentiating some or all of them?

Local vs. Cosmopolitan Dynamics in the Mediterranean

Historical studies are by nature interdisciplinary. Not only do historical processes gradually blur into the opacity of prehistory, but the study of a specific historical period can benefit from data and the hermeneutic frameworks from a variety of other fields: sociology and anthropology, archaeology, philology, climatology, and human biology. Here, largely using modern autosomal and Y chromosome data we explore genetic patterns within a narrow geographic center, namely the Mediterranean, the Black Sea, and adjoining regions of Europe and the Near East. Our overarching model derives much from Ferdinand Braudel's historical analysis of the Mediterranean and the later revisiting of Mediterranean history from Horden's and Purcell's tome, *The Corrupting Sea.*[65] In an interesting way, Braudel's thesis that the Mediterranean is a specific historical/prehistorical region with common long-term (*longue durée*) historical patterns grounded upon a similar climate and geography and Horden and Purcell's model of the Mediterranean as a highly fragmented and diverse series of micro-ecologies that remain relatively isolated and differentiated from each other except during long-range seafaring trade and migration, are complementary. Braudel's model is largely cosmopolitan in scope while Horden's and Purcell's model underscores the local nature of cultural development in the Mediterranean basin. These two extreme perspectives of Mediterranean history fit several fundamental genetic models of human demography, most notably an insular model of islands interconnected with long and short-range migration. The dipolar forces of insular drift coupled with migration lead to striking results in the distribution of genetic markers across geography.

Cosmopolitan Exemplars

Here we discuss post-Neolithic long-range migration from a genetic perspective. Stimulated by extant geographic distributions of particular Y chromosome haplogroups, many defined over a decade ago, attempts were made to associate them with Mediterranean civilizations. For example, haplogroup J2-M172 with a frequency peak in the Levant was proposed to track the Phoenicians.[66] While much-published Y chromosome results involve lineages defined by markers that have coalescent times that predate the Neolithic, recent results from whole Y chromosome sequences indicate that there are major sublineages that approximate Bronze Age demographic events. An example is haplogroup J2a-M410 which has an estimated split time of ~33,000 years ago.[67] This haplogroup has a geographic span ranging from Pakistan, Iran, the Caucasus, Anatolia, and Mediterranean Europe.[68] Modern distributions and phylogenetic substructure suggest that the M410 SNV arose in the Zagros mountains of present day Iran.[69] Sub-haplogroups J2a-M67 and J2a-Z387 form the majority observed in extant Europeans. There are three main subclades within the J2a-M67, namely Z467, Z500 and CTS900.[70] Currently there is only one occurrence of J2a-M410 reported in an ancient DNA context in Europe. It involves a J2a-CTS900 sublineage observed in a late Bronze Age specimen from Hungary, ca. 1200 BCE.[71] The CTS900 sub-lineage is most frequently observed today in the Caucasus.

The L210 sublineage of J2a-Z467 has been observed in western Anatolia, Cyprus, Sicily, and northern Italy, and among Ashkenazi Jews.[72] The coalescent time for L210 among Tuscan samples dates to approximately 200 BCE, indicating a demographic expansion during the Greco-Roman era. Recently ancient DNA analysis of samples from Roman Era Britain, report one sample with an unusual haplogroup given the geographic context. This individual was a member of haplogroup J2 typical of the Middle East indicating the cosmopolitan nature of the Roman Empire, and his autosomal results show his clustering with individuals from the Levant and Arabian Peninsula.[73]

Sub-haplogroup J2a-Z387 is likewise widely distributed across Mediterranean Europe, particularly in southern Italy, Sicily, and Crete and has a coalescent time of 1500 BCE.[74] It may mirror Late Bronze Age maritime population movements seen in the archaeological record. Haplogroup T is similarly widespread throughout the Near East and the Mediterranean and shows similar demography to the above J2 lineages.[75] Haplogroup T has also been reported in a Late Neolithic sample from Germany, 5207–5079 BCE[76]. While this clade also displays a deep coalescent time, its sublineages expand *ca.* 2000–1400 BCE, during the Bronze Age.[77] The presence of the E-M81 lineage that is prominent in north African Berbers has been observed in Sicily.[78] These examples likely reflect maritime commerce throughout the Mediterranean from the Bronze Age to Antiquity and support an aspect of Braudel's thesis that the Mediterranean Sea constitutes a genetic and cultural region united by long-term and widespread interactions over millennia.

Local Exemplars

Not only do we see long-range expansion across the Mediterranean Sea, we also find evidence of local migration/demographic expansions, particularly in the Aegean and the Balkans.

The island of Crete, was colonized at Knossos by immigrants (*ca.* 7000 BCE) who brought the full Neolithic package to the island. Later during the transition between the Final Neolithic and the Early Bronze Age (*ca.* 3100 BCE), a shift in pottery style may signal the arrival of the Minoans to the island. King and colleagues[79] estimated an expansion time of 3100 BCE for two Y chromosome lineages on Crete, J2a-M319 and J2a-M92. This was estimated using highly polymorphic loci in which the alleles vary in sequence length based on the number of Short Tandemly Repeated (STR) nucleotides, such as $(ATG)_n$ or $(GAGT)_n$ repeat elements. Both J2a-M319 and J2a-M92 are concentrated in central and northwest Anatolian modern populations. These results suggest that a founder population from Anatolia may have immigrated to Crete at this time.

More generally, E-V13 is a Y haplogroup that is frequent among extant populations of Greece and the Balkans and relatively rare in other regions of the Mediterranean.[80] Even though E-V13 has a coalescence time of 10,000 BCE, a major expansion, exemplified by various sublineages, occurred *ca.* 1600 BCE, approximating perhaps the arrival of the Greek speakers to Greece. Because of E-V13's relatively recent expansion time, its original localization to Greece and the Balkans and its less frequent presence across other Mediterranean areas, E-V13 may be an ideal marker to trace the Iron Age/Archaic Age colonization of Greeks across the Mediterranean. This was the aim of a modern study of Provence that estimated a 15% to 20% Y chromosome Greek contribution to the population of Provence.[81] In other studies, E-V13 has also been utilized to estimate the mainland Greek contribution to Sicily[82] and the islands of

Cyprus (13%) and Crete (20%).[83] although some have questioned the specificity of E-V13.[84]

Upcoming Possibilities

Shortcomings including insufficient demographic specificity may, at least in some cases, be mitigated by constructing a high-resolution phylogeny, using whole sequences from several modern DNA samples constrained to the same level of haplogroup resolution. The efficacy of this approach was recently supported using haplogroup N samples common in north temperate Eurasia. Each phylogeny revealed more refined branching structures, many short in length effectively narrowing the interval between split times and the subsequent onset of expansions emerging from past bottlenecks. These newly defined sub-haplogroups often also showed distinctive spatial frequency patterns when genotyped in modern populations.[85] The issue of how well Y chromosomes genotyped to just the V13 level of resolution can discretely reflect the Greek expansion to the central Mediterranean at *ca.* 1600 BCE can be similarly investigated by sequencing multiple E-V13 samples. Subsequent high-resolution phylogenic analysis can help assess if any independent clusters of sub-haplogroups, united by short stems, exist whose estimated coalescent times and modern spatial frequency distributions approximate known colonization sites using appropriate modern population samples. It is prudent to keep in mind that the one-haplogroup–one-migration solution path is an overly simplistic model. Although the possibility exists that ancient DNA studies of well-dated attested Greek specimens going forward could find E-V13 lacking in the temporally relevant prehistoric gene pool, other haplogroups with properties coeval with E-V13 could assume the role of tracking the Greek migrations.

Conclusions

The confluence of modern population genomics and ancient DNA results during the past ~4 years has transformed the field of archaeogenetics.[86] The same technological advances allowing the sequencing of tens of thousands of modern genomes for medical science has also made it feasible to obtain reliable ancient DNA results. This new sequencing capability has also brought Y chromosome phylogenetics to an era of deeper understanding. Uncovering the branch structures at highly granular resolution using "whole" sequences from modern DNA samples has provided a more comprehensive contextual and calibrated framework vital to a fuller understanding of precious nuggets of ancient DNA data. The wherewithal to genotype modern population samples using a dozen or so SNVs that represent branch nodes with coalescent times appropriate to explore specific hypotheses concerning human history can expose previously hidden episodes of consequential past demographic events. One striking development along these lines concerns the demic vs. cultural diffusion debate regarding the arrival of agriculture to Europe. While ancient DNA clearly indicates that the first farmers did in fact migrate from Anatolia and the Near East to Europe, there is scant (~1%) presence of this Y chromosome signal remaining in modern continental populations. Notably both ancient and modern DNA results, including Y chromosome haplogroup R1b coalescent times reveal that the arrival of Steppe peoples largely replaced the earlier European Neolithic genetic legacy during the past ≤5000 years. The modern global Y chromosome gene tree combined with time estimates of the branching events, in fact shows that the post-Neolithic large-scale population growth seen in Europe also largely occurred throughout the world. The maturation of substantive knowledge about Y chromosome phylogenetics now provides a versatile experimental system to forge and test hypotheses regarding the history of anatomically modern humans.

Acknowledgements

P.A.U. was supported by SAP grant SP0#115016 to Prof. Carlos. D. Bustamante.

Notes

1. E.g., Cavalli-Sforza et al. 1994.
2. Rosenberg et al. 2005.
3. Ramachandran et al. 2005.
4. Cavalli-Sforza and Feldman 2003.
5. Novembre et al. 2008.
6. Skoglund et al. 2014; Broushaki et al. 2016.
7. Wall and Slatkin 2012.
8. Alexander et al. 2009.
9. Renfrew 2010.
10. Kidd et al. 2012.
11. Byrnes et al. 2014.
12. Behar et al. 2012.
13. Chiaroni et al. 2009.
14. Kivisild 2015.
15. Underhill and Kivisild 2007; Bedoya et al. 2006.
16. Balaresque et al. 2015.
17. Ray and Excoffier 2009.
18. Harpending 2007.
19. Jorde et al. 1998; Kivisild 2015.
20. Dulik et al. 2012.
21. Behar et al. 2012; Poznik et al. 2013.
22. E.g., Hinds et al. 2005.
23. E.g., Underhill et al. 2000; Jobling and Tyler-Smith 2003; Karafet et al. 2008; Hallast et al. 2015.
24. Jobling and Tyler-Smith 2017.

25. The 1000 Genomes Project Consortium 2010; The 1000 Genomes Project et al. 2015.

26. Stoneking and Krause 2011; Kivisild 2017. and above, Chapter 6.

27. Wei et al. 2013.

28. Poznik et al. 2013.

29. Poznik et al. 2013.

30. Karmin et al. 2015; Batini et al. 2015; Balanovsky et al. 2015; Poznik et al. 2016.

31. Poznik et al. 2016.

32. Hallast et al. 2015; Batini et al. 2015; Balanovsky et al. 2015.

33. Poznik et al. 2016.

34. Rosenberg and Nordborg 2002.

35. Karmin et al. 2015; Poznik et al. 2016.

36. Fu et al. 2014.

37. Scholz et al. 2007.

38. Pope and Terrell 2008.

39. Bergström et al. 2016.

40. Karmin et al. 2015.

41. Keinan and Clark 2012.

42. Karmin et al. 2015; Poznik et al. 2016.

43. Green et al. 2010; Prüfer et al. 2014.

44. Kuhlwilm et al. 2016.

45. Haber et al. 2016.

46. Rasmussen et al. 2014, 2015; Schroeder et al. 2015; Fu et al. 2014.

47. Der Sarkissian et al. 2015; Slatkin and Racimo 2016.

48. E.g., Keller et al. 2012; Schroeder et al. 2015; Rasmussen et al. 2014; Mathieson et al. 2015.

49. Seguin-Orlando et al. 2014; Fu et al. 2016.

50. Olalde et al. 2014.

51. Posth et al. 2016.

52. Pala et al. 2012.

53. Wall and Slatkin 2012.

54. Keller et al. 2012.

55. Rootsi et al. 2012; Voskarides et al. 2016.

56. Mathieson et al. 2015; Hofmanová et al. 2016.

57. Lacan et al. 2011.

58. Keller et al. 2012.

59. Sikora et al. 2014; Gamba et al. 2014.

60. Hofmanová et al. 2016.

61. Hofmanová et al. 2016.

62. Haak et al. 2015; Allentoft et al. 2015.

63. Gamba et al. 2014.

64. Goldstein and Chikhi 2002; Soares et al. 2011; Pinhasi et al. 2012.

65. Braudel 1972; Horden and Purcel 2000.

66. Zalloua et al. 2008.

67. Poznik et al. 2016.

68. Chiaroni et al. 2009.

69. Grugni et al. 2012; Poznik et al. 2016.

70. Poznik et al. 2016.

71. Gamba et al. 2014.

72. King R. J., unpublished results; Magoon et al 2013.

73. Martiniano et al. 2016.

74. Poznik et al. 2016.

75. Mendez et al. 2011.

76. Haak et al. 2015.

77. Poznik et al. 2016.

78. Di Gaetano et al. 2009.

79. King et al. 2008.

80. Cruciani et al. 2007.

81. King et al. 2011.

82. Di Gaetano et al. 2009.

83. Voskarides et al. 2016; King et al. 2008.

84. Tofanelli et al. 2016.

85. E.g., Ilumäe et al. 2016.

86. Slatkin and Racimo 2016; Nielsen et al 2017.

References

Alexander D. H., J. Novembre, and K. Lange. 2009. "Fast model-based estimation of ancestry in unrelated individuals." *Genome Research* 19 (9):1655–1664. DOI: 10.1101/gr.094052.109.

Allentoft M. E., et al. 2015. "Population genomics of Bronze Age Eurasia." *Nature* 522 (7555): 167–172. DOI: 10.1038/nature14507.

Balanovsky, O., et al. 2015. "Deep phylogenetic analysis of haplogroup G1 provides estimates of SNP and STR mutation rates on the human Y-chromosome and reveals migrations of Iranic speakers." *PLoS ONE* 10 (4): e0122968. DOI: 10.1371/journal.pone.0122968.

Balaresque, P., et al. 2015. "Y chromosome descent clusters and male differential reproductive success: young lineage expansions dominate Asian pastoral nomadic populations." *European Journal of Human Genetics* 23 (10): 1413–1422. DOI: 10.1038/ejhg.2014.285.

Batini, C., et al. 2015. "Large-scale recent expansion of European patrilineages shown by population resequencing." *Nature Communications* 6: 7152. DOI: 10.1038/ncomms8152.

Bedoya, G., et al. 2006. "Admixture dynamics in Hispanics: a shift in the nuclear genetic ancestry of a South American population isolate." *Proceedings of the National Academy of Sciences of the United States of America* 103 (19): 7234–7239. DOI: 10.1073/pnas.0508716103.

Behar, D. M., et al. 2012. "A 'Copernican' reassessment of the human mitochondrial DNA tree from its root." *American Journal of Human Genetics* 90 (4): 675–684. DOI: 10.1016/j.ajhg.2012.03.002.

Bergström A., et al. 2016. "Deep roots for Aboriginal Australian Y chromosomes." *Current Biology* 26 (6): 809–813. DOI: 10.1016/j.cub.2016.01.028.

Braudel, F. 1972. *The Mediterranean and the Mediterranean World in the Age of Philip II*, trans. S. Reynolds. London: Collins.

Broushaki F., et al. 2016. "Early Neolithic genomes from the eastern Fertile Crescent." *Science* 353 (6298): 499–503. DOI: 10.1126/science.aaf7943.

Byrnes, J., N. M. Myres, and P. A. Underhill. 2014. "Genetic genealogy in the genomic era." In *Forensic DNA Applications: An Interdisciplinary Perspective*, eds. D. Primorac and M. Schanfield. Boca Raton, FL: CRC Press Taylor & Francis Group. DOI: 10.1201/b16512-23.

Cavalli-Sforza, L. L., and M. W. Feldman. 2003. "The application of molecular genetic approaches to the study of human evolution." *Nature Genetics* 33 (Supplement): 266–275. DOI:10.1038/ng1113.

Cavalli-Sforza, L. L., P. Menozzi, and A. Piazza. 1994. *History and Geography of Human Genes*. Princeton: Princeton University Press.

Chiaroni, J., P. A. Underhill, and L. L. Cavalli-Sforza. 2009. "Y chromosome diversity, human expansion, drift, and cultural evolution," *Proceedings of the National Academy of Sciences of the United States of America* 106 (48): 20174–20179. DOI: 10.1073/pnas.0910803106.

Cruciani. F., et al. 2007. "Tracing past human male movements in northern/eastern Africa and western Eurasia: new clues from Y-chromosomal haplogroups E-M78 and J-M12." *Molecular Biology and Evolution* 24 (6): 1300–1311. DOI: 10.1093/molbev/msm049.

Der Sarkissian, C., et al. 2015. "Ancient genomics." *Philosophical Transactions of the Royal Society London B, Biological Sciences* 370: 20130387. DOI: 10.1098/rstb.2013.0387.

Di Gaetano, C., et al. 2009. "Differential Greek and northern Africa migrations to Sicily are supported by genetic evidence from the Y chromosome." *European Journal of Human Genetics* 17: 91–99. DOI: 10.1038/ejhg.2008.120.

Dulik, M. C., et al. 2012. "Mitochondrial DNA and Y chromosome variation provides evidence for a recent common ancestry between Native Americans and Indigenous Altaians." *American Journal of Human Genetics* 90 (2): 229–246. DOI:10.1016/j.ajhg.2011.12.014.

Fu, Q., et al. 2014. "Genome sequence of a 45,000-year-old modern human from western Siberia." *Nature* 514 (7523): 445–450. DOI: 10.1038/nature13810.

Fu, Q., et al. 2016. "The genetic history of Ice Age Europe." *Nature* 534 (7606): 200–205. DOI: 10.1038/nature17993.

Gamba, C., et al. 2014. "Genome flux and stasis in a five millennium transect of European prehistory." *Nature Communications* 5: 5257. DOI: 10.1038/ncomms6257.

Goldstein, D. B., and L. Chikhi. 2002. "Human migrations and population structure: what we know and why it matters." *Annual Review of Genomics and Human Genetics* 3: 129–152. DOI: 10.1146/annurev.genom.3.022502.103200.

Green R. E., et al. 2010. "A draft sequence of the Neandertal genome." *Science* 328 (5979): 710–722. DOI: 10.1126/science.1188021.

Grugni, V., et al. 2012. "Ancient migratory events in the Middle East: new clues from the Y-chromosome variation of modern Iranians." *PLoS ONE* 7 (7): e41252. DOI: 10.1371/journal.pone.0041252.

Haak, W. 2015. "Massive migration from the steppe was a source for Indo-European languages in Europe." *Nature* 522 (7555): 207–211. DOI: 10.1038/nature14317.

Haber, M., et al. 2016. "Ancient DNA and the rewriting of human history: be sparing with Occam's razor." *Genome Biology* 17: 1. DOI: 10.1186/s13059-015-0866-z.

Hallast, P., et al. 2015. "The Y-chromosome tree bursts into leaf: 13,000 high-confidence SNPs covering the majority of known clades." *Molecular Biology and Evolution* 32 (3): 661–673. DOI: 10.1093/molbev/msu327.

Harpending, H. 2007. "Humans: demographic history." In *Encyclopedia of Life Sciences*. Chichester: John Wiley & Sons Ltd. Available at http://www.els.net. DOI: 10.1002/9780470015902.a0005077.pub2.

Hinds, D. A., et al. 2005. "Whole-genome patterns of common DNA variation in three human populations." *Science* 307 (5712): 1072–1079. DOI: 10.1126/science.1105436.

Hofmanová, Z., et al. 2016. "Early farmers from across Europe directly descended from Neolithic Aegeans." *Proceedings of the National Academy of Sciences of the United States of America* 113 (25): 6886–6891. DOI: 10.1073/pnas.1523951113.

Horden, P., and N. Purcell. 2000. *The Corrupting Sea: A Study of Mediterranean History*. Oxford: Wiley-Blackwell.

Ilumäe, A.-M., et al. 2016. "Human Y-chromosome haplogroup N: a non-trivial time-resolved phylogeography that cuts across language families." *American Journal of Human Genetics* 99: 163–173. DOI: 10.1016/j.ajhg.2016.05.025.

Jobling, M. A., and C. Tyler-Smith. 2003. "The human Y chromosome: an evolutionary marker comes of age." *Nature Reviews Genetics* 4: 598–612. DOI: 10.1038/nrg1124.

Jobling, M. A., and C. Tyler-Smith. 2017. "Human Y-chromosome variation in the genome-sequencing era. *Nature Reviews Genetics* 18: 485–497. DOI: 10.1038/nrg.2017.36.

Jorde, L. B., M. Bamshad, and A. R. Rogers. 1998. "Using mitochondrial and nuclear DNA markers to reconstruct human evolution." *BioEssays* 20:126–136. DOI: 10.1002/(SICI)1521-1878(199802)20:2<126:AID-BIES5>3.0.CO;2-R.

Karafet, T. M., et al. 2008. "New binary polymorphisms reshape and increase resolution of the human Y chromosomal haplogroup tree." *Genome Research* 18 (5): 830–838. DOI: 10.1101/gr.7172008.

Karmin, M., et al. 2015. "A recent bottleneck of Y chromosome diversity coincides with a global change in culture." *Genome Research* 25 (4): 459–466. DOI: 10.1101/gr.186684.114.

Keinan, A., and A. G. Clark. 2012. "Recent explosive human population growth has resulted in an excess of rare genetic variants." *Science* 336 (6082): 740–743. DOI: 10.1126/science.1217283.

Keller, A., et al. 2012. "New insights into the Tyrolean Iceman's origin and phenotype as inferred by whole-genome sequencing." *Nature Communications* 3: 698. DOI: 10.1038/ncomms1701.

Kidd, J., et al. 2012. "Population genetic inference from personal genome data: impact of ancestry and admixture on human genomic variation." *American Journal of Human Genetics* 91 (4): 660–671. DOI: 10.1016/j.ajhg.2012.08.025.

King, R. J., et al. 2008. "Differential Y-chromosome Anatolian influences on the Greek and Cretan Neolithic." *Annals of Human Genetics* 72 (2): 205–214. DOI:10.1111/j.1469-1809.2007.00414.x.

King R. J., et al. 2011. "The coming of the Greeks to Provence and Corsica: Y chromosome models of archaic Greek colonization of the western Mediterranean." *BMC Evolutionary Biology* 11: 69. DOI: 10.1186/1471-2148-11-69.

Kivisild, T. 2015. "Maternal ancestry and population history from whole mitochondrial genomes." *Investigative Genetics* 6: 3. DOI: 10.1186/s13323-015-0022-2.

Kivisild, T. 2017. "The study of human Y chromosome variation through ancient DNA." *Human Biology* 136: 529–546. DOI 10.1007/s00439-017-1773-z.

Kuhlwilm M., et al. 2016. "Ancient gene flow from early modern humans into Eastern Neanderthals." *Nature* 530 (7591): 429–433. DOI: 10.1038/nature16544.

Lacan, M., et al. 2011. "Ancient DNA reveals male diffusion through the Neolithic Mediterranean route." *Proceedings of the National Academy of Sciences of the United States of America* 108 (24): 9788–9791. DOI: 10.1073/pnas.1100723108.

Magoon, G. R., et al. 2013. "Generation of high-resolution *a priori* Y-chromosome phylogenies using "next-generation" sequencing data." *Bioarchiv* November 22, 2013 DOI: https://doi.org/10.1101/000802.

Martiniano, R., et al. "Genomic signals of migration and continuity in Britain before the Anglo-Saxons." *Nature Communications* 7: 10326. DOI: 10.1038/ncomms10326.

Mathieson I., et al. 2015. "Genome-wide patterns of selection in 230 ancient Eurasians." *Nature* 528 (7583): 499–503. DOI: 10.1038/nature16152.

Mendez, F. L., et al. 2011. "Increased resolution of Y chromosome haplogroup T defines relationships among populations of the Near East, Europe, and Africa." *Human Biology* 83 (1), 39–53. DOI: 10.3378/027.083.0103.

Nielsen, R., et al. 2017. "Tracing the peopling of the world through genomics." *Nature* 541: 302–310. DOI: 10_1038/nature21347.

Novembre, J., et al. 2008. "Genes mirror geography within Europe." *Nature* 456 (7281): 98–101. DOI: 10.1038/nature07331.

Olalde, I., et al. 2014. "Derived immune and ancestral pigmentation alleles in a 7,000-year-old Mesolithic European." *Nature* 507 (7491): 225–228. DOI: 10.1038/nature12960.

Pala, M., et al. 2012. "Mitochondrial DNA signals of late glacial recolonization of Europe from Near Eastern refugia." *American Journal of Human Genetics* 90 (5), 915–924. DOI: 10.1016/j.ajhg.2012.04.003.

Pinhasi, R., et al. 2012. "The genetic history of Europeans." *Trends in Genetics* 28: 496–505. DOI: 10.1016/j.tig.2012.06.006.

Pope, K. O., and J. E. Terrell. 2008. "Environmental setting of human migrations in the circum-Pacific region." *Journal of Biogeography* 35: 1–21. DOI: 10.1111/j.1365-2699.2007.01797.

Posth, C., et al. 2016. "Pleistocene mitochondrial genomes suggest a single major dispersal of Non-Africans and a late glacial population turnover in Europe." *Current Biology* 26 (6): 827–833. DOI: 10.1016/j.cub.2016.01.037.

Poznik, G. D., et al. 2013. "Sequencing Y chromosomes resolves discrepancy in time to common ancestor of males versus females." *Science* 341 (6145): 562–565. DOI: 10.1126/science.1237619.

Poznik, G. D., et al. 2016. "Punctuated bursts in human male demography inferred from 1,244 worldwide Y chromosome sequences." *Nature Genetics* 48 (6): 593–599. DOI: 10.1038/ng.3559.

Prüfer, K., et al. 2013. "The complete genome sequence of a Neanderthal from the Altai Mountains." *Nature* 505 (7481): 43–9. DOI: 10.1038/nature12886.

Ramachandran, S., et al. 2005. "Support from the relationship of genetic and geographic distance in human populations for a serial founder effect originating in Africa." *Proceedings of the National Academy of Sciences of the United States of America* 102 (44): 15942–15947. DOI: 10.1073/pnas.0507611102.

Rasmussen, M., et al. 2014. "The genome of a late Pleistocene human from a Clovis burial site in western Montana." *Nature* 506 (7487): 225–229. DOI: 10.1038/nature13025.

Rasmussen, M., et al. 2015. "The ancestry and affiliations of Kennewick Man." *Nature* 523 (7561): 455–458. DOI: 10.1038/nature14625.

Ray, N., and L. Excoffier. 2009. "Inferring past demography using spatially explicit population genetic models." *Human Biology* 81 (2–3): 141–157. DOI: 10.3378/027.081.0303.

Renfrew, C. 2010. "Archaeogenetics—towards a 'new synthesis?'" *Current Biology* 20 (4): R162–R165. DOI: 10.1016/j.cub.2009.11.056.

Rootsi, S., et al. 2012. "Distinguishing co-ancestries of European and Caucasian human Y-chromosomes within haplogroup G." *European Journal of Human Genetics* 20 (12): 1275–1282. DOI: 10.1038/ejhg.2012.86.

Rosenberg, N. A., and M. Nordborg. 2002. "Genealogical trees, coalescent theory and the analysis of genetic polymorphisms." *Nature Reviews Genetics* 3 (5): 380–390. DOI: 10.1038/nrg795.

Rosenberg, N. A., et al. 2005. "Clines, clusters, and the effect of study design on the inference of human population structure." *PLoS Genetics* 1 (6): e70. DOI: 10.1371/journal.pgen.0010070.

Scholz, C. A., et al. 2007. "East African megadroughts between 135 and 75 thousand years ago and bearing on early-modern human origins." *Proceedings of the National Academy of Sciences of the United States of America* 104 (42): 16416–16421. DOI: 10.1073/pnas.0703874104.

Schroeder, H., et al. 2015. "Genome-wide ancestry of 17th-century enslaved Africans from the Caribbean." *Proceedings of the National Academy of Sciences of the United States of America* 112 (12): 3669–3673. DOI: 10.1073/pnas.1421784112.

Seguin-Orlando, A., et al. 2014. "Paleogenomics: genomic structure in Europeans dating back at least 36,200 years." *Science* 346 (6213): 1113–1118. DOI: 10.1126/science.aaa0114.

Sikora, M., et al. 2014. "Population genomic analysis of ancient and modern genomes yields new insights into the genetic ancestry of the Tyrolean Iceman and the genetic structure of Europe." *PLoS Genetics* 10 (5): e1004353. DOI: 10.1371/journal.pgen.1004353.

Skoglund, P., et al. 2014. "Investigating population history using temporal genetic differentiation." *Molecular Biology and Evolution* 31 (9): 2516–2527. DOI: 10.1093/molbev/msu192.

Slatkin, M., and F. Racimo, 2016. "Ancient DNA and human history." *Proceedings of the National Academy of Sciences of the United States of America* 113 (23): 6380–6387. DOI. 10.1073/pnas.1524306113.

Soares, P., et al. 2011. "Learning about human population history from ancient and modern genomes." *Nature Review Genetics* 12 (9): 603–614. DOI: 10.1038/nrg3029.

The 1000 Genomes Project Consortium. 2010. "A map of human genome variation from population scale sequencing." *Nature* 467 (7319): 1061–1073. DOI: 10.1038/nature09534.

The 1000 Genomes Project Consortium et al. 2015. "A global reference for human genetic variation." *Nature* 526 (7571): 68–74. DOI: 10.1038/nature15393.

Tofanelli, S., et al. 2016. "The Greeks in the West: genetic signatures of the Hellenic colonization in southern Italy and Sicily." *European Journal of Human Genetics* 24: 429–436. DOI: 10.1038/ejhg.2015.124.

Underhill, P. A., and T. Kivisild. 2007. "Use of Y chromosome and mitochondrial DNA population structure in tracing human migrations." *Annual Review Genetics* 41: 539–564. DOI: 0.1146/annurev.genet.41.110306.130407.

Underhill, P. A., et al. 2000. "Y chromosome sequence variation and the history of human populations." *Nature Genetics* 26 (3): 358–361. DOI: 10.1038/81685.

Voskarides, K., et al. 2016. "Y-chromosome phylogeographic analysis of the Greek-Cypriot population reveals elements consistent with Neolithic and Bronze Age settlements." *Investigative Genetics* 7: 1. DOI: 10.1186/s13323-016-0032-8.

Wall J. D., and Slatkin M. 2012. "Paleopopulation genetics." *Annual Review Genetics* 46: 635–649. DOI:10.1146/annurev-genet-110711-155557.

Wei, W., et al. 2013. "A calibrated human Y-chromosomal phylogeny based on resequencing." *Genome Research* 23 (2): 388–395. DOI: 10.1101/gr.143198.112.

Zalloua, P. A., et al. 2008. "Identifying genetic traces of historical expansions: Phoenician footprints in the Mediterranean." *American Journal of Human Genetics*: 83 (5): 633–642. DOI: 10.1016/j.ajhg.2008.10.012.

NOTE: Page numbers followed by *t* indicate a table. Those followed by *f* indicate a figure.

A NOTE ON THE TYPE

THIS BOOK has been composed in Miller, a Scotch Roman typeface designed by Matthew Carter and first released by Font Bureau in 1997. It resembles Monticello, the typeface developed for The Papers of Thomas Jefferson in the 1940s by C. H. Griffith and P. J. Conkwright and reinterpreted in digital form by Carter in 2003.

Pleasant Jefferson ("P. J.") Conkwright (1905–1986) was Typographer at Princeton University Press from 1939 to 1970. He was an acclaimed book designer and AIGA Medalist.

The ornament used throughout this book was designed by Pierre Simon Fournier (1712–1768) and was a favorite of Conkwright's, used in his design of the *Princeton University Library Chronicle*.